Payne

RECURSIVE FUNCTION THEORY

To Clifford Spector
(December 20, 1930 – July 29, 1961)

PROCEEDINGS OF
SYMPOSIA IN PURE MATHEMATICS
VOLUME V

RECURSIVE FUNCTION THEORY

AMERICAN MATHEMATICAL SOCIETY
PROVIDENCE, RHODE ISLAND
1962

Proceedings of the Fifth Symposium in Pure Mathematics
of the American Mathematical Society

This Symposium was conducted, and the proceedings prepared in part, by the American Mathematical Society under Subcontract IDA-2631-8 to Contract NONR 2631 (00).

Held at the Hotel New Yorker
New York, New York
April 6-7, 1961

First edition, 1962
Second printing with corrections, 1970

International Standard Book Number 0-8218-1405-2
Library of Congress Catalog Number 50-1183

CONTENTS

PREFACE

A symposium on Recursive Function Theory was held on April 6 and 7, 1961 in conjunction with the New York meeting of the American Mathematical Society. This Symposium was sponsored by the American Mathematical Society, the Association for Computing Machinery, and The Association for Symbolic Logic, with financial support from project FOCUS of the Institute for Defense Analyses. The Organizing Committee consisted of Professor S. C. Kleene, Chairman, Professor J. C. E. Dekker, Professor John McCarthy, Professor J. B. Rosser, and Professor J. R. Shoenfield.

Fourteen papers were presented at the Symposium. Thirteen of these are contained in this volume. The paper *Recursive Unsolvability of Post's problem of "Tag"*, presented by Professor Marvin Minsky has appeared in volume 74 (1961) of the Annals of Mathematics. In its place he contributed another paper to this book.

This volume is dedicated to the memory of Professor Clifford Spector who died on July 29, 1961. His paper was put in final form by Professor George Kreisel.

J. C. E. Dekker,
Editor of these Proceedings
Rutgers, The State University

PREFACE

PROVABLY RECURSIVE FUNCTIONALS OF ANALYSIS: A CONSISTENCY PROOF OF ANALYSIS BY AN EXTENSION OF PRINCIPLES FORMULATED IN CURRENT INTUITIONISTIC MATHEMATICS

BY

CLIFFORD SPECTOR[1]

The central problem in Hilbert's program in the foundations of mathematics was to show by elementary methods that from a classical proof of an elementary theorem one can always construct an elementary proof of the same theorem. Our paper is a contribution to a modification of Hilbert's program.

Received by the editor, September 23, 1961.

[1] In his last letter to me, written and posted on 26 July 1961, Spector wrote: 'Enclosed is a typewritten draft with part of the last section omitted...

$$(x)\neg\neg(E\alpha)(y)P(x,\alpha,y)\rightarrow\neg\neg(x)(E\alpha)(y)P(x,\alpha,y)\ .$$

This special case as well as the general case of axiom F can be proved using bar induction. In the special case bar induction at type one is used, which I claim can be reduced to bar induction at type zero. However, the latter reduction may apply only to existential formulas, but I think it is exactly an existential formula that occurs. So the result may be correct, but there is still some work to do.

'I intend to include a § 12.2 giving a proof of the Gödel translation of axiom F. If possible it would be nice to include a 12.3 giving the result above.'

The typescript mentioned in the quotation above, consists of the sections up to and including 12.1 of the present paper, with about half a page (crossed out) of a projected 12.2. This last half page states that the proof of the Gödel translation of axiom F would use a generalization of Hilbert's substitution method as illustrated in the special case of § 12.1. However Spector's notes do not contain any details, so that it is not quite clear how to reconstruct the proof he had in mind. In order to achieve the general aim he intended, we give a different proof in 12.2 which uses bar induction at type zero to derive bar recursion at type one (these notions are defined in 6.2 and 6.3 below), and hence, by Spector's own § 10 below, the Gödel 1958 translation [4] of the special case of axiom F mentioned in his letter. § 12.3 reduces the consistency of classical analysis, with the comprehension axiom restricted to arithmetic and hyper-arithmetic properties, to the special case of axiom F above. It is likely that a reconstruction of the proof intended by Spector would be informative.

At the suggestion of K. Gödel the title was changed, to avoid misunderstandings, from *Provably recursive functionals in analysis; a consistency proof by an extension of intuitionistic principles.*

Spector's typescript did not contain any footnotes. I have added footnotes wherever explanations seem necessary; perhaps more freely than was Spector's custom.

Spector's typescript, being incomplete, did not contain the acknowledgments he undoubtedly wished to make. Research on the present paper was begun in summer 1959 at the Ohio State University under a grant from the National Science Foundation, and continued during the academic year 1960–1961 at the Institute for Advanced Study,

Hilbert intended that "elementary" should mean strictly finitary in the sense of combinatorial properties of symbols. However the successful attempts to apply Hilbert's program to arithmetic have involved an extension of finitistic principles by means of transfinite induction (or recursion), intuitionistic logic, or computable functionals of finite type. We propose to extend Gödel's consistency proof for arithmetic [**4**], which uses computable functionals of finite type, to a consistency proof for analysis by introducing one new principle of definition called bar recursion. Our proof provides a method for eliminating the nonconstructive and impredicative features of classical analysis in favor of the "elementary" principles formalized in Σ_4.

The system Σ_4, defined below, is a logic-free theory of computable functionals of finite type in which all formulas are of the form $u = v$, where u and v are terms of type zero denoting natural numbers. An effective process for computing the value of a closed term of type zero can be obtained by making use of the defining equations of the functionals which occur in the term. It is through an analysis of this computational process that we hope to obtain a major improvement of the results of this paper.

The least constructive feature of Σ_4 is bar recursion which defines a functional by means of a recursion over a well-founded class of functionals. This is the only feature of Σ_4 which is not *obviously* intuitionistic. Whether or not it is an intuitionistic principle is a debatable issue at this stage of the development of intuitionistic mathematics. Bernays and Gödel have both expressed the opinion that on the basis of what the intuitionists have already accepted (e.g. the bar theorem), bar recursion should also be acceptable. Kreisel reports that Heyting told him that bar recursion is intuitionistically acceptable. The author believes that the bar theorem is itself questionable, and that until the bar theorem can be given a suitable foundation, the question of whether bar recursion is intuitionistic is premature.[2]

The principal result of this paper is that if an $\bigwedge\bigvee$-formula (universal function quantifiers followed by existential quantifiers followed by a quantifier-free formula) is provable in classical anaysis, then it can be proved by the elementary methods of Σ_4. Since every arithmetical formula is classically equivalent to an $\bigwedge\bigvee$-formula, it follows that every arithmetical theorem proved by analytic methods can be proved in Σ_4.[3] Furthermore, the statement

---

Princeton, New Jersey, under contract with the Office of Naval Research, administered by the Ohio State University Research Foundation. From paragraph 3 of the introduction below, and from conversations with Spector I know that he valued highly his discussions with P. Bernays and K. Gödel on the subject of the present paper.—G. Kreisel.

[2] This view, that the bar theorem is in need of a suitable foundation, is almost universal, and not inconsistent with that expressed by Bernays and Gödel. For, they presumably mean that any plausible general principle which would establish the bar theorem within a suitable formal framework of constructive mathematics would also establish bar induction of 6.3 and hence bar recursion.

[3] More precisely, in general, it is of course not the arithmetic formula itself which can be proved in Σ_4, but its $\bigwedge\bigvee$ form ($=$ its no-counterexample-interpretation in the sense of [**9**]).

that a number is the Gödel number of a general recursive functional (whose arguments are functions) is also an $\bigwedge\bigvee$-formula [**6**], and therefore the provably recursive functionals are bar recursive.

The essential improvement over Kreisel [**10**] concerns the choice of schemata for generating the functionals employed in the interpretation. Kreisel's classical application of the least number operator (restricted to recursive expressions, but with existence proved classically) is replaced by bar recursion. By further analysis of the computation process of the bar recursive functionals we hope to obtain information of interest both from the point of view of constructivity and of obtaining independence theorems.

1. Formalization of analysis. For economy we write the axioms of analysis so that the classical system can be obtained from the intuitionistic system by adjoining the axiom schema for the law of the excluded middle. Most of what mathematicians call analysis including measure theory can be formalized within the classical system (see [**5**, vol. 2, Supplement IV]). The intuitionistic system is quite weak; in fact it can be interpreted within intuitionistic arithmetic [**11**, para. 14, Remark 14].

1.1. *Propositional calculus.*

A1 $p \to p$.

A2 If q, then $p \to q$.

A3 If p and $p \to q$, then q.

A4 If $p \to q$ and $q \to r$, then $p \to r$.

A5 $p \wedge q \to p$, $q \wedge p \to p$, $p \to p \vee q$, $q \to p \vee q$.

A6 If $p \to r$ and $q \to r$, then $p \vee q \to r$.

A7 If $r \to p$ and $r \to q$ then $r \to p \wedge q$.

A8 If $p \wedge q \to r$, then $p \to (q \to r)$.

A9 If $p \to (q \to r)$, then $p \wedge q \to r$.

A10 $0' = 0 \to p$.

A11 $p \vee (p \to 0' = 0)$ (classical system only).

Let $\neg p$ be an abbreviation for $p \to 0' = 0$.

1.2. We show now that Kleene's axioms [**6**, pp. 82, 101] for the intuitionistic propositional calculus are derivable from A1—A10. The assertions below apply to A1—A10.

(1.2.1) If $p \to q$ and $p \to (q \to r)$, then $p \to r$.

Proof. $p \to p$, $p \to q$, $p \to (p \wedge q)$, $p \to (q \to r)$, $(p \wedge q) \to r$, $p \to r$.

(1.2.2) $$[(p \to q) \wedge (p \to (q \to r)) \cdot \wedge\, p] \to r\,.$$

Proof. Write (1.2.2) as $R \to r$. Then R implies each of its three conjuncts. After three applications of (1.2.1) we obtain $R \to r$.

(1.2.3) $$(p \to q) \to [(p \to (q \to r)) \to (p \to r)]\,.$$

Apply A8 to (1.2.2).

(1.2.4) $p \to (q \to p)$. (A5, A8)

(1.2.5) $p \to (q \to p \wedge q)$. (A1, A8)

(1.2.6) If $p \to (q \to r)$, then $q \to (p \to r)$.

PROOF. Derive $q \wedge p \to p \wedge q$ from A5, A7. Using A7, A8, the problem reduces to: if $p \wedge q \to r$ then $q \wedge p \to r$, which is obtained from transitivity of implication (A4).

(1.2.7) $[(p \to r) \wedge (q \to r)] \to [(p \vee q) \to r]$.

PROOF. Write (1.2.7) as $P \to Q$. Then P implies each of its two conjuncts, and applying (1.2.6) we obtain $p \to (P \to r)$, $q \to (P \to r)$, and therefore $(p \vee q) \to (P \to r)$. One more application of (1.2.6) gives (1.2.7).

(1.2.8) $(p \to r) \to [(q \to r) \to ((p \vee q) \to r)]$. ((1.2.7), (A8)).

(1.2.9) $(p \to 0' = 0) \to (p \to q)$.

PROOF. (1.2.9) reduces to $(p \to 0' = 0) \wedge p \to q$. Using a proof similar to that of (1.2.2), $(p \to 0' = 0) \wedge p \to 0' = 0$, and then applying transitivity (A4) and A10 the proof is completed.

Kleene's axioms (in order) are our (1.2.4), (1.2.3), A3, (1.2.5), A5, (1.2.8), (1.2.3) with $0' = 0$ for r, (1.2.9). Axiom A10 can be derived from A1—A9, B, C, D [**6**].

1.3. *Quantifier rules.* The remaining rules (and axioms) belong to both the classical and intuitionistic systems. See § 2.2 for syntactical notions not explained here.

B1, 2 If $Q \to P(a)$, then $Q \to \bigwedge a P(a)$;
if $P(a) \to Q$, then $\bigvee a P(a) \to Q$,

where a is a variable not occurring free in Q.

B3, 4 $\bigwedge a P(a) \to P(u)$, $P(u) \to \bigvee a P(a)$,

where a is a variable of the same type as the term u and u is free for a in $P(a)$.

1.4. *Equality axioms.* Here and below x, y, z are variables of type (0), i.e. are variables for natural numbers.

C1 $x = x$, $x = y \wedge z = y \to x = z$,

C2 $x = y \to T(x) = T(y)$,

where $T(x)$ is a term of type zero.

C3 $x' = 0 \to 0' = 0$,

C4 $x' = y' \leftrightarrow x = y$.

1.5. *Induction rule, axiom of choice at type zero.*

D If $P(0)$ and $P(x) \rightarrow P(x')$, then $P(x)$.

E $\bigwedge x \bigvee a P(x, a) \rightarrow \bigvee b \bigwedge x P(x, bx)$,

where b is a variable whose argument is of type zero and whose value is of the same type as the variable a.

The classical comprehension axiom in the form

$$\bigvee a \bigwedge x [ax = 0 \leftrightarrow P(x) \cdot \wedge \cdot ax = 1 \leftrightarrow \neg P(x)] \quad (1.5.1)$$

can be derived from the axiom of choice E.

For the purpose of dealing with classical analysis, is suffices to consider only variables of type zero, functions from natural numbers to natural numbers, and functions from natural numbers to functions. However since we will be introducing functionals of higher type later, and since there is not much additional trouble to include them from the very beginning we do so. We do not include the axiom of choice, comprehension axioms, or equality axioms at higher types.

2. Notation.

2.1. *Conventions.* The letters $a - h$, $A - H$, and with subscripts denote variables for functionals of finite type or finite sequences of such variables. The letters i, j, k, m, n denote natural numbers (informal); p, q, r, P, Q, R, denote formulas; s, t, denote type symbols; u, v, w, T denote terms; x, y, z denote variables of type zero; X, Y, Z denote variables for functionals whose values are of type zero. Greek letters denote constants.

2.2. *Definition of term and type symbols.*

(2.2.1) (0) is a type symbol.

(2.2.2) If s and t are type symbols, then so is (st); s is the type of the argument and t is the type of the value.

(2.2.3) Type symbols are obtained only as above.

(2.2.4) For each type symbol there are variables and constants of that type. No variable or constant of one type is a variable or constant of another type. Each variable and constant of a given type is a term of that type.

(2.2.5) If u is a term of type (st) and v is a term of type s, then (uv) is a term of type t.

(2.2.6) 0 is a term of type zero. If u is a term of type zero, so is (u').

(2.2.7) Terms are obtained only as above.

The prime formulas of our formal systems are always of the form $u = v$, where u and v are terms of type zero.

2.3. *Abbreviations.* If u and v are terms of the same type other than (0), then $u = v$ is an abbreviation for

$$(\cdots (ue_1)e_2 \cdots e_n) = (\cdots ((ve_1)e_2) \cdots e_n) , \quad (2.3.1)$$

where $e_1, \cdots, e_n$ are distinct variables chosen so that both sides are of type zero and no e_i occurs in u or in v; $a(b_1, \cdots, b_n)$ is an abbreviation for $ab_1 \cdots b_n$, which in turn is an abbreviation for $(\cdots (ab_1) \cdots b_n)$; $(a_1, \cdots, a_m)b$ is an abbreviation for $(a_1b, \cdots, a_mb)$, where b itself may be a finite sequence of variables. Thus we can represent a function of n variables, and also a finite sequence of functionals whose arguments are of the same type can be treated as a single functional. These abbreviations are sometimes ambiguous, but the meaning is clear in context. In the statements and proofs of many of the theorems we speak of a variable or constant as though it denoted a single functional, whereas a finite sequence of functionals is understood. This simplifies the description of the proofs. Full generality can be obtained by the use of suitable abbreviations.

3. Interpretation. An *interpretation* of a formal system Σ_i in a system Σ_j (in the spirit of Kreisel [**9**]) consists of a finitistically defined function which maps each formula $P(Q)$ of Σ_i to an infinite sequence of formulas $P_n(Q_n)$ $(n = 0, 1, \cdots)$ in Σ_j; and finitistic proofs that:

3.1. If $\vdash_i P$, then, for some n, $\vdash_j P_n$.

3.2. If $\vdash_i P \rightarrow Q$, then $\bigwedge n \bigvee m \bigvee T \vdash_j [P_n(T(b)) \rightarrow Q_m(b)]$, where b is the list of variables occurring free in Q_m but not in Q, $T(b)$ is a list of terms containing only the variables b free which are substituted for the free variables of P_n not occurring free in P.

3.3. If P contains no logical symbols, then P_n is finitistically equivalent to P, for all n.

It is only for the sake of the quantifier-free system Σ_4 $(j = 4)$ that the interpretation of P need consist of more than one formula. When the interpretation of P is always a single formula, the interpretation is called a *translation*.

4. Interpretation of Σ_1 in Σ_2. Let Σ_1 be classical analysis, i.e. the theory of functionals of finite type with axioms A – E. Let Σ_2 be intuitionistic analysis together with the additional axiom schema

F $$\bigwedge x \neg\neg P(x) \rightarrow \neg\neg \bigwedge x P(x) .$$

I.e. Σ_2 is obtained from Σ_1 by replacing the law of the excluded middle by axiom F.[4]

The schema F is chosen not because we believe it is of intuitionistic significance, but to provide a formal system in which classical analysis is easily interpreted, and whose logical basis is intuitionistic.

---

[4] As observed in [**10**, p. 108, 2.421], the Σ_2 translation of (1.5.1) is $\neg\neg(\bigwedge x)[P_1(x) \vee \neg P_1(x)]$(*), where P_1 is the Σ_2 translation of P. (*) follows from F applied to $P_1(x) \vee \neg P_1(x)$ as $P(x)$ since $\neg\neg(A \vee \neg A)$ holds. Conversely, since $(\bigwedge x)[P(x) \vee \neg P(x)] \rightarrow [(\bigwedge x)\neg\neg P(x) \rightarrow (\bigwedge x)P(x)]$, we get $(\bigwedge x)\neg\neg P(x) \rightarrow \neg\neg(\bigwedge x)P(x)$ from $\neg\neg(\bigwedge x)[P(x) \vee \neg P(x)]$, i.e. F follows from the schema (*). (Note that E does not imply intuitionistically its own Σ_2 translation.)—That (*) follows from F can be seen indirectly by [**3**] since (*) is the Σ_2-translation of (1.5.1) and F of E, and, as noted in 1.5 above, (1.5.1) follows from E in classical predicate logic [by taking $a = 0 \leftrightarrow P(x) \cdot \wedge \cdot a = 1 \leftrightarrow \neg P(x)$ for $P(x,a)$ in E]. See footnote 18 for a comparison of E and 1.5.1.—It should be noted that the observation above of [**10**, 2.421], holds also when (1.5.1) is formulated with *free* variables in $P(x)$, while the relative consistency proof in the second half of 2.421 only holds for (1.5.1) without free variables.

The presentation of Σ_1 and the proofs below can be simplified by not using the connectives $\vee$ and $\bigvee$, and by writing the induction schema as a single axiom with a function variable. However we do not take advantage of this fact because our present approach yields a stronger result concerning the interpretation of Σ_2 in Σ_4.

The interpretation of Σ_1 in Σ_2 is given by means of a translation which is intuitionistically equivalent to that of Gentzen [**2**, p. 532], which appears as the ° translation in [**6**, § 81], and also to that of Gödel [**3**], which appears as the ′ translation in [**6**], see proof of [**6**, Theorem 60 (b2)]). The translation is given by thinking of $p \dot{\vee} q$ as an abbreviation for $\neg\neg(p \vee q)$ and $\dot{\bigvee} aP(a)$ for $\neg\neg\bigvee aP(a)$, where for the moment we write the connectives $\vee$, $\bigvee$ of Σ_1 as $\dot{\vee}$, $\dot{\bigvee}$. Gentzen's abbreviations are $\neg(\neg p \wedge \neg q)$ and $\neg\bigvee a \neg P(a)$, which are intuitionistically equivalent to the abbreviations above by virtue of $\neg(p \vee q) \leftrightarrow (\neg p \wedge \neg q)$ and $\neg\bigvee aP(a) \leftrightarrow \bigwedge a \neg P(a)$.

The proof that this translation preserves provability (i.e. 3.1) can be given as in [**6**, § 81] except that the translation of the axiom of choice is proved using F, E, and transitivity of implication. Condition 3.2 reduces to

4.1. if $\vdash_i P \to Q$, then $\vdash_j P_0 \to Q_0$,

which is now a consequence of 3.1. Condition 3.3 also follows from 3.1.

5. Primitive recursive functionals. The primitive recursive functionals are the functionals obtained by means of the schemata PR1—PR6 below. This is the class of functionals which appears in Gödel's 1958 interpretation of intuitionistic arithmetic [**4**]. It is larger than Kleene's class of primitive recursive functionals because, in the recursion schema PR6, the value of the functional ϕ may be of any finite type, whereas Kleene's functionals have values of type zero only. By an oversight Kreisel's schemata [**10**, p. 110] lack a schema for abstraction (lambda operator), which is implicit in our notation. With the inclusion of abstraction, Kreisel's schemata define the same class of functionals as ours, taking into account the difference in the type structure.

The class of bar recursive functionals is obtained by adjoining to PR1—PR6 the schema for bar recursion given in the next section.

We recall that equality between terms of type other than zero is an abbreviation. In each schema, ϕ is the functional to be defined, and θ_1 and θ_2 are previously defined functionals. Formally a new constant is introduced for each particularization of each schema. It is assumed that the types of the functionals in the schemata are arbitrary provided that the equations are well-formed and the conventions of § 2 are not violated. Let $a = (a_1, \cdots, a_m)$, $b = (b_1, \cdots, b_n)$; and $1 \leq i \leq m$ (ba is defined in § 2).

PR1	$\phi a = a_i$	projection
PR2	$\phi cba = c(ba)$	substitution
PR3	$\phi = \theta_1\theta_2$	composition
PR4	$\phi = 0$	zero

PR5 $\phi x = x'$ successor

PR6 $\phi 0 = \theta_1$, $\phi(x') = \theta_2(\phi x, x)$ primitive recursion

Schemata PR1—PR5 can be replaced by the schema for explicit definition

ED $\phi a_1 \cdots a_m = T$,

where T is a term (see § 2.2) whose constants have been previously defined and whose variables are taken from the set $\{a_1, \cdots, a_m\}$. A functional satisfying

DBC $\phi a = [\theta_1 a \text{ if } \theta_3 a = 0;\ \theta_2 a \text{ if } \theta_3 a \neq 0]$

is obtained thus: $\phi_1 0 = \theta_1$, $\phi_1(x') = \theta_2$; $\phi a = \phi_1(\theta_3 a)$. For each type other than zero, there is a functional whose numerical value is 0, and we denote this functional by $\mathbf{O}$. I.e. $\mathbf{O}a_1 \cdots a_m = 0$.

6. Bar recursion and bar induction. Bar recursion is a principle of definition and bar induction a corresponding principle of proof. Bar induction is a generalization of Brouwer's bar theorem [1] to higher types. We discuss bar induction here primarily to point out the relationship between bar recursion and the bar theorem. Since our immediate objective is to obtain a quantifier-free interpretation of analysis, bar recursion rather than induction is appropriate.[5] In the appendix we indicate how bar induction can be used to obtain an interpretation in a system with quantifiers.

6.1. *Representation of finite sequences.* In order to state the two principles we require a method for representing arbitrary finite sequences of functionals of a given type. A convenient representation for the finite sequence $c_0, \cdots, c_{z-1}$ is the pair (z, C), where

(6.1.1) $$Cx = [c_x \text{ if } x < z;\ \mathbf{O} \text{ otherwise}] .$$

We write $\langle c_0, \cdots, c_{z-1}\rangle$ for the functional C defined in (6.1.1). Conversely the pair (z, C) represents the sequence $C0, \cdots, C(z-1)$.

6.2. *Bar recursion.* For ease in reading we omit showing G, H, Y as argu-

[5] The difference is not essential since bar recursion can be derived from bar induction: in fact, the first sentence of § 6.4 below, suitably interpreted, leads to a formal derivation. "$\phi(y, C)$ is defined" may be read *either* as (a) there is a unique ϕ^* with arguments x and C' with the following properties: for $x < y$, ϕ^* is the zero functional (of appropriate type) for all C'; for $x \geqq y$ and all C'_z satisfying $C_z = C'_z$ for $z < y$, ϕ^* satisfies the defining equations for ϕ in § 6.2, *or* as (b) for each y and C the computational process mentioned in the introduction above terminates. [There is an analogy between the justifications of bar recursion and the familiar justifications of definition by induction: (a) corresponds to the usual set theoretic approach which requires the existence of unique functions which satisfy primitive recursive definitions, (b) to the 'metamathematical' approach which shows that primitive recursive functions are *reckonable* from their recursion equations in the sense of [6, p. 295.] A derivation of bar recursion in sense (a) from bar induction is given in 12.2 for the case when C ranges over number theoretic functions with numerical values: bar induction is assumed for type zero only (bar theorem).

ments of ϕ.

$$\phi(x, C) = \begin{cases} G(x, \langle C0, \cdots, C(x-1)\rangle) \text{ if } Y(\langle C0, \cdots, C(x-1)\rangle) < x\,, \\ H[\lambda a\phi(x', \langle C0, \cdots, C(x-1), a\rangle, x, \langle C0, \cdots, C(x-1)\rangle] \text{ otherwise.} \end{cases}$$

Thus $\phi(x, C)$ is defined outright if $Y(\langle C0, \cdots, C(x-1)\rangle) < x$, and in terms of $\lambda a\phi(x', \langle C0, \cdots, C(x-1), a\rangle)$ otherwise.

For use as a formal schema, the rhs should be written in the formal notation for primitive recursive functionals.

6.3. *Bar induction.* The statement of this principle is in a form suggested by Kleene's statement of Brouwer's bar theorem [8]. In the bar theorem a is of type zero and C is a function. In bar induction a can be of any finite type.

Hyp1	$\bigwedge C \bigvee x P(x, C)$	(well-foundedness) .
Hyp2	$P(x, C) \vee \neg P(x, C)$	(decidability of P) .
Hyp3	$P(x, C) \rightarrow Q(x, C)$	(basis) .
Hyp4	$[\bigwedge a Q(x', \langle C0, \cdots, C(x-1), a\rangle)] \rightarrow Q(x, \langle C0, \cdots, C(x-1)\rangle)$ (induction) .	
Conc	$Q(0, C)$	

6.4. *Comparision of bar recursion and the bar theorem.* Each of the five lines of the bar theorem (§ 6.3) has an analogue in terms of bar recursion obtained by writing "$Y(\langle C0, \cdots, C(x-1)\rangle) < x$" for "$P(x, C)$", and "$\phi(x, C)$ is defined" for "$Q(x, C)$". That Y satisfies the well-foundedness condition can be proved intuitionistically as follows. In the computation of YC only a finite number of values of C occur. Hence for x sufficiently large, $Y(\langle C0, \cdots, C(x-1)\rangle) = YC$, and by taking x even larger,

$$Y(\langle C0, \cdots, C(x-1)\rangle) < x\,.$$

The decidability of Y is a known property of intuitionistic functionals.[6]

[6] More precisely, what is assumed here is not merely the computability of values of Y, but a computable bound for the modulus of continuity of Y at the argument C; or, equivalently, that there is a decidable property $P'_Y(x, C)$ which (i) ensures that $\langle C0, \cdots, C(x-1)\rangle$ is secured with respect to Y, and (ii) is well-founded. ["Secured with respect to Y" means that there are approximations (neighborhoods) $U_0, \cdots, U_{x-1}$ of $C0, \cdots, C(x-1)$ such that for all C', satisfying $C'y$ in U_y for $y < x$, $YC = YC'$; e.g. if C is of type (0(0, 0)) (sequence of number theoretic functions)

$$(\bigvee c_0 \cdots c_{x-1})\{(\bigwedge i)_{<x}(\bigwedge y)(y < c_i \rightarrow Ciy = C'iy)\cdot\rightarrow YC = YC'\}.]$$

Evidently, in general, $Y(\langle C0, \cdots, C(x-1)\rangle) < x$ does not ensure $Y(\langle C0, \cdots, Cx\rangle) < x+1$ and, *a fortiori*, not (i). Thus, while for bar recursion, as formulated in 6.2, the well-foundedness of the relation $Y(\langle C0, \cdots, C(x-1)\rangle) < x$, say $P_Y(x, C)$, is sufficient, the proof of this fact seems to require a detour via P'_Y. The essential difference between P_Y and P'_Y is that P_Y is clearly computable in Y, x and C, while P'_Y is not, unless Y is interpreted to range over *non-extensional* functionals (cf. [**11**, para. 12, Remark 10]). Thus, the postulate of the existence of a decidable P'_Y, the so-called 'continuity axiom' for Y, forces the variables to range over non-extensional functionals, e.g. 'representing

The essential difference between the bar theorem and bar recursion is the restriction of the type of a in the theorem to type zero. On the other hand, the recursion principle is more restrictive in the sense that it does not involve species (P and) Q, but only computable functionals.

7. Gödel translation of Σ_2 into Σ_3. Let Σ_3 contain as formulas the $\bigvee\bigwedge$-formulas of Σ_2. We use Σ_3 only as an intermediate point in describing the interpretation of Σ_2 in Σ_4, and we will not concern ourselves with proof-theoretic properties of Σ_3.[7]

Gödel's translation [4, p. 285] will be thought of as a step by step process in which the connective with maximum scope is considered last. Each connective other than implication generates one step. For our purposes we find it convenient to describe the translation of implication in two steps.

A formula without quantifiers is its own translation. Let p, q, r be formulas of Σ_2, and let $\bigvee a \bigwedge b P(a, b)$, $\bigvee c \bigwedge d Q(c, d)$, $\bigvee e \bigwedge f R(e, f)$ be their translations, where each of $a - f$ is finite list of variables, possibly empty, and P, Q, R are quantifier-free formulas. Then the translation of

$$p \wedge q \text{ is } \bigvee ac \bigwedge bd[P(a, b) \wedge Q(c, d)] ,$$

$$p \vee q \text{ is } \bigvee xac \bigwedge bd[x = 0 \wedge P(a, b) \cdot \vee \cdot x = 1 \wedge Q(c, d)] ,$$

$$(\bigvee g)p \text{ is } \bigvee ga \bigwedge b P(a, b) ,$$

$$(\bigwedge g)p \text{ is } \bigvee A \bigwedge gb P(Ag, b) ,$$

The translation of $p \rightarrow q$ is given in two steps beginning with

7.1. $$\bigvee a \bigwedge b P(a, b) \rightarrow \bigvee c \bigwedge d Q(c, d) ,$$

7.2. $$\bigwedge a \bigvee c \bigwedge d \bigvee b [P(a, b) \rightarrow Q(c, d)] ,$$

7.3. $$\bigvee CB \bigwedge ad[P(a, Bad) \rightarrow Q(Ca, d)] .$$

functions' of [**10**, p. 115], with obvious modifications, or algorithms for computing the functionals, as in [**1**, p. 63, line 17]. On the other hand bar recursion is satisfiable both by (classes of) extensional and (of) non-extensional functionals. Hence, e.g., there is the possibility of (i) a computational process for bar recursion functionals which uses only the values of auxiliary functions at a finite number of arguments and not their representing functions (which embody the continuity properties of the auxiliary functions), and (ii) independence results which exploit the fact that the bar recursion schema is satisfied by classes of functionals which do not satisfy bar induction (with the continuity axiom), such as the partial recursive functionals defined by Kleene's schemata S1–S9 on all countable arguments in the sense of [**7**, pp. 91–95]. However, from the point of view of constructive foundations, it is likely that the bar induction schema and the non-extensional interpretation of the higher type variables are more fundamental.

[7] §§ 7–9 verify in detail the properties of the ($\Sigma_2 \rightarrow \Sigma_3$) translation stated in [**4**] and illustrated in [**10**, para. 3.3]. The verification is based on the formalization of intuitionistic logic in 1.1 above instead of following Gödel's own suggestion in [**4**, p. 286] for a formalization particularly suited to the present purpose. Specifically, some of the rules of 1.1 are redundant; e.g. A2. [For, by A5, $q \wedge p \rightarrow q$; hence, by A8, $q \rightarrow (p \rightarrow q)$; hence by A3, if q then $p \rightarrow q$]. Also, A7 may be replaced by Perm [**4**, p. 286]. A more important difference is that Gödel uses only intensional equality with the corresponding inference rule [**4**, p. 283, footnote 3], while below G6 is formulated in § 8.2 (G6 does not seem to be used: it would not be intuitionistically acceptable.)

As a special case we consider $\neg p$, i.e. $p \to 0' = 0$.

The two steps are:

7.4. $$\bigwedge a \bigvee b[P(a, b) \to 0' = 0] ,$$

7.5. $$\bigvee B \bigwedge a \neg P(a, Ba) .$$

For an example we compare the translations of

7.6. $$p \to (q \to r) ,$$

7.7. $$(p \wedge q) \to r .$$

The translation of 7.6 is obtained thus:

7.8. $$\bigvee a \bigwedge b P(a, b) \to \bigwedge c \bigvee e \bigwedge f \bigvee d[Q(c, d) \to R(e, f)] ,$$

7.9. $$\bigvee a \bigwedge b P(a, b) \to \bigvee ED \bigwedge cf [Q(c, Dcf) \to R(Ec, f)] ,$$

7.10. $$\bigwedge a \bigvee ED \bigwedge cf \bigvee b[P(a, b) \to \cdot\, Q(c, Dcf) \to R(Ec, f)] ,$$

7.11. $$\bigvee EDB \bigwedge acf (P(a, Bacf) \to \cdot\, Q(c, Dacf) \to R(Eac, f)] .$$

The translation of 7.7 is obtained with less pain.

The final translation is:

7.12. $$\bigvee EBD \bigwedge acf [P(a, Bacf) \wedge Q(c, Dacf) \cdot \to R(Eac, f)] .$$

8. Σ_4. The system Σ_4 contains no logical connectives. Its terms are formed as in § 2.2. All formulas are of the form $u = v$, where u and v are terms of type zero. The constants of Σ_4 denote bar functionals. The role of the propositional connectives is taken over by a set of constants called *propositional functions.* The axioms and rules are given below. There are deliberate redundancies.

8.1. *Propositional functions.* The following equations are taken as axioms.[8]

(8.1.1) $\operatorname{imp}(0, y) = y$, $\operatorname{imp}(x', y) = 0$, $\operatorname{imp}(x, 0) = 0$.

(8.1.2) $\operatorname{con}(0, y) = y$, $\operatorname{con}(x, 0) = x$, $\operatorname{con}(x', y') = 0'$.

(8.1.3) $\operatorname{dis}(0, y) = 0$, $\operatorname{dis}(x, 0) = 0$, $\operatorname{dis}(x', y') = 0'$.

(8.1.4) $\operatorname{eq}(0, y) = y$, $\operatorname{eq}(x, 0) = x$, $\operatorname{eq}(x', y') = \operatorname{e}(x, y)$.

Abbreviations: $u_1 = u_2 \to v_1 = v_2$ is an abbreviation for

$$\operatorname{imp}(\operatorname{eq}(u_1, u_2), \operatorname{eq}(v_1, v_2)) = 0 ,$$

and similarly for $\vee$, $\wedge$.

8.2. *Axioms and rules of inference.*

G1. The defining equations for the bar recursive functionals (including the propositional functions) are axioms.

G2. Substitution of a term u for a variable a of the same type: if $P(a)$ then $P(u)$.

G3. Equality axioms, C.

[8] The elimination of propositional connectives (applied to decidable relations) by number theoretic functions is well-known. It is mentioned (in the present context) in [4, p. 284, line 6].

G4. Induction, D.

G5. If $\mathrm{eq}(u, v) = 0$, then $u = v$, where u and v are terms of type zero.

G6. Weak extensionality: if $P \to ue = ve$, then $P \to T(u) = T(v)$, where e is a list of distinct variables which do not occur in u, v, P such that ue and ve are of type zero. The usual form of extensionality cannot be expressed in this system.

8.3. *Derivable theorems and rules.*

(8.3.1) $\mathrm{imp}(x, x) = 0$ (by induction on x).

(8.3.2) $\mathrm{eq}(x, x) = 0$ (induction). Hence if $u = v$, then $\mathrm{eq}(u, v) = 0$.

(8.3.3) Proof by cases: if $P(0)$ and $P(x')$ then $P(x)$.

(8.3.4) Modus ponens.

(8.3.5) Every substitution instance of a propositional tautology is provable (use proof by cases).

9. Principal theorem: interpretation of Σ_2 in Σ_4. Let p be a formula in Σ_2, and let $\bigvee a \bigwedge b P'(a, b)$ be its Gödel translation in Σ_3. Then the interpretation of p in Σ_4 is given by the infinite sequence of formulas

$$P(\phi_n, b) \qquad (n = 0, 1, \cdots),$$

where $\phi_0, \phi_1, \cdots$ is an enumeration of the constants in Σ_4 of appropriate type, and P is obtained from P' by expressing $=, \to, \wedge, \vee$ in terms of eq, imp, con, dis.

It suffices to prove the following two statements.

9.1. If $\vdash_2 p$, then for some n, $\vdash_4 P(\phi_n, b)$.

9.2. If $\vdash_2 [p \to q]$, then for every m, there are n and k such that

$$\vdash_4 [P(\alpha_m, \beta_k b) \to Q(\gamma_n, b)] .$$

The α, β, γ enumerate the constants of appropriate types. Condition 3.3 is satisfied because, if p contains no quantifiers, then the translation of p is obtained by expressing the propositional connectives of p in terms of the propositional functions and is therefore finitistically equivalent to p.

We can dispose of 9.2 immediately by showing that it is a consequence of 9.1. Assume $\vdash_2 [p \to q]$. We treat the case that p and q have no free variables. The contrary case poses no difficulties. From 9.1 it follows that there are constants β, γ such that

9.3. $$\vdash_4 [P(a, \beta a d) \to Q(\gamma a, d)] .$$

We obtain 9.2 by setting $\beta_k = \beta\alpha_m$, $\gamma_n = \gamma\alpha_m$.

The proof of 9.1 involves treating each of the axioms and rules of Σ_2 and will be given in the next section. The essential difference between our proof of 9.1 and Gödel [4] is the addition of axiom F.

Proof of 9.1. We shall consider each of the axioms and rules of Σ_2 in turn, and show that in each case the desired constants can be found. Axiom F is considered in § 10.

In order to avoid unnecessarily complicated formulas we treat only the case that no free variables occur other than those explicitly shown. The only difficulty which arises when free variables are present occurs when one of the hypotheses of a rule of inference contains free variables not occurring in the conclusion and not explicitly shown. This difficulty is overcome by substituting the zero functional of appropriate type for such variables, making use of the substitution rule in Σ_4.

A1. Trivial.

A2. Let γ_1 be given such that $\vdash_4 Q(\gamma_1, d)$. We wish to find β, γ such that

$$\vdash_4 P(a, \beta ad) \to Q(\gamma a, d) .$$

This is accomplished by setting $\gamma a = \gamma_1$. When free variables are present γ and γ_1 have these free variables for arguments. The purpose of the remark above is to guarantee that each such argument of γ_1 is an argument of γ, for otherwise γ would not be well-defined. Henceforth we omit "$\vdash_4$".

A3. We are given

A3.1 $$P(\alpha, b) ,$$

A3.2 $$P(a, \beta ad) \to Q(\gamma a, d) ,$$

and we wish to find γ_1 such that $Q(\gamma_1, d)$. Substitute α for a in A3.2 and βad for b in A3.1, and apply modus ponens in Σ_4. The result is $Q(\gamma\alpha, d)$. Hence we set $\gamma_1 = \gamma\alpha$.

A4. We are given

A4.1 $$P(a, \beta ad) \to Q(\gamma a, d) ,$$

A4.2 $$Q(c, \delta cf) \to R(\varepsilon c, f) .$$

We wish to find β_1 and ε_1 such that

A4.3 $$P(a, \beta_1 af) \to R(\varepsilon_1 a, f) .$$

Substitute γa for c in A4.2, $\delta(\gamma a, f)$ for d in A4.1, and apply transitivity of implication in Σ_4 to the resulting two formulas. The result is

A4.4 $$P[a, \beta(a, \delta(\gamma a, f))] \to R(\varepsilon(\gamma a), f)] .$$

Hence we set

A4.5 $$\beta_1 af = \beta(a, \delta(\gamma a, f)) , \qquad \varepsilon_1 a = \varepsilon(\gamma a) .$$

A5. As a typical case consider $p \to (p \vee q)$. The next to the last step in the Gödel translation is

A5.1 $$\bigwedge a \bigvee xec \bigwedge fd \bigvee b[P(a, b) \to [x = 0 \wedge P(e, f) \cdot \vee \cdot x = 1 \wedge Q(c, d)]] .$$

The desired functionals are obtained by what amounts to setting $x = 0$, $e = a$, $c = \mathbf{0}$, $b = f$.

A6. We are given

A6.1 $$P(a, \beta_1 af) \to R(\varepsilon_1 a, f) ,$$

A6.2 $$Q(c, \delta_1 cf) \to R(\varepsilon_2 c, f) .$$

We wish to obtain β, δ, ε such that

A6.3 $$[x = 0 \wedge P(a, \beta xacf) \cdot \vee \cdot x = 1 \wedge Q(c, \delta xacf)] \to R(\varepsilon xac, f) .$$

This is accomplished by setting

A6.4 $$\beta 0acf = \beta_1 af, \quad \delta(x')acf = \delta_1 cf, \quad \varepsilon 0ad = \varepsilon_1 a , \quad \varepsilon(x')ac = \varepsilon_2 c .$$

A7. Similar to A6.

A8, A9. See the example at the end of § 7.

A10. Trivial.

B1. If $p \rightarrow q(e)$, then $p \rightarrow \bigwedge e q(e)$. We are given

B1.1 $$P(a, \beta ead) \rightarrow Q(e, \gamma ea, d) ,$$

and the formula we wish to derive is the same as B1.1 except that e occurs as a right-most instead of a left-most argument.

B2. Is similar to B1.

B3. $\bigwedge e p(e) \rightarrow p(T)$, where T is a term free for e in $p(e)$. The next to the last step in the Gödel translation of this formula is

B3.1 $$\bigwedge A \bigvee c \bigwedge d \bigvee eb[P(e, Ae, b) \rightarrow P(T, c, d)] .$$

The solution is obtained by what amounts to setting $e = T$, $c = AT$, $b = d$. The restriction on T guarantees that no variable in T occurs in c or in d. Furthermore, when free variables are present, each variable which occurs (free) in T occurs free in B3. Since e and c are permitted to depend on the free variables of B3, they may depend on the free variables of T, and therefore the solution is well-defined.

B4 is similar to B3. The axioms of group C are their own interpretations and are axioms in Σ_4.

D. If $p(0)$ and $p(x) \rightarrow p(x')$, then $p(x)$. We are given

D.1 $$P(0, \alpha, b) ,$$

D.2 $$P(x, a, \beta xad) \rightarrow P(x', \gamma xa, d) .$$

We wish to obtain α_1 such that

D.3 $$P(x, \alpha_1 x, b) .$$

Hence we set

D.4 $$\alpha_1 0 = \alpha , \quad \alpha_1(x') = \gamma(x, \alpha_1 x) .$$

Then

D.5 $$P(0, \alpha_1 0, b) , \quad P(x, \alpha_1 x, \beta(x, \alpha_1 x, b)) \rightarrow P(x', \alpha_1(x'), b) .$$

The problem reduces to the following lemma.

Lemma. *In Σ_4, if $Q(0, b)$ and $Q(x, \beta xb) \rightarrow Q(x', b)$, then $Q(x, b)$.*

Proof. Assume the premises are derivable and define

D.6 $$pd(0) = 0 , \quad pd(x') = x ; \quad x \dot{-} 0 = x , \quad x \dot{-} y' = pd(x \dot{-} y) ;$$

D.7 $\delta 0xb = b\,, \qquad \delta(y', x, b) = \beta(x \dot{-} y', \delta yxb)\,;$

D.8[9] $x \leqq y \leftrightarrow x \dot{-} y = 0\,; \qquad 0 < x \leftrightarrow (pdx)' = x\,;$

D.9 $R(x, y, b) \leftrightarrow [y \leqq x \rightarrow Q(y, \delta(x \dot{-} y, x, b)]\,.$

Then

D.10 $0 < y \leqq x \rightarrow \delta(x \dot{-} y, x, b) = \beta(x, \delta(x \dot{-} y', x, b))\,,$

and by induction on y, $R(x, y, b)$. Finally,

D.11 $R(x, x, b) \leftrightarrow Q(x, b)\,.$

10. The interpretation of F is provable in Σ_4. This is the only point where we make use of bar recursion, and we use it in the following restricted form, where the parameter G_0 is not exhibited as an argument of ϕ for greater readability.

$$\text{BR} \qquad \phi z C x = \begin{cases} Cx & \text{if } x < z\,, \\ \mathbf{0} & \text{if } x \geqq z \wedge Y(\langle C0, \cdots, C(z-1)\rangle) < z\,, \\ \phi(z', \langle C0, \cdots, C(z-1), a_0 \rangle, x) & \text{otherwise}\,, \end{cases}$$

where

$$a_0 = G_0(z, \lambda a \phi(z', \langle C0, \cdots, C(z-1), a\rangle)\,,$$

and by convention, $\phi(z, C) = \lambda x \phi(z, C, x)$.

In this form, bar recursion is used to solve for an infinite system of unknowns $C = (C0, C1, \cdots)$. A condition R and a functional Y are given, so to speak, and we wish to find C such that $R(YC, C)$. The condition R is

---

[9] The lemma is correct, but the proof has some minor defects. Thus D 10 is misstated, presumably for

D′10 $y < x \rightarrow \delta(x \dot{-} y, x, b) = \beta(y, \delta(x \dot{-} y', x, b))\,,$

the second half of D 8 is a theorem of Σ_4 (with D 6) and not a definition, and it is tacitly assumed that certain properties of the *predecessor*, *cut-off subtraction* and *order* are formally provable in Σ_4 from D 6 and D 8 for pd, $\dot{-}$, $\leqq$. It is simpler to replace D 8 by D′8: $y < x \leftrightarrow x \dot{-} y \neq 0$ (cf. [5, vol. 1, p. 309, *l*.7]), write $y < x \vee y = x$ as $y \leqq x$, and derive the following formulae (line references being to p. 305 of [5, vol. 1], but we do not use recursion equations for addition). D′9 $x \neq 0 \rightarrow (pdx)' = x$ (*l*.2). D′10. $pdx \neq 0 \rightarrow x \neq 0$. D′11. $x \neq 0 \rightarrow x \dot{-} pdx = 0'$ (*l*.8). D′12. $x \dot{-} x = 0$. D′13. $y' \leqq x \rightarrow y < x$ ($y' = x \rightarrow y = pdx$, apply D′11; $y' < x \rightarrow y < x$ (*l*.19)). D′14. $y < x \rightarrow x \dot{-} y = (x \dot{-} y')'$ (since $y < x \rightarrow x \dot{-} y \neq 0$, $x \dot{-} y = (pd(x \dot{-} y))'$ by D′9; $pd(x \dot{-} y) = x \dot{-} y'$ by D6). D′15. $y \leqq x \rightarrow x \dot{-} (x \dot{-} y) = y$ (for $y = 0$ and $y = x$ by D′12; so it is sufficient to show $z' < x \rightarrow x \dot{-} (x \dot{-} z') = z'$: for $z = 0$, since $0' < x \rightarrow x \neq 0$ and $x \dot{-} 0' = pdx$, by D′11; if $z' < x$, $pd(x \dot{-} (x \dot{-} z'')) = x \dot{-} (x \dot{-} z'')' = x \dot{-} (x \dot{-} z')$ by D′14, $= z'$ by induction hypothesis; so, by D′10, $x \dot{-} (x \dot{-} z'') \neq 0$, and, by D′9, $x \dot{-} (x \dot{-} z'') = z''$). D′10 follows from D′15 by replacing y in D7 by $x \dot{-} y'$ when y' is replaced by $x \dot{-} y$ by D′14. $R(x,y,b) \rightarrow R(x,y',b)$: by D′13 it is sufficient to show $y < x \rightarrow Q(y, \delta(x \dot{-} y, x,b)) \rightarrow Q(y', \delta(x \dot{-} y', x,b))$, i.e., by D′15, $y < x \rightarrow \cdot Q(y, \beta(y, \delta(x \dot{-} y', x,b)) \rightarrow Q(y', \delta(x \dot{-} y'), x,b)$; but this follows from the second premise of the lemma. $R(x, x, b) \rightarrow Q(x, b)$ since $x \leqq x$ by D′8, $x \dot{-} x = 0$ by D′12, and $\delta(0,x,b) = b$.

informal.[10]

If the range of Y were finite, say $YC \leqq n$, then we could proceed by setting $Cx = \mathbf{0}$ for $x > n$, solve for Cn in terms of $C0, \cdots, C(n-1)$ so as to satisfy $R(n, C)$, solve for $C(n-1)$ in terms of $C0, \cdots, C(n-2)$ to satisfy $R(n-1, C)$, etc. until we obtain C such that $R(x, C)$ holds for all $x \leqq n$. If the range of Y is not finite, the process must be modified, and we no longer obtain $R(x, C)$ for all x in the range of Y.

In the general case the solution is given by means of the functional ϕ defined in BR. The reason for introducing bar recursion here is to reduce the problem of solving for infinitely many unknowns to solving for a single unknown. The functional G_0 is intended to give the solution to the one-dimensional problem. The functional ϕ at the argument $(z, \langle C0, \cdots, C(z-1)\rangle)$ expresses C in terms of the given $C0, \cdots, C(z-1)$. The basis step occurs when $Y(\langle C0, \cdots, C(z-1)\rangle) < z$, and we set $Cx = \mathbf{0}$ for $x \geqq z$. On the other hand, if $Y(\langle C0, \cdots, C(z-1)\rangle) \geqq z$, then we assume $\phi(z', \langle C0, \cdots, C(z-1), a\rangle)$ expresses C in terms of the given $C0, \cdots, C(z-1)$ and variable $a (= Cz)$ and we wish to find a value for a such that $C = \phi(z', \langle C0, \cdots, C(z-1), a\rangle)$ satisfies $R(z, C)$. It is presumed that G_0 supplies the desired value for a.

In the sequence of lemmas below we show that the final value of $C = \phi(0, \mathbf{O})$ is a solution to $R(YC, C)$, assuming that G_0 gives the solution to the one-dimensional problem. The proofs are informal, but are written in such a way that they can be converted to formal proofs in Σ_4 without excessive effort. Elementary properties of primitive recursive functionals such as obtained in [**6**] and in [**5**, vol. 1, §7], will be taken for granted.

---

[10] The (heuristic) observations in the following two paragraphs do not indicate under what hypotheses on R the present method provides a C satisfying $R(YC, C)$. It may possibly be helpful to start with the problem in hand, namely 10.4 below. Consider first the special case when Y in 10.4 has the constant value m_0; it is then sufficient to find, for given A and D, functionals B and C such that $P[m_0, A0B, B(A0B)] \to P(m_0, C_{m_0}, DC)$; to do this, one solves the more general problem

$$(\bigvee B_*)(\bigwedge a)(\bigvee C_*)\,[P(m_0, a, B_*a) \to P(m_0, C_*m_0, DC_*)]\ ,$$

where B_* may depend on m_0, but not on a (and, in fact, does not even depend on P). Put $B_*b = D(\langle m_0 \mid b\rangle)$ for all b where $\langle m \mid b\rangle$ denotes the functional with natural numbers as arguments, $= b$ at m, and otherwise $= \mathbf{0}$. Then $C_* = \langle m_0 \mid a\rangle$ solves this problem. If $a = A0B$ we have a solution of the original problem. Note that C_* does not depend on P.

For the general case of 10.4, consider $P(m, a, Ba) \to P(m, Cm, DC)$ and suppose E_m is the 'generalization' of $\langle m \mid b\rangle$, i.e., E_m expresses C (to be found) in terms of Cm: $C = E_m(Cm)$ (i). Then, as in the special case, we put $B_m a = D(E_m a)$ for all a and m. We need a value n such that $YC = n$ (ii), and $Cn = AnB_n$, $= A(n, \lambda a D(E_n a))$ by definition of B_n. If further $Cn = G(n, E_n)$ (iii), then $Cn = AnB_n$ is obviously satisfied provided $G(m, E') = A(m, \lambda a\, D(E'a))$ for all m and E', as in 10.9 below. Now, Lemma 1 below shows that (i) is satisfied by C_0 when $E_m a = \phi(m', \langle C_0 O, \cdots, C_0(m-1), a\rangle)$ and ϕ is defined by BR, and Lemma 3 shows that (ii) and (iii) are satisfied by $n = YC_0$, all this for *arbitrary* G_0 in BR. So we are free to define G_0 by 10.9 as required. It seems worth while to search for a more systematic treatment of the general case where the tentative introduction of E_m and G is motivated by general theorems.

LEMMA 1. *Let* $C_0 = \phi(0, \mathbf{O})$, *where* ϕ *is defined as in* BR. *Then*

$$C_0 = \phi(x, \langle C_0 0, \cdots, C_0(z-1)\rangle)$$

for all z.

PROOF. By induction on z. When $z = 0$ the conclusion is the same as the hypothesis. Assume, as hypothesis of the induction, that

$$C_0 = \phi(z, \langle C_0 0, \cdots, C_0(z-1)\rangle) .$$

Case 1. $Y(\langle C_0 0, \cdots, C_0(z-1)\rangle) < z$. Then

$$C_0 = \langle C_0 0, \cdots, C_0(z-1)\rangle = \langle C_0 0, \cdots, C_0 z\rangle .$$

Therefore, $Y(\langle C_0 0, \cdots, C_0 z\rangle) < z < z'$. Hence

$$\phi(z', \langle C_0 0, \cdots, C_0 z\rangle) = \langle C_0 0, \cdots, C_0 z\rangle = C_0 .$$

Case 2: otherwise. Then $C_0 = \phi(z', \langle C_0 0, \cdots, C_0(z-1), a_0\rangle)$. Hence, $C_0 z = a_0$, and therefore $C_0 = \phi(z', \langle C0, \cdots, C_0 z\rangle)$.

LEMMA 2. *If* C_0 *is defined as in Lemma* 1 *and* $n = YC_0$, *then*

$$Y(\langle C_0 0, \cdots, C_0(n-1)\rangle) \geqq n .$$

PROOF. Suppose the conclusion is false. Then, by the previous lemma, we would have $C_0 = \langle C_0 0, \cdots, C_0(n-1)\rangle$, from which we obtain a contradiction: $n = YC_0 = Y(\langle C_0 0, \cdots, C_0(n-1)\rangle) < n$.

LEMMA 3. *With* C_0 *and* n *as above*, $C_0 n = G_0(n, \lambda a \phi(n', \langle C_0 0, \cdots, C_0(n-1), a\rangle))$.

PROOF. By the previous lemmas, $C_0 = \phi(n', \langle C_0 0, \cdots, C_0(n-1), a_0\rangle)$. The conclusion follows from BR.

LEMMA 4. *The interpretation of* $\bigwedge x \neg\neg p(x) \cdot \rightarrow \cdot \neg\neg \bigwedge y p(y)$ *is provable in* Σ_4. *I.e. one of the formulas in the infinite sequence is provable.*

PROOF. Beginning with

10.1. $$\bigwedge x \neg\neg \bigvee a \bigwedge b P(x, a, b) \rightarrow \neg\neg \bigwedge y \bigvee c \bigwedge d P(y, c, d) ,$$

the Gödel translation is obtained as follows, with the last step omitted:

10.2. $$\bigwedge x B \bigvee a P(x, a, ba) \rightarrow \neg\neg \bigvee C \bigwedge y d P(y, Cy, d) ,$$

10.3. $$\bigvee A \bigwedge x B P[x, AxB, B(AxB)] \rightarrow \bigwedge Y D \bigvee C P[YC, C(YC), DC] ,$$

10.4. $$\bigwedge A Y D \bigvee x B C\{P[x, AxB, B(AxB)] \rightarrow P[YC, C(YC), DC]\} .$$

We wish to obtain x, B, C as functionals of A, Y, D.[11] The C and Y play the same role as in the previous lemmas. The condition R mentioned above is defined by

10.5. $$R(n, C) \leftrightarrow \bigvee x B\{P[x, AxB, B(AxB)] \rightarrow P(n, Cn, DC)\} .$$

[11] Note for reference: if the type of a in 10.1 above is s, then the type of C is $(0, s)$ (sequence of objects, each of type s), and the type of Y is $((0, s), 0)$ (numerical valued operation defined on sequences of objects of type s). These types are independent of b.

Let A, Y, D_0 be fixed, and let ϕ be defined as in BR with G_0 yet to be determined. We obtain the desired reduction by expressing the problem in terms of G_0 in place of C_0. Let

10.6. $$E_m = \lambda a \phi(m', \langle C_0 0, \cdots, C_0(m-1), a\rangle) .$$

On the basis of Lemmas 3 and 1, we have

10.7. $$C_0 n = G_0(n, E_n) , \quad C_0 = E_n[G_0(n, E_n)] , \quad (n = YC_0) .$$

The obvious choice for x is $n = YC_0$. It suffices to find B_m $(m = 0, 1, \cdots)$ and G_0 such that, for each m,

10.8. $$P[m, AmB_m, B_m(AmB_m)] \to P[m, G_0 m E_m, D(E_m(G_0 m E_m))] .$$

For if we set $m = YC_0$ and apply 10.7 to 10.8 we obtain the quantifier-free part of 10.4.

Using bar recursion we have reduced the problem of satisfying 10.4 to that of finding G_0 and B_m $(m = 0, 1, \cdots)$ such that 10.8 holds for the given A, D, and all m, where E is defined, via 10.6 and BR, in terms of G_0. A solution to 10.8 is given by

10.9. $$G_0 m E' = A(m, \lambda a D(E'a)) \qquad \text{(all } m, E') ,$$

10.10. $$B_m a = D(E_m a) \qquad \text{(all } m, a) ,$$

where E in 10.10 is defined in terms of the G_0 of 10.9 as described above. This yields

10.11. $$x = YC_0 , \qquad AxB = C_0(YC_0) , \qquad B(AxB) = DC_0 ,$$

and thus the solution to 10.4 is given independently of P.

11. Principal result. Let $\bigwedge a \bigvee b P(a, b)$ be a provable closed $\bigwedge\bigvee$-formula in analysis (Σ_1 or Σ_2). Then there is a constant (bar functional) ϕ in Σ_4 such that $\vdash_4 P(a, \phi a)$.

This result is obtained by applying the interpretation of Σ_1 in Σ_2 and of Σ_2 in Σ_4.

COROLLARY. *A recursive ordering which can be proved to be a well-ordering in classical analysis is a provable well-ordering in* Σ_4. (*The well-foundedness property of an ordering* $<$ *is proved in* Σ_4 *by exhibiting a functional* ϕ *such that* $\vdash_4 \neg a((\phi a)') < a(\phi a)$, *where* ϕa *is of type zero.*)

12. Appendix. *Interpretations by means of bar induction.* In this section we shall indicate briefly and informally how bar induction can be used in place of bar recursion to obtain an interpretation of Σ_2.[12] In particular we

[12] More precisely, the derivation of bar recursion from bar induction, indicated in § 6.4 and footnote 5, is elaborated in the appendix (§ 12.2). But the principal interest here lies in the particular results, namely in proving, for given $P(x)$ in the form $(\bigvee a)(\bigwedge b)P(x, a, b)$, the Σ_4 translation of axiom F applied to $P(x)$ by means of bar induction at as low a type as possible. Bar induction of type zero was explicitly asserted by Brouwer.

shall show how a proof of the interpretation of axiom F for prenex[13] formulas in arithmetic can be given using bar induction at type zero (Brouwer's bar theorem). The interpretation is given by the Gödel translation of Σ_2 into Σ_3. For our present purposes, the axioms of Σ_3 are those of intuitionistic analysis together with bar induction and continuity axioms (partially described below).

We state without proof that the translation of each axiom and rule of Σ_2 excluding axiom F is respectively a theorem or derived rule of Σ_3. The proof is similar to that given above with Σ_4 in place of Σ_3. The only problem which remains is to show that the translation of F is provable in Σ_3. However before we consider this problem we shall consider a simpler problem proposed by Kreisel in [**10**, § 5.22].

12.1. Kreisel asks if there is an intuitionistic proof of the Gödel translation of

(12.1.1) $$\neg\neg \bigwedge x \bigvee y \bigwedge z[R(x, y) \vee \neg R(x, z)] ,$$ [14]

where R is decidable, i.e. $\bigwedge xy[R(x, y) \vee \neg R(x, y)]$. In fact it was this problem that, at Kreisel's suggestion, the author solved before considering the problem of this paper. We answer Kreisel's question affirmatively by proving (12.1.1) in Σ_3 with bar induction only at type zero. This restricted Σ_3 is equivalent to Kreisel's unpublished formalization of the intuitionistic theory of functionals of finite type.

Because of the presence of the axiom of choice, it suffices to prove the next to the last step in the translation of (12.1.1), which is

(12.1.2) $$\bigwedge XZ \bigvee b\{R[Xb, b(Xb)] \vee \neg R[Xb, Zb]\} ,$$

where b is of type ((0)(0)), i.e. is a function from natural numbers to natural numbers. Kreisel correctly conjectured that the ε-substitution method of [**5**, vol. 2], could be used to obtain an intuitionistic proof of (12.1.2).

The substitution method can be described as follows. Let X, Z, and R be

---

[13] In § 12.3, we reduce the case of arbitrary arithmetic formulae to that of prenex ones by use of elementary intuitionistic predicate logic (although, of course, not every formula has an *equivalent* prenex form).

[14] This is equivalent to the comprehension axiom (1.5.1) applied to the formula $(\bigvee y)R(x, y)$ for $P(x)$. It is clear, that if the Σ_4-translation of (12.1.1) is written in the form

(*) $$(\bigvee B)(\bigwedge XZ)\{R[X\bar{B}, \bar{B}(X\bar{B})] \vee \neg R[X\bar{B}, Z\bar{B}]\} ,$$

where $\bar{B}$ denotes $B(X, Z)$ and the type of $\bar{B}$ is ((0), (0)) (this is the $\bigvee\bigwedge$ form of (12.1.2)) there is no primitive recursive functional B which satisfies (*) if $(\bigwedge x)(\bigvee y)(\bigwedge z)[R(x, y) \vee \neg R(x, z)]$ is not recursively satisfiable. For, every primitive recursive functional is general recursive; apply [**10**, 7.2 and 4.32]. In contrast, (*) is satisfied for countable X and Z by a functional B defined by means of Kleene's schemata S1-S9 restricted to countable arguments as in [**7**, pp. 91–95]. This was shown (classically) in [**10**] and follows also from the fact that bar recursion functionals are definable by Kleene's schemata restricted to countable arguments. Note finally that for each pair of primitive recursive functionals (= ordinal recursive functionals of order $< \varepsilon_0$ in the sense of [**9**]) X_0, Z_0, there is (intuitionistically) a primitive recursive b_0 satisfying 12. 1.2; this is shown in Theorem III on p.50 of [**9**], i.e., condition (γ) for the no-counterexample-interpretation.

given. Let b^0 be any function (of type ((0)(0))). We shall define a process for obtaining functions $b^1, b^2, \cdots$ so that if

$$R[Xb^n, b^n(Xb^n)] \vee \neg R[Xb^n, Zb^n] \tag{12.1.3}$$

the process terminates, and if b^n does not satisfy (12.1.3), then b^{n+1} is obtained as follows. Let b^{n+1} coincide with b^n except at the argument Xb^n and set

$$b^{n+1}(Xb^n) = Zb^n . \tag{12.1.4}$$

We observe that if $Xb^m = Xb^n$ for $m < n$ the process terminates. For assume n is chosen as small as possible. Then $Xb^m = Xb^n$, $b^n(Xb^n) = Zb^m$, and therefore

$$R[Xb^n, b^n(Xb^n)] \leftrightarrow R[Xb^m, Zb^m] . \tag{12.1.5}$$

The right side is true since the construction did not terminate at stage m, etc. Thus, for each x, the value of $b^n x$ changes at most once, and it suffices to prove that there exist $m \neq n$ such that $Xb^m = Xb^n$.

For a classical proof which uses the continuity of X, we extend the sequence $b^0, \cdots, b^n$ in case the process terminates by repeating the last term infinitely often. Then the sequence $b^0, b^1, \cdots$ approaches a limit b. By the continuity of X, an intuitionistic axiom, Xb^n becomes constant for n sufficiently large, and therefore the process terminates. The only point at which we have strayed from intuitionistic reasoning is the assertion that the sequence b^n converges. We would like to prove

$$\bigwedge x \bigvee k \bigwedge m(m \geqq k \rightarrow b^m x = b^k x) \tag{12.1.6}$$

which cannot be done simply on the assumption that each value changes at most once.

For an intuitionistic proof we let b^0 be a function which is zero except possibly at a finite number of arguments. Then each b^n has this property, and we can represent b^n by the finite sequence

$$b^n 0, \cdots, b^n(j-1) ,$$

where j is chosen so that $b^n x = 0$ for all $x \geqq j$. The process can be described in terms of these finite sequences. Note that b^0 has infinitely many representations all of which are permitted. In determining the representation for b^{n+1}, either we use a sequence of the same length as for b^n or we extend the length so as to include the new nonzero value.

The continuity axiom as applied to X states in intuitive terms that for every b there is a k such that Xb can be computed on the basis of $b0, \cdots, b(k-1)$ alone, in which case we say that the finite sequence $\langle b0, \cdots, b(k-1)\rangle$ is *secured*. In applying bar induction (Brouwer's bar theorem) we set

(12.1.7) $P(k, b) \leftrightarrow [\langle b0, \cdots, b(k-1)\rangle$ is secured$]$,

(12.1.8) $Q(k, b) \leftrightarrow [$for every finite sequence $\langle c0, \cdots, c(m-1)\rangle$, the process applied to $\langle b0, \cdots, b(k-1), c0, \cdots, c(m-1)\rangle$ as the initial sequence either terminates or, for some n, $Xb^n < k]$.

The first two hypotheses of the principle of bar induction (§ 6.3) are consequences of the continuity axiom, which asserts that every functional is continuous in Kreisel's terminology [10] (is countable in Kleene's terminology [7]). To show that the next two hypotheses are satisfied we consider two cases.

Case 1. $\langle b0, \cdots, b(k-1)\rangle$ is secured. We wish to show that $Q(k, b)$. If $Xb^0 < k$ we are through. If $Xb^0 \geqq k$, then $Xb^1 = Xb^0$ and the process terminates.

Case 2. $\langle b0, \cdots, b(k-1)\rangle$ is not secured. Let $c0, \cdots, c(m-1)$ be given, and assume $\bigwedge yQ(k', \langle b0, \cdots, b(k-1), y\rangle)$. We want to derive

$$Q(k, \langle b0, \cdots, b(k-1)\rangle) .$$

We can assume without loss of generality that $m > 0$, for if $m = 0$, we can set $c0 = 0$ and change m from 0 to 1. By hypothesis, $Q(k', \langle b0, \cdots, b(k-1), c0\rangle)$. Hence the process applied to $b0, \cdots, b(k-1), c0, \cdots, c(m-1)$ as the initial sequence either terminates or, for some n, $Xb^n \leqq k$. This reduces the problem to the case that the process has not terminated at stage n, $Xb^n = k$, and for all $i < n$, $Xb^i > k$. Then the first k components of the finite sequence for b^{n+1} are $b0, \cdots, b(k-1)$.

We now consider the process from b^{n+1} on. If $Xb^{n+i} = k$ in the continuation, the process terminates. Applying the hypothesis a second time but with b^{n+1} as the starting point, the process either terminates, or, for some i, $Xb^{n+i} < k$, which concludes Case 2.[15]

12.2. As usual a formula is called Σ_1^1 (not to be confused with the system $\Sigma_1!$) if it consists of an existential (number theoretic) function quantifier, followed by a universal numerical quantifier with quantifier-free scope.

THEOREM. *The* Σ_4 *(Gödel,* 1958) *translation of axiom* F *applied to* Σ_1^1 *formulae is provable in intuitionistic analysis with the axiom of choice and bar induction at type zero together with the continuity axiom for functionals of type* ((0, 0), 0).

On account of footnote 11 and § 10 it is sufficient to derive the particular case of BR used in § 10, when C is of type (0, (0, 0)) (sequences of number theoretic functions with numerical values), in other words, bar recursion of type one.

The reduction to type zero follows trivially from the familiar homomorphism between the sets of all number theoretic functions and of all sequences of such functions. The standard homomorphisms are primitive recursive (p.r.) in the sense of section 5: to the sequence of functions C corresponds the function C^*, where

$$C^*(x) = C_{x_1}(x_2) ,$$

when $x = \langle x_1, x_2\rangle$ is one of the usual p.r. numberings of pairs of natural numbers. There is a p.r. functional Y^* of Y such that, for all C, $Y^*C^* = YC$.

We denote $1 + \max(y_1 : y < x)$ by $\bar{x}_1$.

By the axiom of continuity there is a decidable P_{Y^*} such that $(\bigwedge C^*)(\bigvee x)P_{Y^*}(x, C^*)$

[15] As mentioned in footnote 1, Spector intended to prove the Σ_4 translation of axiom F by generalizing the argument of § 12.1, thereby providing an alternative approach to that of § 10. Instead, we use that section in 12.2 below.

and $\{P_{Y^*}(x, C^*) \wedge (\bigwedge y)[y < x \rightarrow C_1^*(y) = C^*(y)]\} \rightarrow Y^*C_1^* = Y^*C^*$, i.e.

$$\langle C^*0, \cdots, C^*(x-1)\rangle$$

is Y^*-secured.

(Here $=$ denotes numerical equality since C^* is numerical valued.)

To apply bar induction of 6.3, we put (using 'C^*' instead of the 'C' of 6.3):

(i) $$P(x, C^*)\colon P_{Y^*}(x, C^*) \wedge Y^*(\langle C^*0, \cdots, C^*(x-1)\rangle) = Y^*C^* < \bar{x}_1$$

and

(ii) $Q(x, C^*)$: there is a unique $\phi^{(x)}(u, C_1)$ (or, more explicitly, a unique $\phi_{\langle C^*0, \cdots, C^*(x-1)\rangle}$ with arguments u and C_1) with the following properties:

For all u and C_1,

(a) $u < \bar{x}_1 \rightarrow \phi^{(x)}(u, C_1) = \mathbf{0}$, and $(\bigvee v)[v < x \wedge C_1^*(v) \neq C^*(v)] \rightarrow \phi^{(x)}(u, C_1) = \mathbf{0}$;

(b) $u \geqq \bar{x}_1 \wedge (\bigwedge v)[v < x \rightarrow C_1^*(v) = C^*(v)]$ implies:

(bi) $\phi^x(u, C_1) = G(u, \langle C_1 0, \cdots, C_1(u-1)\rangle)$ if $Y(\langle C_1 0, \cdots, C_1(u-1)\rangle) < u$,

and otherwise

(bii) $\phi^x(u, C_1) = H[\lambda c \phi^x(u', \langle C_1 0, \cdots, C_1(u-1), c\rangle), u, \langle C_1 0, \cdots, C_1(u-1)\rangle]$.

Hyp $1-2$ are evidently satisfied (cf. § 6.4), Hyp 3 is satisfied because $P(x, C^*)$ implies that $\phi^{(x)}(u, C_1)$ is defined outright by (a) or (bi). To establish Hyp 4, we assume, for (numerical) a,

$$(\bigwedge a) Q(x', \langle C^*0, \cdots, C^*(x-1), a\rangle)\,.$$

By the axiom of choice at type 0, we have $\phi^{(x')}(u, C_1; a)$ satisfying conditions (a) and (b) above with x' in place of x, i.e. for all u and C_1 :

(a′) $u < (\bar{x}')_1 \rightarrow \phi^{(x')}(u, C_1; a) = \mathbf{0}$,

$$\{(\bigvee v)[v < x \wedge C_1^*(v) \neq C^*(v)] \vee C_1^*x \neq a\} \rightarrow \phi^{(x')}(u, C_1; a) = \mathbf{0}$$

(b′) $u \geqq (\bar{x}')_1 \wedge C_1^*(x) = a \wedge (\bigwedge v)[v < x \rightarrow C_1^*(v) = C^*(v)]$

implies

(bi′) $\phi^{(x')}(u, C_1; a) = G(u, \langle C_1 0, \cdots, C_1(u-1)\rangle)$ if $Y(\langle C_1 0, \cdots, C_1(u-1)\rangle) < u$

and otherwise

(bii′) $\phi^{(x')}(u, C_1; a)$

$$= H[\lambda c \phi^{(x')}(u', \langle C_1 0, \cdots, C_1(u-1), c\rangle; a), u, \langle C_1 0, \cdots, C_1(u-1)\rangle]\,.$$

Then the following $\phi(u, C_1)$ is the unique ϕ satisfying $Q(x, C^*)$: Definition by cases.

Case 1. If $u < \bar{x}_1$ or $(\bigvee v)[v < x \wedge C_1^*(v) \neq C^*(v)]$ then $\phi(u, C_1) = \mathbf{0}$. [This is required outright by condition (a) in $Q(x, C^*)$.] For the remaining cases note that if $(\bigvee v)[v < x \wedge C_1^*(v) \neq C^*(v)]$ the premise of (b′) cannot be satisfied, and $\phi^{(x')}(u, C_1; a) = \mathbf{0}$ for all a. Cases 2 or 3 apply only if Case 1 does not.

Case 2. $\bar{x}_1 = (\bar{x}')_1$. Then $u \geqq (\bar{x}')_1 \leftrightarrow u \geqq \bar{x}_1$, and we put

$$\phi(u, C_1) = \phi^{(x')}(u, C_1; C_1^*x)\,.$$

[Then, with $a = C_1^*x$, the premises of (b′) and (b) are equivalent.]

Case 3. $\bar{x}_1 < (\bar{x}')_1$. We use the property of the usual numberings of pairs that in this case $(\bar{x}')_1 = \bar{x}_1 + 1$, and $\bar{x}_1 = x_1$ where, as above, $x = \langle x_1, x_2 \rangle$.

Subcase 3.1. If $u > \bar{x}_1$, i.e. $u \geqq (\bar{x}')_1$, we put $\phi(u, C_1) = \phi^{(x')}(u, C_1; C_1^* x)$, as in Case 2.

Subcase 3.2. If $u = \bar{x}_1$

$$\phi(u, C_1) = G(u, \langle C_1 0, \cdots, C_1(u-1) \rangle) \quad \text{if} \quad Y(\langle C_1 0, \cdots, C_1(u-1) \rangle) < u$$

and otherwise

$$\phi(u, C_1) = H[\lambda c \phi^{(x')}(u', \langle C_1 0, \cdots, C_1(u-1) c \rangle; c x_2), u, \langle C_1 0, \cdots, C_1(u-1) \rangle] .$$

[By Subcase 3.1 $\phi(u', C_2)$ is equated, for all C_2, to $\phi^{(x')}(u', C_2; C_2^* x)$: if C_2 denotes $\langle C_1 0, \cdots, C_1(\bar{x}_1 - 1), c \rangle$ then, in the present case, $C_2^* x = c x_2$.]

This establishes Hyp 4, and $Q(0, \mathbf{0})$ asserts the existence of a unique function satisfying the bar recursion schema.

As is to be expected, a derivation of bar recursion in the 'computational' sense (b) of footnote 5 is less impredicative: more formally speaking, bar induction is applied to a predicate $Q(x, C^*)$ without alternating function quantifiers. Note that though we say that bar induction is applied at type zero, i.e. C^* ranges over sequences of *numbers*, the species Q is defined by means of quantification over objects of higher type.

12.3. *Restricted forms of the comprehension schema* (1.5.1). In this section we shall derive (Lemma 2) the Σ_2 translation of the comprehension schema for 'hyperarithmetic' $P(x)$ in Σ_2 by applying F to Σ_1^1 formulae (on the basis of a suitable definition[16] of 'hyperarithmetic'), and then show in the same subsystem of Σ_2 (Lemma 3) that all arithmetic formulae are hyperarithmetic on this definition.

Definition. $P(x)$ is called hyperarithmetic if there are quantifier-free $S_1(x, y, \alpha)$ and $S_2(x, z, \beta)$ (x, y, z number variables, α, β function variables) such that, for all x,

$$(\bigvee \alpha)(\bigwedge y) S_1(x, y, \alpha) \rightarrow P(x) ,$$
$$(\bigvee \beta)(\bigwedge z) S_2(x, y, \beta) \rightarrow \neg P(x)$$
$$\neg\neg(\bigvee t)(\bigvee \alpha)(\bigvee \beta)(\bigwedge y)(\bigwedge z)\{[t = 0 \rightarrow S_1(x, y, \alpha)] \wedge [t \neq 0 \rightarrow S_2(x, z, \beta)]\} .$$

The following lemma shows that the Σ_2 translation, described in § 4, of one of the standard classical definitions of *hyperarithmetic* (12.3.1) is covered by the (intuitionistic) definition of *hyperarithmetic* just given. For the present purpose it is of course desirable to consider a weak definition of *hyperarithmetic* in order to make Lemma 2 and the theorem below as strong as possible.

Lemma 1. *If the* Σ_2 (*Gödel* [3]) *translation of*

(12.3.1) $$(\bigwedge x)[(\bigvee \alpha)(\bigwedge y) S_1(x, y, \alpha) \leftrightarrow (\bigwedge \beta)(\bigvee z) \neg S_2(x, z, \beta)]$$

is (*intuitionistically*) *provable, then* $\neg\neg(\bigvee \alpha)(\bigwedge y S_1(x, y, \alpha)$,

$$\neg(\bigvee \alpha)(\bigwedge y) S_1(x, y, \alpha), \neg\neg(\bigvee \beta)(\bigwedge z) S_2(x, z, \beta), \neg(\bigvee \beta)(\bigwedge z) S_2(x, z, \beta)$$

[16] This definition is a slight modification of a definition given in a letter of Spector of 21 July 1961. The original form led to the difficulties mentioned in footnote 1.

are all hyperarithmetic, $\neg\neg(\bigvee\alpha)(\bigwedge y)S_1(x, y, \alpha)$ *being itself the* Σ_2 *translation of* $(\bigvee\alpha)(\bigwedge y)S_1(x, y, \alpha)$.

Let $P(x)$ denote $\neg\neg(\bigvee\alpha)(\bigwedge y)S_1(x, y, \alpha)$.

Evidently $(\bigvee\alpha)(\bigwedge y)S_1(x, y, \alpha) \to P(x)$.

The translation of (12.3.1) is equivalent to the conjunction of

(12.3.2) $$\neg\neg(\bigvee\alpha)(\bigwedge y)S_1(x, y, \alpha) \to (\bigwedge\beta)\neg(\bigwedge z)S_2(x, z, \beta)$$

and

(12.3.3) $$(\bigwedge\beta)\neg(\bigwedge z)S_2(x, z, \beta) \to \neg\neg(\bigvee\alpha)(\bigwedge y)S_1(x, y, \alpha) .$$

Since $(\bigvee\beta)(\bigwedge z)S_2(x, z, \beta) \to \neg(\bigwedge\beta)\neg(\bigwedge z)S_2(x, z, \beta)$, by contraposing (12.3.2),

$$(\bigvee\beta)(\bigwedge z)S_2(x, z, \beta) \to \neg P(x) .$$

Denote $(\bigvee\alpha)(\bigwedge y)S_1(x, y, \alpha)$ by A, $(\bigvee\beta)(\bigwedge z)S_2(x, z, \beta)$ by B: by (12.3.3), $\neg B \to \neg\neg A$, hence, by propositional inference, $\neg\neg(A \vee B)$, hence by [**10**, p. 112, 3.52], $\neg\neg(\bigvee t)[(t = 0 \to A) \wedge (t \neq 0 \to B)]$, and thus the third condition in the definition of hyperarithmetic $P(x)$.

The treatment of the other cases is parallel.

Lemma 2. *If* $P(x)$ *is hyperarithmetic,* $\neg\neg(\bigwedge x)[P(x) \vee \neg P(x)]$ *is provable in* Σ_2 *with axiom* F *applied to a* Σ_1^1 *formula.*

By usual contraction of quantifiers,

(12.3.4) $$(\bigvee t)(\bigvee\alpha)(\bigvee\beta)(\bigwedge y)(\bigwedge z)\{[t = 0 \to S_1(x, y, \alpha)] \wedge [t \neq 0 \to S_2(x, z, \beta)]\}$$

is equivalent to a Σ_1^1 formula. Thus, by axiom F applied to (12.3.4)

$$\neg\neg(\bigwedge x)\{(\bigvee\alpha)(\bigwedge y)S_1(x, y, \alpha) \vee (\bigvee\beta)(\bigwedge z)S_2(x, z, \beta)]\} ,\quad \text{i.e.}$$

$$\neg\neg(\bigwedge x)(A \vee B) .$$

But, since $\{[A \to P(x)] \wedge [B \to \neg P(x)]\} \to \{(A \vee B) \to [P(x) \vee \neg P(x)]\}$, by the first two conditions on hyperarithmetic predicates, we have

$$\neg\neg(\bigwedge x)[P(x) \vee \neg P(x)] .$$

Lemma 3. *Any arithmetic predicate* $A(x)$ *can be proved to be hyperarithmetic in* Σ_2 *with axiom* F *applied to* Σ_1^1 *formulae.*

Note that for the immediate purpose (of a Σ_4 translation of the arithmetic comprehension axiom) the following weaker result is sufficient.

Lemma 3′. *Any arithmetic predicate* $B(x)$ *without existence or disjunction symbols can be proved to be hyperarithmetic in* Σ_2 *with axiom* F *applied to* Σ_1^1 *formulae.*

Proof of Lemma 3′. By applying the axiom of choice E to Σ_1^1 formulae both $B(x)$ and $\neg B(x)$ can be proved (classically) in Σ_1 to be equivalent to Σ_1^1 formulae (via prenex normal forms), say $(\bigvee\alpha)(\bigwedge y)S_1(x, y, \alpha)$ and $(\bigvee\beta)(\bigwedge z)S_2(x, z, \beta)$. By the Σ_2 translation (para 4 above) and Lemma 1 of the present section, $B^\circ(x)$ can be proved to be hyperarithmetic in Σ_2, with F applied to Σ_1^1 formulae only. But $B^\circ(x)$ is $B(x)$ itself.

We now prove Lemma 3 for arbitrary arithmetic $A(x)$.

We associate with each A a 'strongest' prenex formula A_P, and a prenex formula A_P^* which is very nearly the prenex form of $\neg(A_P)$, such that (i) intuitionistically, $A_P \to A$, $A_P^* \to \neg A$, and (ii) in classical arithmetic, $\neg A_P \leftrightarrow A_P^*$. By the axiom of choice E applied to Σ_1^1 formulae, one gets Σ_1^1 formulae A_1 and A_1^*, $A_1 \leftrightarrow A_P$, $A_1^* \leftrightarrow A_P^*$ (both intuitionistically and classically) and, since F is the Σ_2 translation of E, a classical proof that A_P is hyperarithmetic in the sense of (12.1.3). By Lemma 1 we then get Lemma 3. The detour via prenex forms is used to avoid variables of higher type which occur in the Σ_4 (Gödel, 1958) translation of arithmetic formulae with iterated implications. We keep types low in order to apply F only to Σ_1^1 formulae.

Definitions of A_P and A_P^*. Let $^\circ A_P$, $^\circ A_P^*$ denote the quantifier-free parts of A_P, A_P^* respectively, and (QA), (QA^*) their prefixes.

All transformations below are classical equivalences. We prove, by induction, $A_P \to A$, $A_P^* \to \neg A$.

(i) If A is quantifier-free, A_P is A, A_P^* is $\neg A$.

(ii) If A is $\neg B$, A_P is B_P^*, A_P^* is B_P, and so, if $B_P \to B$ and $B_P^* \to \neg B$, $A_P \leftrightarrow B_P^* \to \neg B \leftrightarrow A$ and $A_P^* \leftrightarrow B_P \to B \to \neg A$.

If A is $B \wedge C$, $B \vee C$, $B \to C$, suppose that the sets of bound variables of B_P, B_P^*, C_P and C_P^* are pairwise disjoint (which can always be achieved by renaming of variables).

(iii) If A is $B \wedge C$, A_P is $(QB)(QC)(^\circ B_P \wedge {}^\circ C_P)$ and A_P^* is

$$(\bigvee t)(QB^*)(QC^*)[(t = 0 \to {}^\circ B_P^*) \wedge (t \neq 0 \to {}^\circ C_P^*)] .$$

Then $A_P \to A$ is clear, $A_P^* \to (B_P^* \vee C_P^*) \to (\neg B \vee \neg C) \to \neg(B \wedge C) \leftrightarrow \neg A$.

(iv) If A is $B \vee C$, A_P is $(\bigvee t)(QB)(QC)[(t = 0 \to {}^\circ B_P) \wedge (t \neq 0 \to {}^\circ C_P)]$, A_P^* is $(QB^*)(QC^*)(^\circ B_P^* \wedge {}^\circ C_P^*)$. $A_P \to (B_P \vee C_P) \to (B \vee C) \leftrightarrow A$, and A_P^* is $(\neg B \wedge \neg C)_P$, hence, by (ii) and (iii), $A_P^* \to (\neg B) \wedge (\neg C) \leftrightarrow \neg A$.

(v) If A is $B \to C$, A_P is $(C \vee \neg B)_P$, A_P^* is $(B \wedge \neg C)_P$ and $A_P \to A$ follows from $(C \vee \neg B) \to (B \to C)$, $A_P^* \to \neg A$ from $(B \wedge \neg C) \to \neg(B \to C)$.

The proofs of $A_P \to A$, $A_P^* \to \neg A$ are clear for:

(vi) If A is $(\bigwedge x)B(x)$, A_P is $(\bigwedge x)[B(x)]_P$, A_P^* is $(\bigvee x)[B(x)]_P^*$ and

(vii) If A is $(\bigvee x)B(x)$, A_P is $(\bigvee x)[B(x)]_P$, A_P^* is $(\bigwedge x)[B(x)]_P^*$.

By the axiom of choice at type zero, applied only to Σ_1^1 formulae, any prenex formula is equivalent to a Σ_1^1 formula, say (12.3.4):

$$A_P \leftrightarrow (\bigvee \alpha)(\bigwedge y)S(x, y, \alpha) , \qquad A_P^* \leftrightarrow (\bigvee \beta)(\bigwedge z)S_2(x, z, \beta) .$$

Since in classical arithmetic $\neg A_P \leftrightarrow A_P^*$, and since F applied to Σ_1^1 formulae is the Σ_2 translation of axiom E in 1.4, applied to Σ_1^1 formulae $(\bigvee a)P(x, a)$, we have a proof in Σ_2, with F applied to Σ_1^1 formulae, of the Σ_2 translation of

$$(\bigwedge x)[(\bigvee \alpha)(\bigwedge y)S_1(x, y, \alpha) \leftrightarrow \neg(\bigvee \beta)(\bigwedge z)S_2(x, z, \beta)] .$$

By Lemma 1, $\neg\neg[(\bigvee \alpha)(\bigwedge y)S_1(x, y, \alpha) \vee (\bigvee \beta)(\bigwedge z)S_2(x, z, \beta)]$, and so, by axiom F, applied to a Σ_1^1-formula

(12.3.6) $$\neg\neg(\bigwedge x)[(\bigvee \alpha)(\bigwedge y)S_1(x, y, \alpha) \vee (\bigvee \beta)(\bigwedge z)S_2(x, z, \beta)] .$$

By (12.3.5), $A_P \to A$ and $A_P^* \to \neg A$,

$$\neg\neg(\bigwedge x)[A(x) \vee \neg A(x)]\ .$$

DEFINITION. The *hyperarithmetic comprehension schema* (for Σ_1) consists of all instances of the form

$$(\bigwedge x)\{[(\bigvee \alpha)(\bigwedge y)S_1(x, y, \alpha) \to P(x)] \wedge [(\bigvee \beta)(\bigwedge z)S_2(x, z, \beta) \to \neg P(x)] \wedge [(\bigvee \alpha)(\bigwedge y)S_1(x, y, \alpha) \vee (\bigvee \beta)(\bigwedge z)S_2(x, z, \beta)]\} \to (1.5.1)\ ,$$

where S_1 and S_2 denote quantifier-free expressions of Σ_1 and $P(x)$ denotes an arbitrary formula of Σ_1 (in which x occurs free and a not); x, y are number variables, α, β, a (of 1.5.1) are function variables of type (0, 0).

THEOREM. *The Σ_4 translation of the comprehension axiom* (1.5.1) *restricted to hyperarithmetic (and, in particular, arithmetic) formulae $P(x)$ and the axiom of choice* E *applied to Σ_1^1 formulae are provable by means of bar induction at type zero.*

By Lemmas 2 and 3 and the theorem of the preceding § 12.2.

From the point of view of consistency proofs the theorem above, applied to arithmetic formulae $P(x)$ in (1.5.1), is inferior to the familar finitist relative consistency proof of that system relative to first order arithmetic.[17] In fact, for any finite subsystem of this (predicative) extension of first order arithmetic the consistency can be proved in Gödel's T [4] itself. On the other hand, our proof of the theorem is less elementary because it uses Brouwer's bar theorem; or more formally, footnote 14 shows that our proof cannot be carried out in T for any (finite) subsystem which includes (1.5.1) with $(\bigvee y)R(x, y)$ for $P(x)$, where $(\bigvee y)R(x, y)$ is recursively enumerable but not recursive, simply because the Σ_4 translation of such an instance of (1.5.1) does not hold in T. However, for the full hyperarithmetic case of (1.5.1) in the form

$$P \text{ is hyperarithmetic} \to (\bigvee \alpha)(\bigwedge x)\{[\alpha(x) = 0 \leftrightarrow P(x)] \wedge [\alpha(x) \neq 0 \leftrightarrow \neg P(x)]\}$$

(the 'first' impredicative case) the present proof is the first consistency proof on accepted intuitionistic principles. The interest of this result lies in the fact that in this part of analysis one can carry out not only the standard theory of Lebesgue-measure, but also of Borel sets (the axiom of choice of type zero being included[18]).

---

[17] More precisely, the finitist consistency proof applies to the system where *both* 1.5.1 *and* induction (axiom D) are restricted to arithmetic $P(x)$ (containing no *bound* variables of higher type). In the *Theorem* above, D is not so restricted. The present method does not allow a convenient separation between restricted and unrestricted induction. This explains the need for less elementary principles noted in the text. The argument in [5, vol. 2, pp. 366-367] proves the consistency of first order arithmetic by means of restricted 1.5.1 and unrestricted induction.

[18] Comparison of classical analysis based (i) on E (as in Spector's Σ_1) and (ii) on 1.5.1 (as in [10]). As noted in 1.5, (i) includes (ii). But even for special cases of E which are derivable from 1.5.1 in Σ_1 (without E), E is more economical in that the type of bar induction needed to prove the Σ_4 translation of E (or: the corresponding F) is lower. E.g., E applied to Σ_1^1 formulae is derivable from 1.5.1; if $(\bigvee z)R(\alpha,z,x)$, R recursive,

REFERENCES

1. L. E. J. Brouwer, *Über Definitionsbereiche von Funktionen,* Math. Ann. vol. 97 (1927) pp. 60–76.

2. Gerhard Gentzen, *Die Widerspruchsfreiheit der reinen Zahlentheorie,* Math. Ann. vol. 112 (1936) pp. 493–565.

3. Kurt Gödel, *Zur intuitionistischen Arithmetik und Zahlentheorie,* Ergebnisse eines math. Koll., Heft 4 for 1931–1932, (1933) pp. 34–38.

4. ———, *Über eine bisher noch nicht benützte Erweiterung des finiten Standpunktes,* Dialectica vol. 12 (1958) pp. 280–287.

5. D. Hilbert and P. Bernays, *Grundlagen der Mathematik,* Berlin, Springer, vol. 1, 1934, xii+471 pp.; vol. 2, 1939, xii+498 pp.

6. S. C. Kleene, *Introduction to metamathematics,* Amsterdam, North-Holland Publishing Co.; Groningen, P. Nordhoff Ltd., and Princeton, D. Van Nostrand Inc.; 1952, x+550 pp.

7. ———, *Countable functionals, Constructivity in mathematics,* Amsterdam, North-Holland Publishing Co., 1959, pp. 81–100.

8. ———, *Foundations of intuitionistic mathematics,* Studies in Logic, to appear.

9. G. Kreisel, *On the interpretation of non-finitist proofs.* I and II, J. Symb. Logic vol. 16 (1951) pp. 241–267, and vol. 17 (1952) pp. 43–58.

10. ———, *Interpretation of classical analysis by means of constructive functionals of finite types, Constructivity in mathematics,* Amsterdam, North-Holland Publishing Co., 1959, pp. 101–128. *Errata,* cf. footnote 4 and § 5 above.

11. ———, *On weak completeness of intuitionistic predicate logic,* J. Symb. Logic vol. 25 (1960), to appear.

INSTITUTE FOR ADVANCED STUDY
PRINCETON, NEW JERSEY

POSTSCRIPT

This important paper was written by Clifford Spector during his stay at the Institute for Advanced Study in 1960–61 under a grant from the Office of Naval Research. The discussions P. Bernays and I had with Spector (see footnote 1) took place after the main result (contained in § 10 of the paper) had been established already. However it ought to be mentioned that during the time Spector first established this result he was in close contact with Kreisel. It was Spector's express intention to give to Kreisel a good deal of credit for his work. Originally a joint publication by Spector and Kreisel was envisaged. This plan was dropped because Spector had taken over the elaboration by himself and because the version of the proof which was to be published was due to Spector. Also Spector alone, at that time, was working on an extension of the result in the direction of stricter constructivity which he hoped to include in his paper. K. Gödel

then $(\bigwedge z)R(\rho,z,x)$ where ρ is defined as follows: order finite sequences of natural numbers by finite differences, and let μ^* be the 'least element' operator for this ordering; $\bar{\rho}(n) = \mu^* m(\bigvee \beta)(\bigwedge z)[\bar{\beta}(n) = m \wedge R(\beta,z,x)]$. But to prove the existence of ρ, one (apparently) has to apply 1.5.1 to a $P(x)$ which is a Boolean combination of $\sum_1^1$ formulas, while we have a Σ_4 translation of 1.5.1 by bar induction of *lowest* type only for $P(x)$ which are hyperarithmetic. On the other hand, the *Theorem* of 12.2 proves the Σ_4 translation of E applied to $\sum_1^1$ formulae by means of the bar theorem itself (bar induction of lowest type).

REPRESENTABILITY OF SETS IN FORMAL SYSTEMS

BY

ANDRZEJ MOSTOWSKI

The aim of this paper is to advocate a method (due in principle to Gödel [2], though never elaborated by him in details) to present in a uniform way the theories of recursive, hyperarithmetical and related families of sets. The gist of the method is to define these families using the notion of representability in suitable formalized theories. The techniques worked out by Kleene and other writers yield probably better results when one wants to discuss properties of a single family; the writer believes however that the method developed below is very helpful when one wants to discuss common properties of these families and to detect reasons of their affinities.

The method will be presented for families consisting of sets of integers and sets of functions. An extension to higher types has not yet been tried. I do not know whether it would meet essential difficulties.

The writer had planned to entitle the paper "Kleene's theories as I see them." Although the final title is more conservative, the influence of Kleene's work on the present paper should be obvious to every reader even moderately acquainted with the literature.

CHAPTER I. GENERAL THEORY OF REPRESENTABILITY

I.1. **Formal systems.** In Chapters I and II we shall deal with formal systems having a common language and differing from each other by the notion of consequence. The common language of these systems is that of second order arithmetic [3] with constants for both types of objects (integers and functions). Latin l.c. letters will be used for variables and constants of type 0 (integers) and Greek l.c. letters for variables and constants of type 1 (functions). We use the first letters of the alphabet for constants and the last ones for variables. Numerals are denoted by $\delta_0, \delta_1, \cdots$.

We denote by Ax the set of axioms consisting of the usual axioms for the propositional and functional calculus with identity, of Peano's axioms for arithmetic, of the so-called pseudo-definitions, i.e., axioms of the form $(E\xi)(x)[\xi(x) = 0 \equiv \Phi]$, where Φ is a formula in which the variable ξ is not free, cf. [3], and finally of the special form of the axiom of choice which allows one to permute the functional and the numerical quantifiers, cf. [11, p. 217]. If X is a set of formulas then $Cn_0(X)$ denotes the set of formulas which can be obtained from $Ax \cup X$ by the usual rules of proof.

We assume that in each formal system which will be considered in our theory there is defined a function of consequence Cn_S acting on sets of formulas and yielding such sets. Further we assume that this function satisfies the

Received by the editor April 6, 1961

following axioms:[1]

(I) $X \subseteq Cn_S(X)$.

(II) $X \subseteq Y$ *implies* $Cn_S(X) \subseteq Cn_S(Y)$.

(III) $Cn_0(Cn_S(X)) \subseteq Cn_S(X)$ *and* $Cn_S(Cn_0(X)) \subseteq Cn_S(X)$.

(IV) *If* Ψ *is a closed formula and* $\Phi \in Cn_S(X \cup \{\Psi\})$, *then* $\Psi \supset \Phi \in Cn_S(X)$.

(V) *There are infinitely many inessential constants of both types.*

A constant a or α is inessential for S if for every set X of formulas none of which contains a (or α) the condition $\Phi \in Cn_S(X)$ implies $\Phi' \in Cn_S(X)$ where Φ' results from Φ by a substitution for a (or α) of a free variable of the appropriate type which does not occur in Φ.

Note that $Cn_0(Cn_S(X)) = Cn_S(Cn_0(X)) = Cn_S(X)$ by (II), (III), and the obvious properties of Cn_0.

In some theorems we shall assume that Cn_S is an idempotent operation. These theorems are marked by an asterisk.

If Z is a set of formulas then $\mathrm{Ext}_S(Z)$ denotes the system S' with the function of consequence defined thus: $Cn_{S'}(X) = Cn_S(X \cup Z)$.

I.1.1. *If there are infinitely many constants (of both types) which are inessential for S and do not occur in formulas of Z, then the system* $\mathrm{Ext}_S(Z)$ *satisfies* (A)–(E).

Let λ be a functional variable or a functional constant and φ a function from integers to integers. We denote by $D_\varphi(\lambda)$ the set of formulas $\lambda(\delta_n) = \delta_{\varphi(n)}$ and call this set the diagram of φ. To maintain the symmetry between both types we denote by $D_n(l)$ the formula $l = \delta_n$; here l is a numerical constant or a numerical variable.

I.2. **Representable sets and functions.** Let ω be the set of integers $\geqq 0$, ω^ω the set of all mappings from ω into ω. Elements of ω will be briefly called numbers and elements of ω^ω functions. We denote functions by the letters φ, ψ, ϑ, $\cdots$. We denote by $R_{k,l}$ the Cartesian product $\omega^\omega \times \cdots \times \omega^\omega \times \omega \times \cdots \times \omega = (\omega^\omega)^k \times \omega^l$; elements of $R_{k,l}$ are denoted by German letters $\mathfrak{p}, \mathfrak{q}, \cdots$. Thus $\mathfrak{p}$ is a sequence $(\varphi_1, \cdots, \varphi_k, n_1, \cdots, n_l)$ consisting of k functions and l numbers.

Let the German l.c. letters $\mathfrak{a}, \mathfrak{b}, \cdots, \mathfrak{v}, \mathfrak{w}, \cdots$ denote sequences consisting of k inessential constants (or variables) of type 1 and l inessential constants (or variables) of type 0. Such sequences are called briefly k, l sequences and we shall write $\mathfrak{a} = (\alpha_1, \cdots, \alpha_k, a_1, \cdots, a_l)$ and similarly for other letters. We put

$$D_{\mathfrak{q}}(\mathfrak{a}) = D_{\varphi_1}(\alpha_1) \cup \cdots \cup D_{\varphi_k}(\alpha_k) \cup D_{n_1}(a_1) \cup \cdots \cup D_{n_l}(a_l) .$$

A set $A \subseteq R_{k,l}$ is weakly represented in S by a formula Φ if Φ has k free variables of type 1, l free variables of type 0 and

[1] These are esentially Tarski's axioms [**17**]; we did not include all of Tarski's axioms in order to have a wider range of applications.

(1) $$\mathfrak{q} \in A \equiv \Phi(\mathfrak{a}) \in Cn_S(D_{\mathfrak{q}}(\mathfrak{a}));$$

here $\Phi(\mathfrak{a})$ is the formula obtained from Φ by a substitution of the constants $\mathfrak{a}$ for the free variables of Φ. The family of weakly representable subsets of $R_{k,l}$ is denoted by $\mathscr{R}_{k,l}(S)$ or briefly by $\mathscr{R}_{k,l}$ when S is fixed.

If besides (1) the equivalence

$$\mathfrak{q} \text{ non} \in A \equiv \sim \Phi(\mathfrak{a}) \in Cn_S(D_{\mathfrak{q}}(\mathfrak{a}))$$

holds, then we say that A is strongly represented in S by Φ. The family of strongly representable subsets of $R_{k,l}$ is denoted by $\mathscr{R}^*_{k,l}(S)$ or briefly by $\mathscr{R}^*_{k,l}$.

It is worth while to remark that families $\mathscr{R}_{k,l}$ and $\mathscr{R}^*_{k,l}$ may well coincide, e.g., if S is a complete system.

A mapping $f\colon R_{k,l} \to R_{m,n}$ is represented in S by a formula Φ with $k + m$ free variables of type 1 and $l + n$ free variables of type 0 if for every $\mathfrak{q}$ in $R_{k,l}$

(2) $$(E!\,\mathfrak{w})\,\Phi(\mathfrak{a}, \mathfrak{w}) \in Cn_S(D_{\mathfrak{q}}(\mathfrak{a})),$$

(3) $$D_{f(\mathfrak{q})}(\mathfrak{b}) \subseteq Cn_S(D_{\mathfrak{q}}(\mathfrak{a}) \cup \{\Phi(\mathfrak{a}, \mathfrak{b})\});$$

here $\mathfrak{v}$ and $\mathfrak{a}$ are k, l sequences of variables and of inessential constants and $\mathfrak{w}$ and $\mathfrak{b}$ are m, n sequences. The family of representable mappings of $R_{k,l}$ into $R_{m,n}$ will be denoted by $\mathscr{F}_{k,l;m,n}(S)$ or briefly by $\mathscr{F}_{k,l;m,n}$.

If $m = 0$, i.e., if f maps $R_{k,l}$ into ω or into a Cartesian product of finitely many copies of ω, we can replace (3) by

(3′) $$\Phi(\mathfrak{a}, \mathfrak{b}) \in Cn_S(D_{\mathfrak{q}}(\mathfrak{a}) \cup D_{f(\mathfrak{q})}(\mathfrak{b}))$$

and obtain an equivalent condition. Since (2) and (3′) imply (for $m = 0$)

(3″) $$\sim \Phi(\mathfrak{a}, \mathfrak{b}) \in Cn_S(D_{\mathfrak{q}}(\mathfrak{a}) \cup D_{\mathfrak{n}}(\mathfrak{b})) \qquad \text{for } \mathfrak{n} \neq f(\mathfrak{q})$$

we infer that

I.2.1. *If* $f \in \mathscr{F}_{k,l;0,n}$ *then the relation* $f(\mathfrak{q}) = \mathfrak{n}$ *is strongly representable. If* $f\colon R_{k,l} \to R_{0,n}$ *and the relation* $f(\mathfrak{q}) = \mathfrak{n}$ *is weakly representable, then* $f \in \mathscr{F}_{k,l;0,n}$.

No similar theorem holds if $m \neq 0$.

I.3. **Properties of strongly representable sets and of representable functions.** We list below a series of elementary theorems whose proofs can be obtained immediately from the definitions.

I.3.1. $\mathscr{R}^*_{k,l}$ *is a Boolean algebra of sets.*

I.3.2. *If* $A \in \mathscr{R}^*_{k,l}$, *then* $A \times \omega \in \mathscr{R}^*_{k,l+1}$ *and* $A \times \omega^\omega \in \mathscr{R}^*_{k+1,l}$. *Every set which arises from A by a "permutation of axes" belongs to* $\mathscr{R}^*_{k,l}$ *along with A.*

I.3.3. *If* $A \in \mathscr{R}^*_{k,l+1}$, *then the set* $A^n = \{\mathfrak{q}\colon (\mathfrak{q}, n) \in A\}$ *belongs to* $\mathscr{R}^*_{k,l}$.

A theorem similar to 3.3 would be false for the operation $A^\varphi = \{\mathfrak{q}\colon (\varphi, \mathfrak{q}) \in A\}$; we only have a weaker result

I.3.4. *If* $A \in \mathscr{R}^*_{k+1,l}(S)$ *and* α *is a constant inessential for* S *then* $A^\varphi \in \mathscr{R}^*_{k,l}(\mathrm{Ext}_S(D_\varphi(\alpha)))$.

I.3.5*. *If* $f \in \mathscr{F}_{k,l;m,n}$ *and* $A \in \mathscr{R}^*_{m,n}$, *then* $f^{-1}(A) \in \mathscr{R}^*_{k,l}$.

The notion of a recursive mapping $f: R_{k,l} \to R_{m,n}$ will be defined formally in §II.1. It will be seen there that for $k = m = 0$, $n = 1$, this notion coincides with the usual notion of a recursive function with l arguments.

I.3.6. *If f is a recursive mapping of $R_{k,l}$ into $R_{m,n}$ and $A \in \mathscr{R}^*_{m,n}$, then $f^{-1}(A) \in \mathscr{R}^*_{k,l}$.*

Note that no assumption of idempotency of Cn_S was needed in Theorem 3.6.

I.3.7. *$\mathscr{F}_{k,l;m,n}$ contains all recursive mappings; permutations and identifications of variables do not lead outside the family of representable functions.*

I.3.8. *Superposition of two functions one of which is representable and the other recursive leads to representable functions.*

I.3.9*. *The family of representable functions is closed with respect to superpositions.*

I.3.10. *$A \in \mathscr{R}^*_{m,n}$ if and only if the characteristic function of A belongs to $\mathscr{F}_{m,n;0,1}$.*

I.3.11. *If $A \in \mathscr{R}^*_{k,l+1}$ and if for every $\mathfrak{q}$ in $R_{k,l}$ there is an n such that $(\mathfrak{q}, n) \in A$, then the function $\min_n [(\mathfrak{q},n) \in A]$ belongs to $\mathscr{F}_{k,l;0,1}$.*

As the last theorem we note that the family of strongly representable sets is not affected by extensions of S. More exactly

I.3.12. *If X is a consistent set of formulas, then $\mathscr{R}^*_{k,l}(S) = \mathscr{R}^*_{k,l}(\mathrm{Ext}_S(X))$.*

I.4. **Properties of weakly representable sets.**

I.4.1. *If $A, B \in \mathscr{R}_{k,l}$, then $A \cap B \in \mathscr{R}_{k,l}$.*

I.4.2. *If $A \in \mathscr{R}_{k,l}$ and $B \in \mathscr{R}^*_{k,l}$, then $A \cup B \in \mathscr{R}_{k,l}$.*

It is not known whether the union of two weakly representable sets is always weakly representable. Most probably this is not the case, but no counter-example is known at present.

Theorems I.3.2, I.3.3 and I.3.4 remain true for weakly representable sets. Theorems I.3.5*, I.3.6, are probably false if $\mathscr{R}^*$ is replaced by $\mathscr{R}$. The following weak form of 3.5* survives:

I.4.3. *If $A \in \mathscr{R}_{0,l}$ and $f \in \mathscr{F}_{0,1;0,l}$, then $f^{-1}(A) \in \mathscr{R}_{0,1}$.*

The following example shows that Theorem I.3.12 does not hold when $\mathscr{R}^*$ is replaced by $\mathscr{R}$. Let Cn_S be the function Cn_β defined in [**10**] and Π a formula such that the set $Cn_S(\{\Pi\})$ be consistent and complete (see note p. 48). In this case the family $\mathscr{R}_{0,1}(\mathrm{Ext}_S(\{\Pi\}))$ is a Boolean algebra because it coincides with $\mathscr{R}^*_{0,1}(\mathrm{Ext}_S(\{\Pi\}))$. In [**10**] it has been shown that the family $\mathscr{R}_{0,1}(S)$ coincides with the family Π^1_2 and hence is not closed under complementation.

We note a weaker theorem (which we may note in passing, holds not only for $\mathscr{R}_{k,l}$ but for $\mathscr{R}^*_{k,l}$ as well).

I.4.4. *If Π is a closed formula, then $\mathscr{R}_{k,l}(\mathrm{Ext}_S(\{\Pi\})) \subseteq \mathscr{R}_{k,l}(S)$.*

Indeed, if Φ represents A in $\mathrm{Ext}_S(\{\Pi\})$, then $\Pi \supset \Phi$ represents A in S.

It is remarkable that under special, but not too narrow assumptions the analogue of Theorem I.3.12 can be proved for the family $\mathscr{R}_{k,l}$. We shall discuss this phenomenon (discoved for recursive sets by Ehrenfeucht and Feferman) in later sections.

I.5. **Universal functions.** Let $e_0, e_1, \cdots$ (or more exactly $e_0^{(k,l)}, e_1^{(k,l)}, \cdots$) be an enumeration of the Gödel numbers of formulas with k free functional variables and l free number variables. The formula with the Gödel number e_n is denoted by $\breve{e}_n$. The sequence e_n is primitive recursive and logical operations on formulas (including substitutions) correspond to primitive recursive operations on integers e_n. Put

$$U(n) = U^{(k,l)}(n) = \{\mathfrak{q} : \breve{e}_n(\mathfrak{a}) \in Cn_S(D_{\mathfrak{q}}(\mathfrak{a}))\} .$$

I.5.1. *$\mathscr{R}_{k,l}$ coincides with the family of sets $U^{(k,l)}(n)$, $n = 0, 1, \cdots$.*

Thus $U^{(k,l)}$ is a universal function for the family $\mathscr{R}_{k,l}$.

Theorem I.5.1 provides us in the usual way with examples of nonrepresentable sets.

I.5.2. *The set $\{(\mathfrak{q}, n) : (\mathfrak{q}, n) \notin U^{(k,l+1)}(n)\}$ is not weakly representable.*

If one wants an example of a subset of $\mathscr{R}_{k+1,0}$ which is not weakly representable one can take the set

$$\{(\varphi, \varphi_1, \cdots, \varphi_k) : (\varphi, \varphi_1, \cdots, \varphi_k) \notin U^{(k+1,0)}(\varphi(0))\} .$$

I.5.3. *The set $\{(\mathfrak{q}, p) : \mathfrak{q} \in U^{(k,l+1)}(p)\}$ is not strongly representable.*

Let T be the set of Gödel numbers of closed formulas which are provable in S, i.e., which belong to $Cn_S(0)$.

I.5.4. *There is a recursive function of two variables $g(n, m) = g_n(m)$ such that $U^{(0,1)}(n) = g_n^{-1}(T)$.*

I.5.5. $T \notin \mathscr{R}^*_{0,1}$, $\omega - T \notin \mathscr{R}_{0,1}$.

Theorems I.4.1 — I.4.3 and the analogues of Theorems I.3.2 — I.3.4 for weakly representable sets have strengthened versions showing that if the operations mentioned in these theorems are performed on sets $U(p)$, $U(q)$, $\cdots$ then the result is a set $U(\chi(p, q, \cdots))$ where the function χ is recursive. E.g. the strengthened version of Theorem I.4.1 reads:

$$U(p) \cap U(q) = U(\chi(p, q)), \text{ where } \chi(p, q) = \min_r [e_r = \ulcorner\breve{e}_p \,\&\, \breve{e}_q\urcorner] .$$

Let $A \subseteq R_{k,l+1}$, $s \leqq l$, $\mathbf{m} = (n_{s+1}, \cdots, n_l)$ and let $A_{\mathbf{m},e}$ be the set

$$\{(\varphi_1, \cdots, \varphi_k, n_1, \cdots, n_s) : (\varphi_1, \cdots, \varphi_k, n_1, \cdots, n_s, \cdots, n_l, e) \in A\} .$$

If $A = U^{(k,l+1)}(r)$, then $A_{\mathbf{m},e} = U^{(k,s)}(r')$, where r' depends on $\mathbf{m}$, e, and r. Using the strengthened versions of Theorems I.4.1—I.4.3 and of the analogues of Theorems I.3.2 — I.3.4 for the family $\mathscr{R}$ and repeating the proof of Kleene [4] we obtain the recursion theorem (or, in Myhill's terminology, the fixed point theorem).

I.5.6. *There is a recursive function $E(\mathbf{m}, r)$ such that if $A = U^{(k,l+1)}(r)$, then $A_{\mathbf{m},E(\mathbf{m},r)} = U^{(k,s)}(E(\mathbf{m}, r))$.*

I.6*. **Degrees of representability.** In the whole §6 we assume that $Cn_S(Cn_S(X)) = Cn_S(X)$.

Let φ, ψ be functions, i.e. elements of $R_{1,0}$. We say that the degree of representability of φ (in S) is not higher than that of ψ (symbolically $\varphi \leqq_S \psi$), if φ is representable in $\mathrm{Ext}_S(D_\psi(\alpha))$, where α is an inessential constant for S:

$$\varphi \leqq_S \psi \equiv \varphi \in \mathscr{F}_{0,1;0,1}(\mathrm{Ext}_S(D_\psi(\alpha))) .$$

This definition is an obvious adaptation of the definition due to Kleene-

Post [6] to the more general situation considered here.

Similarly as for recursive degrees we define $\varphi \approx_S \psi$ as $(\varphi \leqq_S \psi)$ & $(\psi \leqq_S \varphi)$.

I.6.1*. *The relation* $\leqq_S$ *is reflexive and transitive; the relation* $\approx_S$ *is an equivalence relation.*

The equivalence classes under $\approx_S$ we call degrees of representability. We show similarly as in [6] that

I.6.2*. *Degrees form an upper semi-lattice whose minimal element is* $\mathscr{F}_{0,1;0,1}$.

I.6.3*. *If* $\varphi \in \mathscr{F}_{0,1;0,1}$ *and* $\psi(n) = \vartheta(\varphi(n)), \zeta(n) = \varphi(\vartheta(n))$, *then* $\psi \leqq_S \vartheta$ *and* $\zeta \leqq_S \vartheta$, *but in general* ψ non $\approx_S \zeta$.

In the next two theorems we denote by χ_A the characteristic function of a set A and put $\tau = \chi_T$ (cf. I.5.4).

I.6.4*. *If* $A \in \mathscr{R}_{0,1}$, *then* $\chi_A \leqq_S \tau$.

Indeed, $\chi_A(n)$ can be represented as $\tau(\varphi(n))$, where φ is recursive.

I.6.5*. *If* T' *is the set of Gödel numbers of formulas provable in* $\mathrm{Ext}_S(D_\tau(\alpha))$, *where* α *is an inessential constant of* S, *then* $\tau <_S \chi_{T'}$ *and* $\chi_A <_S \chi_{T'}$ *for every* A *in* $\mathscr{R}_{0,1}$.

The theorem is proved by showing that $\tau \approx_S \chi_{T'}$ would imply that T' is strongly representable in $S' = \mathrm{Ext}_S(D_\tau((\alpha))$ which contradicts I.5.5 (with S replaced by S'). The formula $\tau \leqq_S \chi_{T'}$ is a consequence of the following result: for arbitrary φ, if T^* is the set of Gödel numbers of formulas provable in $\mathrm{Ext}_S(D_\varphi(\alpha))$, then $\chi_{T^*} \leqq_S \varphi$.

Theorems I.6.4* and I.6.5* generalize the basic properties of the "jump operation" of [6].

I.7. **Separability and decidability.** We call two subsets A, B of $R_{k,l}$ separable in S if there is a C in $\mathscr{R}^*_{k,l}$ such that $A \subseteq C$ and $B \cap C = 0$.

I.7.1. *T and the set T^R of Gödel numbers of formulas which are refutable in S are not separable in S.*

The proof is identical with the proof of 2.5.B in [3].

S is called S-undecidable if $T \notin \mathscr{R}^*_{0,1}$; it is called essentially S-undecidable if no consistent system $\mathrm{Ext}_S(X)$ is S-decidable. From I.7.1 we obtain the result that, under the assumptions made in §I.1,

I.7.2. *S is essentially S-undecidable.*

I.8. **Properties** (A), (C), **and** (S). Most of the formal systems encountered in practice enjoy one or more of the following fundamental properties:

(A) *If* $A, B \in \mathscr{R}_{k,l}$, *then* $A \cup B \in \mathscr{R}_{k,l}$, $k, l = 0, 1, 2, \cdots$.

(C) *If X is a consistent weakly representable set of formulas then the set*

$$M_{k,l}(X) = \{(\mathfrak{q}, p) : \breve{e}_p(\mathfrak{a}) \in Cn_S(D_{\mathfrak{q}}(\mathfrak{a}) \cup X)\}$$

is weakly representable, i.e., $M_{k,l}(X) \in \mathscr{R}_{k,l+1}(S)$, $k, l = 0, 1, 2, \cdots$.

The phrase "weakly representable set of formulas" means of course a set such that the set of the Gödel numbers of its elements is weakly representable.

A special case of (C) in which we assume $X = 0$ is called (C_0). Sets $M_{k,l}(0)$ are denoted simply by $M_{k,l}$.

(S) *For every* A, B in $\mathscr{R}_{k,l}$ *there is a formula* Φ *with k free functional variables and l free number variables such that*

$$\Phi(\mathfrak{a}) \in Cn_S(D_{\mathfrak{q}}(\mathfrak{a})),\ if\ \mathfrak{q} \in A - B\,,$$
$$\sim \Phi(\mathfrak{a}) \in Cn_S(D_{\mathfrak{q}}(\mathfrak{a})),\ if\ \mathfrak{q} \in B - A\,,$$
$$\Phi(\mathfrak{a}) \in Cn_S(D_{\mathfrak{q}}(\mathfrak{a}))\ or \sim \Phi(\mathfrak{a}) \in Cn_S(D_{\mathfrak{q}}(\mathfrak{a})),\ if\ \mathfrak{q} \in A \cup B\,.$$

Φ is called a separating formula for A and B.

Condition (S) was formulated for the first time (in a slightly weaker form and for a special system S) by Shepherdson [**15**]. Condition (C) was used implicitly by several authors.

I.8.1. *Each of the properties* (A), (C), (S) *is preserved when one passes from S to a system* $\mathrm{Ext}_S(D_\varphi(\alpha))$, *where α is an inessential constant and φ a function.*

I.8.2. (C_0) *implies that* $\{(\mathfrak{q}, p) : \mathfrak{q} \in U^{(k,l)}(p)\} \in \mathscr{R}_{k,l+1}(S)$ *and* $T \in \mathscr{R}_{0,1}(S) - \mathscr{R}^*_{0,1}(S)$.

Proof. The first part is identical with (C_0). In view of I.5.5 it is sufficient to prove that $T \in \mathscr{R}_{0,1}$. Let φ be a recursive function such that $e^{(0,1)}_{\varphi(\ulcorner\Phi\urcorner)} = \ulcorner\Phi \,\&\, (x = x)\urcorner$ for every closed formula Φ. Since[2]

$$\ulcorner\Phi\urcorner \in T \equiv \breve{e}^{(0,1)}_{\varphi(\ulcorner\Phi\urcorner)}(a) \in Cn_S(D_0(a)) \equiv (0, \varphi(\ulcorner\Phi\urcorner)) \in M_{0,1}$$

we infer by 3.6 that $T \in \mathscr{R}_{0,1}(S)$.

I.8.3. (S) *implies that $\mathscr{R}^*_{k,l}$ coincides with the family of sets $A \subseteq R_{k,l}$ such that $A \in \mathscr{R}_{k,l}$* and *$R_{k,l} - A \in \mathscr{R}_{k,l}$.*

Obviously every A in $\mathscr{R}^*_{k,l}$ belongs to $\mathscr{R}_{k,l}$ together with its complement (cf. I.3.1). If A and $R_{k,l} - A$ are in $\mathscr{R}_{k,l}$ then every separating formula for these sets strongly represents A in S.

I.8.4. (C) *and* (S) *imply that S is essentially incomplete*, i.e., that for every weakly representable consistent set X of formulas the system $S' = \mathrm{Ext}_S(X)$ is incomplete.

Proof. Condition (C) implies that the set T' of Gödel numbers of formulas provable in S' is weakly representable in S along with X. If S' were complete, then $T' \in \mathscr{R}_{0,1}(S)$ would imply $\omega - T' \in \mathscr{R}_{0,1}(S)$, whence by I.8.3 we would obtain $T' \in \mathscr{R}^*_{0,1}(S)$, whence by I.3.12 $T' \in \mathscr{R}^*_{0,1}(S')$. Since this contradicts I.5.5 we infer that S' is incomplete.

Theorem I.8.4 gives an abstract form of the incompleteness theorem of Rosser [**14**].

I.8.5. (C_0) *and* (S) *imply that the theorem of reduction* [7] *holds in $\mathscr{R}_{k,l}(S)$, $k, l = 0, 1, \cdots$. Hence the second separation principle holds for weakly representable sets and the first separation principle holds for complements of weakly representable sets.*

Proof. If $A, B \in \mathscr{R}_{k,l}(S)$ and Φ is a separating formula for these sets then the reduction of $A \cup B$ is effected by taking

$$A_1 = A \cap \{\mathfrak{q} : \Phi(\mathfrak{a}) \in Cn_S(D_{\mathfrak{q}}(\mathfrak{a}))\}\,,$$
$$B_1 = B \cap \{\mathfrak{q} : \sim \Phi(\mathfrak{a}) \in Cn_S(D_{\mathfrak{q}}(\mathfrak{a}))\}\,.$$

I.8.6. (C_0), (A), *and* (S) *imply that for every pair A, B of disjoint sets in $\mathscr{R}_{k,l}(S)$ there is a formula Θ such that Θ weakly represents A and $\sim\Theta$ weakly represents B.*

---

[2] $\ulcorner\Phi\urcorner$ denotes the Gödel number of the formula Φ.

PROOF. Let ν and σ be recursive functions such that $\nu(n) = \ulcorner\sim\breve{e}_n^{(k,l)}\urcorner$ and $\breve{e}_{\sigma(n)}^{(k,l)} = Sb(x/\delta_n)\,\breve{e}_n^{(k,l+1)}$. Sets

$$A^* = \{(\mathfrak{q}, n) : \mathfrak{q} \in A \vee (\mathfrak{q}, \sigma(\nu(n))) \in M_{k,l}\},$$
$$B^* = \{(\mathfrak{q}, n) : \mathfrak{q} \in B \vee (\mathfrak{q}, \sigma(n)) \in M_{k,l}\}$$

are weakly representable in S according to (A) and (C_0). Let $\breve{e}_q^{(k,l+1)}$ be a separating formula for A^* and B^*; we can assume that the last of its number variables is x. If Θ arises from $\breve{e}_q^{(k,l+1)}$ by the substitution of δ_q for x, then Θ has the desired properties.

Theorem I.8.6 gives an abstract form of a theorem due to Putnam and Smullyan [**13**]; the idea of the proof given here is due to Shepherdson [**15**].

I.8.7. (C), (A), *and* (S) *imply that if X is a weakly representable consistent set of formulas, then for every pair of disjoint sets A, B in $\mathscr{R}_{k,l}(S)$ there is a formula Θ such that Θ weakly represents A and $\sim\Theta$ weakly represents B in* $\mathrm{Ext}_S(X)$.

The proof is similar to that of I.8.6; the only difference is that we replace $M_{k,l}$ by $M_{k,l}(X)$.

I.8.8. (C), (A), *and* (S) *imply that* $\mathscr{R}_{k,l}(S) = \mathscr{R}_{k,l}(\mathrm{Ext}_S(X))$ *for every consistent weakly representable set X of formulas.*

PROOF. The inclusion $\mathscr{R}_{k,l}(S) \subseteq \mathscr{R}_{k,l}(\mathrm{Ext}_S(X))$ follows from I.8.7. If $A \in \mathscr{R}_{k,l}(\mathrm{Ext}_S(X))$ and $\breve{e}_p$ represents A in $\mathrm{Ext}_S(X)$, then $\mathfrak{q} \in A \equiv (\mathfrak{q}, p) \in M_{k,l}(X)$ whence $A \in \mathscr{R}_{k,l}(S)$, since, by (C), $M_{k,l}(X)$ belongs to $\mathscr{R}_{k,l}(S)$.

Theorem I.8.8 represents an abstract form of a theorem discovered by Ehrenfeucht and Feferman [**1**].

It is an open question whether the operation Ext preserves any of the properties (A), (C). For the property (S) the answer is negative as is obvious from the observation that I.8.3 is false for the system S_β discussed in [**10**]; cf. the remark following Theorem I.4.4. However, the property (S) is preserved under finite extensions:

I.8.9. *If S has the property* (S) *then so does the system* $S' = \mathrm{Ext}_S(\{\Pi\})$, *where Π is an arbitrary closed formula.*

PROOF. If $A, B \in \mathscr{R}_{k,l}(S')$, then $A, B \in \mathscr{R}_{k,l}(S)$, cf. I.4.4. Let Φ be a separating formula for A and B in S. We easily show that it is a separating formula for A and B in S'.

CHAPTER II. APPLICATIONS OF THE GENERAL THEORY

II.1. **System** S_0. The function of consequence for this system is simply the function Cn_0. Thus, apart from the existence of inessential constants, S_0 is identical with the system A of [**3**].

II.1.1. *S_0 satisfies axioms* (I) — (V) *and the condition* $Cn_0(Cn_0(X)) = Cn_0(X)$.

The notions of recursive and recursively enumerable subsets of $R_{0,1}$ are known. Subsets of $R_{1,0}$ are called recursive or recursively enumerable if they are unions of recursive (recursively enumerable) sets of neighbourhoods in the Baire space $R_{1,0}$ under the usual numbering of neighbourhoods. These definitions can be generalized in an obvious way to subsets of $R_{k,l}$ for arbitrary k, l. With these definitions we have:

II.1.2. *$\mathscr{R}_{k,l}(S_0)$ and $\mathscr{R}^*_{k,l}(S_0)$ coincide with the families of recursively enumerable and of recursive subsets of $R_{k,l}$.*

II.1.3. *$\mathscr{R}_{k,l}(\mathrm{Ext}_{S_0}(D_\varphi(\alpha)))$ and $\mathscr{R}^*_{k,l}(\mathrm{Ext}_{S_0}(D_\varphi(\alpha)))$ coincide with the families of sets which are recursively enumerable (recursive) in φ.*

II.1.4. *The family $\mathscr{F}_{1,0;0,1}(S_0)$ consists of the mappings f: $R_{1,0} \to \omega$ such that $f(\varphi)$ is a recursive functional in the sense of* [4]; *the family $\mathscr{F}_{1,0;1,0}(S_0)$ consists of the mappings f: $R_{1,0} \to R_{1,0}$ with the following property: there is a recursive functional F with one functional and one numerical variable such that $f(\varphi) = \psi \equiv (n)[\psi(n) = F(\varphi, n)]$.*

The characterization given in II.1.4 can easily be extended to functions in $\mathscr{F}_{k,l;m,n}$ for arbitrary k, l, m, n. Functions of this family will be called recursive.

II.1.5. *S_0 satisfies conditions* (A), (C), *and* (S).

Proof. (A) is obvious from the properties of recursively enumerable sets. (C) follows from the possibility of expressing the relation of provability in S_0 by an existential statement whose initial quantifier binds a numerical variable and has as its scope a formula which defines a recursive relation. (S) is proved as follows.

Let A, B be recursively enumerable sets, let Γ, Δ be formulas which represent them in S_0 and let Π be a formula with the free variables $x, y, \mathfrak{v}$ such that Π strongly represents in S_0 the following relation P:

m is the Gödel number of a formal proof of $\mathfrak{p}$ from $D_{\mathfrak{q}}(\mathfrak{a})$.

Hence $\Pi(\delta_m, \delta_p, \mathfrak{a}) \in Cn_0(D_{\mathfrak{q}}(\mathfrak{a}))$ if $P(m, p, \mathfrak{q})$, and $\sim \Pi(\delta_m, \delta_p, \mathfrak{a}) \in Cn_0(D_{\mathfrak{q}}(\mathfrak{a}))$ if non $P(m, p, \mathfrak{q})$. Repeating the proof of Rosser [**14**] we show that the formula

$$(Ex)\{\Pi(x, \delta_{\ulcorner\Gamma\urcorner}, \mathfrak{v}) \,\&\, (x')[x' < x \supset {\sim}\Pi(x', \delta_{\ulcorner\Delta\urcorner}, \mathfrak{v})]\}$$

is separating for A and B.

II.1.6. *Systems $\mathrm{Ext}_{S_0}(D_\varphi(\alpha))$, where α is an inessential constant of S_0, satisfy conditions* (A), (C), (S).

This theorem which results from I.8.1 and II.1.5 explains why sets recursive or recursively enumerable in arbitrarily given functions have properties similar to absolutely recursive and recursively enumerable sets.

System S_0 is known to possess various properties which do not follow from the general theory of Chapter I (e.g. the existence of effectively inseparable recursively enumerable sets). We shall not deal here with these properties since our aim is to show how much can be already obtained from the general assumptions made in Chapter I and not to develop the theory of recursive sets and their generalizations.

II.2. **Systems S_π.** These are systems obtained from S_0 by the repeated use of the rule ω. The precise definition runs as follows: For a set X of formulas we put

$$F(X) = \{\Phi : (Eq)[((x)[\tilde{e}_q^{(0,1)} \supset \Phi] \in Cn_0(0)) \,\&\, (n)(Sb(x/\delta_n)\tilde{e}_q^{(0,1)} \in X)\};$$
$$Cn_{\pi+1}(X) = F(Cn_\pi(X)),\quad Cn_\lambda(X) = \bigcup_{\pi<\lambda} Cn_\pi(X).$$

Here λ and π are ordinals and λ is a limit ordinal. We define S_π as the system whose function of consequence is Cn_π. By an easy induction on π we obtain

II.2.1. *S_π satisfies conditions* (I) — (V).

II.3. **Constructive definition of systems S_π.** Before we can discuss further properties of systems S_π we must introduce some definitions.

Let φ be an arbitrary function and α a constant inessential for S_0. We denote by R^φ a ternary relation which is recursive in φ and universal for the family of relations (i.e. subsets of $R_{0,2}$) primitive recursive in φ. Let W^φ be the set of e in ω such that R_e^φ is a nonempty well-ordering relation. The order type of R_e^φ is denoted by $|e|^\varphi$. From [**9**] it is known that

$$\{|e|^\varphi : e \in W^\varphi\} = \{\pi : \pi < \omega_1^\varphi\} = \omega_1^\varphi ,$$

where ω_1^φ is the first ordinal not constructible in φ.

Along with the functions Cn_π we shall consider auxiliary functions $\overline{Cn}_e^\varphi$, where e is in W^φ. Whereas the definition of Cn_π would be unacceptable to a constructivistically minded mathematician, the definition of $\overline{Cn}_e^\varphi$ would be almost acceptable for him.

$\Phi \in \overline{Cn}_e^\varphi(X)$ if and only if there is a function $\mathfrak{F}$ whose domain coincides with the field of R_e^φ, whose range consists of sets of formulas and which satisfies the following conditions:

(a) if n_0 is the minimum of R_e^φ, then $\mathfrak{F}(n_0) = Cn_0(X)$;

(b) if n is the successor of n_1 in the ordering R_e^φ, then $\mathfrak{F}(n) = F(\mathfrak{F}(n_1))$;

(c) if n is a limit element of the field of R_e^φ, then $F(n) = \bigcup F(j)$ with summation extended over j such that $R_e^\varphi(j, n)$ and $j \neq n$;

(d) $\Phi \in \bigcup \mathfrak{F}(j)$, where the summation is extended over the field of R_e^φ.

If $e \in W^\varphi$, then there is exactly one function $\mathfrak{F}$ satisfying (a) — (d) for each set X.

II.3.1. $\overline{Cn}_e^\varphi(X) = Cn_{|e|^\varphi}(X)$ *for e in W^φ.*

PROOF. By an easy induction on $|e|^\varphi$.

Let P be a formula which strongly represents R^φ in $\mathrm{Ext}_{S_0}(D_\varphi(\alpha))$; thus P has 3 free numerical variables and one inessential functional constant α. Let $\Delta(\alpha, t)$ be the formula

$$\begin{gathered}(x, y, z)(\{P(\alpha, t, x, x) \,\&\, P(\alpha, t, y, y) \equiv [P(\alpha, t, x, y) \vee P(\alpha, t, y, x)]\} \\ \&\, [P(\alpha, t, x, y) \,\&\, P(\alpha, t, y, x) \supset (x = y)] \,\&\, [P(\alpha, t, x, y) \,\&\, \\ P(\alpha, t, y, z) \supset P(\alpha, t, x, z)] \,\&\, (\beta)(Ex)\{P(\alpha, t, \beta(x+1), \beta(x)) \supset \\ [\beta(x) = \beta(x+1)]\}) .\end{gathered}$$

This formula is obtained by expressing in the language of S_0 the usual definition of well-ordering.

II.3.2. *If $e \in W^\varphi$ and $|e|^\varphi \geqq 1$, then $\Delta(\alpha, \delta_e) \in Cn_{|e|^\varphi+1}(D_\varphi(\alpha))$.*

The proof proceeds by induction on $|e|^\varphi$ and uses the fact that in S_0 we can prove a formula expressing the well-known set-theoretical theorem saying that a relation well-orders its field if and only if every segment of the field is well-ordered.

Kreisel and the author (independently of each other) have shown that there are integers e in W^φ such that $\Delta(\alpha, \delta_e)$ non $\in Cn_\pi(D_\varphi(\alpha))$ for $\pi < |e|^\varphi$. Kreisel calls such integers and the corresponding ordinals $|e|^\varphi$ "autonomous". Their existence shows that Theorem II.3.3 cannot be strengthened by replacing $|e|^\varphi$ by a smaller ordinal.

II.3.3. *There are formulas* $\Gamma(t, x, y, \mathfrak{v}, \xi)$ *and* $\Lambda(t, y, \mathfrak{v}, \xi)$ *such that for every formula* Φ *and every* e *in* W^φ *the following conditions are satisfied*:

(i) *if* $|e|^\varphi \geqq 1$ *and* $R_e^\varphi(f, f)$, *then*

$$\Phi \in \overline{Cn}_f^\varphi(D_\mathfrak{q}(\mathfrak{a})) \equiv \Gamma(\delta_e, \delta_f, \delta_{\ulcorner\Phi\urcorner}, \mathfrak{a}, \alpha) \in Cn_{|e|^\varphi}(D_\mathfrak{q}(\mathfrak{a}) \cup D_\varphi(\alpha)) ,$$

(ii) *if* $|e|^\varphi \geqq 1$, *then*

$$\Phi \in \overline{Cn}_e^\varphi(D_\mathfrak{q}(\mathfrak{a})) \equiv \Lambda(\delta_e, \delta_{\ulcorner\Phi\urcorner}, \mathfrak{a}, \alpha) \in Cn_{|e|^\varphi}(D_\mathfrak{q}(\mathfrak{a}) \cup D_\varphi(\alpha)) ,$$

(iii) *if* $|e|^\varphi \geqq 1$, *then*

$$\Phi \notin \overline{Cn}_e^\varphi(D_\mathfrak{q}(\mathfrak{a})) \equiv \sim\Lambda(\delta_e, \delta_{\ulcorner\Phi\urcorner}, \mathfrak{a}, \alpha) \in Cn_{|e|^\varphi+1}(D_\mathfrak{q}(\mathfrak{a}) \cup D_\varphi(\alpha)) ,$$

(iv) Λ *is satisfied in the standard model of* S_0 *under the interpretation of* $t, y, \mathfrak{v}, \xi$ as $e, \ulcorner\Phi\urcorner, \mathfrak{q}, \varphi$ *if and only if* $\Phi \in \overline{Cn}_e^\varphi(D_\mathfrak{q}(\mathfrak{a}))$.

PROOF (IN OUTLINE). To construct the formula Γ we express in the language of S_0 the arithmetized definition of the relation $\Phi \in \overline{Cn}_f^\varphi(D_\mathfrak{q}(\mathfrak{a}))$. The formula which we obtain in this way has the form

$$(\eta)[Z_1(t, \xi, \eta) \,\&\, Z_2(t, \xi, \eta, \mathfrak{v}) \supset \eta(2^x(2y + 1)) = 0] ,$$

where Z_1 and Z_2 can be described as follows:

$Z_1(t, \xi, \eta)$ is the formula

$$(u, v)[\eta(2^u(2v + 1)) = 0 \supset P(\xi, t, u, u)]$$

(i.e. Z_1 "says" that $\eta(2^u(2v + 1)) = 0$ implies that u is in the field of R_t^ξ).

$Z_2(t, \xi, \eta, \mathfrak{v})$ is a conjunction of three formulas each of which gives necessary and sufficient conditions for the vanishing of $\eta(2^u(2v + 1))$ in cases (a) when u is the minimum of R_t^ξ; (b) when u is the successor of an element u_1; (c) when u is a limit element. The formulas describe (in the language of S) the three situations described in points (a), (b), (c) of the definition of $\overline{Cn}_e^\varphi(X)$.

Having constructed the formula Γ we show that

(1°) Γ is satisfied in the standard model of S_0 with $t, x, y, \xi, \mathfrak{a}$ interpreted as $e, f, \ulcorner\Phi\urcorner, \varphi, \mathfrak{q}$ if and only if $\Phi \in \overline{Cn}_f^\varphi(D_\mathfrak{q}(\mathfrak{a}))$.

(2°) If f is in the field of R_e^φ, if τ_f is the order type of the segment of this field determined by the successor of f and if $\Phi \in \overline{Cn}_f^\varphi(D_\mathfrak{q}(\mathfrak{a}))$, then

$$\Gamma(\delta_e, \delta_f, \delta_{\ulcorner\Phi\urcorner}, \mathfrak{a}, \alpha) \in Cn_{\max(1, \tau_f)}(D_\mathfrak{q}(\mathfrak{a}) \cup D_\varphi(\alpha)) .$$

The proof of (1°) is straightforward and the proof of (2°) proceeds by induction on τ_f.

The implication from right to left in (i) results from (1°) and the implication from left to right from (2°).[3]

[3] It is rather significant that we proved this theorem by means of semantical considerations. Probably a syntactical proof would allow us to obtain much stronger results and in particular to characterize the family $\mathscr{R}_{\kappa, \iota}(\mathrm{Exts}_\pi(X))$ with an arbitrary X. We do not know, however, whether a purely syntactical proof exists.

(ii) results from (i) by taking as Λ the formula $(Ex)\Gamma(t, x, y, \mathfrak{v}, \xi)$.

(iii) is proved similarly as (ii), but uses the lemma

$$\Delta(\xi, t) \supset (E!\,\eta)[Z_1(t, \xi, \eta) \,\&\, Z_2(t, \xi, \eta, \mathfrak{v})] \in Cn_0(0)$$

and Theorem II.3.2. (To explain why the lower index in (iii) is $|e|^{\varphi} + 1$ and not simply $|e|^{\varphi}$ as in (ii) we remark that $\sim\Lambda$ begins with a general quantifier and so we must apply once more the rule ω in order to prove this formula.)

(iv) is a direct corollary from (ii) and (iii).

The theorems of this section have nonrelativised versions which we obtain by taking as φ e.g. the constant function 0. When referring to these nonrelativised versions we shall simply omit the index φ and the constant α (or the variable ξ) in the formulas Δ, Γ, and Λ and in the symbol ω_1^{φ}.

II.4. **Properties of** S_π, $\pi < \omega_1$.

II.4.1. *Systems* S_π *for* $1 \leqq \pi < \omega_1$ *satisfy conditions* (A), (C_0) *and* (S).

Proof. Let e be an integer such that $|e| = \pi$.

(A) If Φ, Ψ weakly represent sets $A, B \subseteq R_{k,l}$, then the formula

$$\Lambda(\delta_e, \delta_{\ulcorner \Phi(\mathfrak{a}) \urcorner}, \mathfrak{a}) \vee \Lambda(\delta_e, \delta_{\ulcorner \Psi(\mathfrak{a}) \urcorner}, \mathfrak{a})$$

represents $A \cup B$ in S_π.

(C_0) From the nonrelativised version of II.3.3 and from II.3.1

$$(\mathfrak{q}, p) \in M_{k,l} \equiv \breve{e}_p^{(k,l)}(\mathfrak{a}) \in Cn_\pi(D_{\mathfrak{q}}(\mathfrak{a})) \equiv$$
$$\Lambda(\delta_e, \delta_{g(p)}, \mathfrak{a}) \in Cn_\pi(D_{\mathfrak{q}}(\mathfrak{a})),$$

where $g(p) = \ulcorner \breve{e}_p^{(k,l)}(\mathfrak{a}) \urcorner$. These equivalences prove that if Θ represents g in S_0 then the formula $(Ex)[\Theta(a, x) \,\&\, \Lambda(\delta_e, x, \mathfrak{a})]$ represents $M_{k,l}$ in S_π.

(S) Let $A, B \subseteq R_{k,l}$, let Φ, Ψ represent A, B in S_π and let Θ be the formula

$$(Ex)\{P(\delta_e, x, x) \,\&\, \Gamma(\delta_e, x, \delta_{\ulcorner \Phi(\mathfrak{a}) \urcorner}, \mathfrak{a}) \,\&$$
$$(x')[P(\delta_e, x', x) \,\&\, (x' \neq x) \supset \sim\Gamma(\delta_e, x', \delta_{\ulcorner \Psi(\mathfrak{a}) \urcorner}, \mathfrak{a})]\}.$$

Using an argument similar to that of Rosser [**14**] we show that Θ is a separating formula; the difference between this proof and that of Rosser is that our proof uses the well-ordering R_e instead of $\leqq$.

II.4.2. *Systems* S_π, $1 \leqq \pi < \omega_1$ *do not satisfy conditon* (C).

Proof. $Cn_\pi(0)$ is representable (weakly) in S_π. By I.8.8 condition (C) would imply that $\mathscr{R}_{k,l}(S_\pi) = \mathscr{R}_{k,l}(\mathrm{Ext}_{S_\pi}(Cn_\pi(0))$ which is false, because $\mathrm{Ext}_{S_\pi}(Cn_\pi(0)) = Cn_{\pi.2}(0)$.

II.4.3. *The family* $\mathscr{R}_{0,1}(S_\pi)$, $\pi < \omega_1$, *consists of hyperarithmetic sets*; *every hyperarithmetic set of integers belongs to one of the families* $\mathscr{R}_{0,1}(S_\pi)$, $\pi < \omega_1$.

Proof. The first part follows from the observation that the set of Gödel numbers of formulas in $Cn_\pi(0)$ is hyperarithmetic (as we easily see from the uniqueness of the function $\mathfrak{F}$ used in the definition of $\overline{Cn_e^{\varphi}}$). The second part is proved by showing that for each e in O the set H_e of Kleene [**5**] is weakly representable in $S_{|e|}$.

Theorem II.4.3 reduces the theory of hyperarithmetic sets of integers to the theory of sets representable in systems S_π, $\pi < \omega_1$.

II.5. **Properties of the system S_Ω.**

II.5.1. *The family $\mathscr{R}_{k,l}(S_\Omega)$ consists of Π_1^1 sets, i.e., of sets* $\{\mathfrak{q} : (\varphi)R(\varphi, \mathfrak{q})\}$, *where R is arithmetic.*

PROOF. Evaluating the predicate $\breve{n}(\mathfrak{a}) \in Cn_\Omega(D_\mathfrak{q}(\mathfrak{a}))$ we show easily that it is of type Π_1^1, cf. [14]. Hence so are weakly representable sets. Weak representability of the Π_1^1 sets is an immediate consequence of Orey's theorem [12] (strictly speaking the proof given by Orey is applicable only to sets of integers, but a generalization to the more general case presents no difficulties).

II.5.2. *S_Ω satisfies the condition $Cn_\Omega(Cn_\Omega(X)) = Cn_\Omega(X)$ for every set of formulas.*

PROOF. $Cn_\Omega(X)$ is the smallest set containing X and closed with respect to the rules of proof of S_0 and to the rule ω.

In II.5.4 we shall show that Ω cannot be replaced in II.5.2 by any smaller ordinal. The proof is based on the following important theorem discovered by Shoenfield (unpublished) and Spector [16]:

II.5.3. *If X is a set of formulas and χ the characteristic function of the set of its Gödel numbers, then $Cn_\Omega(X) = Cn_{\omega_1^\chi}(X)$.*

II.5.4. $Cn_\Omega(D_\varphi(\alpha)) = Cn_{\omega_1^\varphi}(D_\varphi(\alpha)) \neq Cn_\pi(D_\varphi(\alpha))$ for $\pi < \omega_1^\varphi$.

PROOF. The equation is an immediate consequence of II.5.3. In order to prove the second part we consider formulas $\Delta(\delta_e, \alpha)$ defined in §II.3. The set of those of the formulas $\Delta(\delta_e, \alpha)$ which are true in the standard model when α is interpreted as φ is not hyperarithmetic in φ, since otherwise so would be the set W^φ which is known to be false [5]. Hence there is no $\pi < \omega_1^\varphi$ such that $\Delta(\delta_e, \alpha) \in Cn_\pi(D_\varphi(\alpha))$ for all e in W^φ. On the other hand, $\Delta(\delta_e, \alpha) \in Cn_\Omega(D_\varphi(\alpha))$ by II.3.2. This shows that the sequence $Cn_\pi(D_\varphi(\alpha))$ is strictly increasing for $\pi < \omega_1^\varphi$.

II.5.5. *There is a formula Ξ such that for each formula Φ*

$$\Phi \in Cn_\Omega(D_\mathfrak{q}(\mathfrak{a})) \equiv \Xi(\delta_{\ulcorner\Phi\urcorner}, \mathfrak{a}) \in Cn_\Omega(D_\mathfrak{q}(\mathfrak{a})) .$$

PROOF. For simplicity's sake we assume that $\mathfrak{q}$ consists of but one function φ. Take as Ξ the formula

$$(Et)[\Delta(\xi, t) \,\&\, \Lambda(t, y, \xi, \xi)] .$$

If $\Xi(\delta_{\ulcorner\Phi\urcorner}, \alpha) \in Cn_\Omega(D_\varphi(\alpha))$, then Ξ is true in the standard model when α is interpreted by φ and hence there is an e such that $\Delta(\delta_e, \alpha)$ and $\Lambda(\delta_e, \delta_{\ulcorner\Phi\urcorner}, \alpha, \alpha)$ are true under the same interpretation of α. Hence e is in W^φ and (cf. II.3.3 (iv)) $\Phi \in \overline{Cn}_e^\varphi(D_\varphi(\alpha)) \subseteq Cn_\Omega(D_\varphi(\alpha))$.[3]

If $\Phi \in Cn_\Omega(D_\varphi(\alpha))$, then by II.5.4 there is a $\pi < \omega_1^\varphi$ such that $\Phi \in Cn_\pi(D_\varphi(\alpha))$ and hence $\Phi \in \overline{Cn}_e^\varphi(D_\varphi(\alpha))$ for an e in W^φ. Using II.3.2 and II.3.3 (ii) we obtain $\Xi(\delta_{\ulcorner\Phi\urcorner}, \alpha) \in Cn_\Omega(D_\varphi(\alpha))$.

II.5.6 *S_Ω satisfies conditions* (A), (C), *and* (S).

(A) If Φ, Ψ weakly represent A, B in S_Ω then the formula $\Xi(\delta_{\ulcorner\Phi(\mathfrak{a})\urcorner}, \mathfrak{a}) \vee \Xi(\delta_{\ulcorner\Psi(\mathfrak{a})\urcorner}, \mathfrak{a})$ weakly represents $A \cup B$ in S_Ω.

(C) The predicate

$$\breve{e}_p^{(k,l)}(\mathfrak{a}) \in Cn_\Omega(D_\mathfrak{q}(\mathfrak{a}) \cup X))$$

is of type Π_1^1, cf. [**14**]. Hence $M_{k,l}(X)$ is of type Π_1^1 and hence it is weakly representable in S_Ω.

(S) We limit ourselves to the case when A and B are subsets of $R_{1,0}$. Let Φ, Ψ weakly represent A, B in S_Ω. Then the following formula Θ

$$(Et)\{\Delta(\xi, t) \,\&\, \Lambda(t, \delta_{\ulcorner\Phi(\alpha)\urcorner}, \xi, \xi) \,\&\, (t')[P(\xi, t, t', t') \supset \sim\Lambda(t', \delta_{\ulcorner\Psi(\alpha)\urcorner}, \xi, \xi)]\}$$

is a separating formula for A and B. To prove, e.g., that $\varphi \in A - B$ implies $\Theta(\alpha) \in Cn_\Omega(D_\varphi(\alpha))$ we choose according to II.5.4 an integer e in W^φ such that $\Phi(\alpha) \in \overline{Cn}_e^\varphi(D_\varphi(\alpha))$ and then proceed similary as in Rosser's proof [**14**] replacing everywhere the less-than relation by R_e^φ.

II.6. **Shoenfield's rule.** This rule[4] is analogous to the rule ω. We say that Φ is provable from a set X of formulas by means of the Shoenfield rule if there is a formula Ψ with exactly one free variable x such that

$$(x)[\Delta(x) \supset \Psi(x)] \supset \Phi \in Cn_0(0)\,,$$
$$\Psi(\delta_e) \in X \textit{ for every } e \textit{ in } W\,.$$

(W is here the (nonrelativised) set of integers such that R_e is a well-ordering and Δ is a formal definition of W; cf. §II.3.)

Let Cn_π^Σ be a function analogous to Cn_π, but based on the Shoenfield rule instead of the rule ω. Let Σ_Ω be the system based on Cn_Ω^Σ as the function of consequence.

II.6.1. $Cn_{\pi+1}^\Sigma(X) = \mathrm{Ext}_{S_{\pi+1}}(X \cup Z_0)$, *where Z_0 is the set of formulas $\sim\Delta(\delta_e)$ with e non$\in W$.*

We obtain a proof of this theorem by showing that (a) every application of the rule ω can be replaced by an application of the Shoenfield rule; (b) $Z_0 \subseteqq Cn_1^\Sigma(0)$; (c) if $\Psi(\delta_e) \in Cn_{S_\pi}(X \cup Z_0)$ for all e in W, then $\Delta(\delta_n) \supset \Psi(\delta_n) \in Cn_{S_\pi}(X \cup Z_0)$ for all n.

Theorem II.6.1 reduces the Shoenfield rule to the ordinary ω rule. As a corollary we obtain

II.6.2. *Σ_Ω satisfies conditions* (I) — (V) *and the equation $Cn_\Omega^\Sigma(Cn_\Omega^\Sigma(X)) = Cn_\Omega^\Sigma(X)$.*

II.6.3. *Sets weakly representable in Σ_Ω coincide with sets of type Π_1^1 relatively to W; sets strongly representable in Σ_Ω coincide with sets hyperarithmetic relatively to W.*

The second part follows from the first. From the evaluation of the predicate $\Phi(\mathfrak{a}) \in Cn_\Omega(D_{\mathfrak{q}}(\mathfrak{a}) \cup Z_0)$ which is of type Π_1^1 relatively to Z_0 it follows that sets weakly representable in Σ_Ω are of type Π_1^1 relatively to W. It remains to show that if A is of this type, then it is weakly representable in Σ_Ω.

Let us assume that there is a relation Q recursive in W such that $\mathfrak{q} \in A \equiv (\psi)(Ep)Q(\bar\psi(p), \mathfrak{q})$. If Φ strongly represents the relation $Q(\bar\psi(p), \mathfrak{q})$ in $\mathrm{Ext}_{S_0}(D_{\chi_W}(\alpha))$, then the formula Ψ: $(\xi)(Ex)\Phi(\xi, x, \mathfrak{v}, \alpha)$ has the following property: $\mathfrak{q} \in A$ if and only if Ψ is satisfied in an arbitrary ω-model of S_0 containing χ_W and $\mathfrak{q}$ under the interpretation of $\mathfrak{v}$ as $\mathfrak{q}$ and α as χ_W. It follows

[4] Its use was suggested by Shoenfield in a conversation.

that $\mathfrak{q} \in A \equiv \Psi(\mathfrak{a}, \alpha) \in Cn_\Omega(D_{\mathfrak{q}}(\mathfrak{a}) \cup D_{\chi_W}(\alpha))$. Now let $X(\alpha)$ be the formula $(x)\{\alpha(x) \leqq 1 \,\&\, [\alpha(x) = 0 \equiv \Delta(x)]\}$. In view of II.3.2 $D_{\chi_W}(\alpha) \subseteq Cn_\Omega(Z_0 \cup \{X(\alpha)\})$ and hence $\mathfrak{q} \in A$ implies $\Psi(\mathfrak{a}, \alpha) \in Cn_\Omega(D_{\mathfrak{q}}(\mathfrak{a}) \cup Z_0 \cup \{X(\alpha)\})$, whence $(\xi)[X(\xi) \supset \Psi(\mathfrak{a}, \xi)] \in Cn_\Omega^\Sigma(D_{\mathfrak{q}}(\mathfrak{a}))$. Conversely, if this condition is satisfied, then $\Psi(\mathfrak{a}, \alpha)$ is satisfied in the standard model under the interpretation of α as χ_W and of $\mathfrak{a}$ as $\mathfrak{q}$, whence $\mathfrak{q} \in A$. Hence A is weakly representable in Σ_Ω.

II.6.4. *Σ_Ω has properties* (A), (C), *and* (S).

(A) The union of two sets of type Π_1^1 in W is of the same type.

(C) If X is a set of formulas such that the set of its Gödel numbers is a Π_1^1 set in W, then the predicate $\breve{e}_p^{(k,l)}(\mathfrak{a}) \in Cn_\Omega(X \cup Z_0 \cup D_{\mathfrak{q}}(\mathfrak{a}))$ is of the same type.

(S) Since $\mathrm{Ext}_{S\Omega}(D_{\chi_W}(\alpha))$ satisfies (S), the same is true of $\mathrm{Ext}_{S\Omega}(D_{\chi_W}(\alpha) \cup \{X(\alpha)\})$, where $X(\alpha)$ is the formula used in the proof of II.6.2 (cf. I.8.9). Since, as can easily be shown,

$$\Phi(\mathfrak{a}, \alpha) \in Cn_\Omega(X \cup D_{\chi_W}(\alpha) \cup \{X(\alpha)\}) \equiv$$
$$(\xi)[X(\xi) \supset \Phi(\mathfrak{a}, \xi)] \in Cn_\Omega(X \cup Z_0)\,,$$

we infer that $\mathscr{R}_{k,l}(\mathrm{Ext}_{\Sigma\Omega}(X)) = \mathscr{R}_{k,l}(\mathrm{Ext}_{S\Omega}(D_{\chi_W}(\alpha) \cup \{X(\alpha)\}))$ which proves the theorem.

Chapter III. Systems with nondenumerable sets of constants

In Chapter III we shall try to obtain some parts of the "classical" theory of projective sets within the frame of the theory of representability. Systems described previously are not suitable for this purpose because the classical theory of projective sets treats each function (i.e. each point in the Baire zero space) as an individually given very simple entity, whereas in the theory set forth previously a function can only be described by an infinite set of formulas.

Following an idea of Kreisel [8] we shall consider systems whose language contains $2^{\aleph_0}$ constants serving to denote individual functions. We call these systems infinitistic.

The language of an infinitistic system contains the same variables, constants, functors, and predicates as the language of S_0. In addition it contains $2^{\aleph_0}$ constants γ_φ, where φ runs over the set of all functions. Logical symbols available in the infinitistic systems are the same as in the finitistic ones. The rules of forming expressions are similar in both systems, the expressions $\gamma_\varphi(x)$ and $\gamma_\varphi(\delta_n)$ being treated as terms denoting numbers.

Similarly as in the finitistic case each infinitistic system is characterized by a function of consequence satisfying conditions (I) — (V) and the additional condition $\gamma_\varphi(\delta_n) = \delta_{\varphi(n)} \in Cn_S(0)$ for arbitrary φ and n.

In the present paper we shall not discuss consequences of these assumptions in the general case, but shall limit ourselves to some particular cases.

III.1. **Arithmetization.** Since formulas of the infinitistic systems may contain constants γ_φ, we cannot use integers to arithmetize these systems. Instead we shall use a mapping of formulas on functions. A mapping of this kind can be obtained as follows.

Let us add to the language of S_0 an infinite number of functional variables $\eta_0, \eta_1, \cdots$ which cannot be bound by quantifiers. A formula Φ of an infinitistic system can be obtained from a formula Φ_0 of the system S_0 (extended by the symbols η_i) by substituting functional constants γ_φ for all the symbols η which occur in Φ_0. Let $\eta_{j_0}, \eta_{j_1}, \cdots, \eta_{j_{k-1}}$ be all these symbols and assume that Φ is obtained from Φ_0 by substituting γ_{φ_i} for η_{j_i}, $i = 0, 1, \cdots, k-1$. Put

$$\varphi(0) = \ulcorner\Phi_0\urcorner\,, \qquad \varphi(kn+s+1) = \varphi_{j_s}(n)\,, \qquad 0 \leqq s < k, n = 0, 1, 2, \cdots$$

We take φ as a "functional Gödel number" of Φ. Since Φ_0 is not uniquely determined by Φ, every formula of the infinitistic system may have (and in general does have) many functional Gödel numbers.

Our definition correlates Gödel numbers only with formulas which contain at least one functional constant. To simplify our exposition we shall exclude other formulas from further consideration, replacing, if necessary, Φ by Φ & $(\gamma_\varphi = \gamma_\varphi)$.

III.1.1. *There is an arithmetically defined function f: $R_{1,0} \to R_{1,0}$ such that if φ is a Gödel number of a formula Φ, then $f(\varphi)$ is a Gödel number of $\sim\Phi$. Similarly for other connectives of the propositional calculus, for quantifiers and for the operation $Sb(\xi_n/\gamma_\varphi)\Phi_1$.*

III.1.2. *Gödel numbers of the axioms of the propositional and predicate calculus, of the axioms of S_0 and of the axioms $\gamma_\varphi(\delta_n) = \delta_{\varphi(n)}$ constitute a set which is open in the space $R_{1,0}$.*

III.1.3. *For every rule of proof the relation: the formula with the Gödel number φ_1 arises from a formula with the Gödel number φ_2 (or from formulas with the Gödel numbers φ_2, φ_3) is arithmetically definable.*

III.2. **System S_0^∞.** Let $Cn_0^\infty(X)$ be the smallest set containing X, all the axioms of the propositional and functional calculi and of the second order arithmetic, all the axioms which have the form $\gamma_\varphi(\delta_n) = \delta_{\varphi(n)}$, and closed with respect to the usual rules of proof. In the axiom schemata (e.g., of the propositional calculus) the schematic letters $\Phi, \Psi, \cdots$ are to be replaced by arbitrary formulas of the infinitistic system. Let S_0^∞ be the system in which Cn_0^∞ is the function of consequence.

A proof in S_0^∞ is a finite sequence of formulas. Via arithmetization we can enumerate proofs in S_0^∞ in such a way that a single function be the Gödel number of a proof. The set of Gödel numbers of proofs is arithmetically definable. From this remark it follows that:

III.2.1 *Sets weakly representable in* $\mathrm{Ext}_{S_0^\infty}(X)$ *are analytic in the set of the Gödel numbers of formulas which belong to X.*

In case $X = 0$ we have a stronger result:

III.2.2. *Every set which is weakly representable in S_0^∞ and contained in $R_{k,l}$ is open in the usual topology of $R_{k,l}$. The same is true for systems* $\mathrm{Ext}_{S_0^\infty}(X)$ *under the assumption that the set of Gödel numbers of formulas which belong to X is open in $R_{1,0}$.*

The proof follows from the observation that small changes of the functions φ for which γ_φ occurs in a formal proof change the given proof again in a formal proof.

III.2.3. *Every open set* $A \subseteq R_{k,l}$ *is weakly representable in* S_0^∞. *If* X *is* ω-*consistent, then* A *is weakly representable in* $\mathrm{Ext}_{S_0^\infty}(X)$.

The proof is obtained by expressing in S_0^∞ the definition of an open set as a union of denumerably many neighbourhoods. In the formula obtained in this way a constant γ_φ occurs, where φ is a function enumerating the neighbourhoods in question.

It is an open problem whether S_0^∞ satisfies conditions (C) and (S); condition (A) is obviously satisfied.

III.3. **Systems** S_π^∞.[5] Using the same ω-rule as in Chapter II we define as in §II.2 functions Cn_π^∞; these functions yield sets of formulas of the infinitistic system when applied to such sets. Let S_π^∞ be the system with Cn_π^∞ as the function of consequence and let T_π^∞ be the set of Gödel numbers of the formulas provable in S_π^∞.

Every proof in S_π^∞ can be represented as a graph. Let G be a directed denumerable graph with one initial vertex V_0 and such that every vertex has either none or denumerably many immediate successors. We assume that G is well-founded. It is then easy to correlate with G an ordinal called the height of G. A normal covering of G is a mapping of its vertices onto pairs (ψ, φ) such that the following conditions are satisfied: (1) ψ is the Gödel number of a closed formula of the form $(x_k)\Phi$ and φ is the Gödel number of a proof in $\mathrm{Ext}_{S_0^\infty}(\{(x_k)\Phi\})$; (2) if V' immediately succeeds V in G and if these vertices are mapped onto pairs (ψ', φ'), (ψ, φ), where ψ is the Gödel number of $(x_k)\Phi$ and ψ' the Gödel number of $(x_m)\Phi'$, then there is an integer n such that φ' is the Gödel number of a proof of $Sb(x_m/\delta_n)\Phi$; (3) under the same assumptions as in (2) there is for every integer n a vertex V'' which immediately succeeds V in G and is such that the pair (ψ'', φ'') onto which V'' is mapped has as its second member the Gödel number of a proof of $Sb(x_m/\delta_n)\Phi$; (4) in the pair which is the image of a vertex without successors the first member is the Gödel number of a formula provable in S_0^∞.

If V_0 is mapped onto the pair (ψ_0, φ_0), where φ_0 is the Gödel number of a proof of Φ, then we say that G together with its normal covering is a graph of a proof of Φ. From these definitions we easily obtain

III.3.1 *If* $\pi < \Omega$ *then* Φ *is provable in* S_π^∞ *if and only if there is a graph* G *of the height* π *such that* G *is a graph of a proof of* Φ.

This theorem together with well-known facts concerning analytic sets yields the following three corollaries:

III.3.2. *The set* T_π^∞ *is analytic for each* $\pi < \Omega$.

III.3.3. *Sets which are weakly representable in* S_π^∞ *are analytic for* $\pi < \Omega$.

III.3.4. *Sets which are strongly representable in* S_π^∞ *are borelian for* $\pi < \Omega$.

III.3.5. *Every analytic set is weakly representable in* S_1^∞.

We shall sketch the proof for analytic subsets of $R_{1,0}$. If A is such a set then there is an open relation Q such that

$$\varphi \in A \equiv (E\psi)(n, p)\, Q(\overline{\psi}(n), p, \varphi)\,.$$

[5] Results stated in this and the next section were obtained by the author in collaboration with Mr. L. Szczerba.

Let Ψ weakly represent Q in S_0^∞ and let $X(\xi, \zeta)$ be a formula such that $X(\gamma_\varphi, \gamma_{\bar\varphi}) \in Cn_1^\infty(0)$ and $(E!\,\xi)X(\gamma_\varphi, \xi) \in Cn_1^\infty(0)$ (cf. [10], where this formula is denoted by "ζ is $\bar\xi$"). With these notations we easily prove that the formula (with one free variable τ)

$$(E\xi, \zeta)(x, y)[X(\xi, \zeta) \,\&\, \Psi(\zeta(x), y, \tau)]$$

weakly represents A in S_1^∞.

Let B_π be the set of functions φ such that the relation $\varphi(2^m(2n+1)) = 0$ is a well-ordering of type π and put $B_\Omega = \bigcup_{\pi<\Omega} B_\pi$. Let Bord($\xi$) be a formula obtained by expressing in the language of S_0 the definition of the set B_Ω (cf. [10]).

III.3.6. *If* $\varphi \in B_\pi$, *then* Bord$(\gamma_\varphi) \in Cn_{\pi+1}^\infty(0)$ *for* $\pi < \Omega$.

The proof proceeds by induction on π and uses the same remark which we mentioned in connection with the proof of II.3.2.

III.3.7. $T_\pi^\infty \neq T_{\pi+1}^\infty$ *for* $\pi < \Omega$.

Proof. Otherwise all formulas Bord(γ_φ), where $\varphi \in B_\Omega$ would be provable in an S_π^∞ which would prove that the set B_Ω is analytic contrary to the well-known theorem stating that this set is exactly CA. Cf. [7].

III.3.8. *Every Borel set is strongly representable in a suitable system* S_π^∞ *with* $\pi < \Omega$.

In the proof of this theorem we use notions of Lusin's theory of sieves. Every Borel set[6] A is determined by a closed sieve W_r (cf. [7, p. 391]) and the number of constituents of the complement of A is at most denumerable. We can assume W_1 to be the whole space so that every φ determines a function ϑ such that $\vartheta(n) \leqq 1$ for $n = 0, 1, 2, \cdots$ and the relation $\vartheta(2^m(2n+1)) = 0$ is nonempty and orders the set of indices n satisfying $\varphi \in W_{r_n}$.

Writing the definition of ϑ in the language of S_0^∞ we obtain a formula $\Theta(\xi, \zeta)$ satisfying the conditions

$$(E!\,\zeta)\Theta(\gamma_\varphi, \zeta) \in Cn_\omega^\infty(0)\,;$$

if φ *determines* ϑ *in the above sense, then* $\Theta(\gamma_\varphi, \gamma_\vartheta) \in Cn_\omega^\infty(0)$.

Using these properties of Θ we can show that the formula $(E\zeta)[\Theta(\xi, \zeta)\,\&$ $\sim$ Bord$(\zeta)]$ strongly represents A in S_π^∞, where π is any infinite ordinal which exceeds the indices of all the constituents of the complement of A.

Theorems III.3.5 and III.3.8 reveal an essential difference between the systems S_π ($\pi < \omega_1$) and S_π^∞ ($\pi < \Omega$). Whereas the families $\mathscr{R}_{k,l}(S_\pi)$ strictly increase with π for $\pi < \omega_1$, the families $\mathscr{R}_{k,l}(S_\pi^\infty)$ are constant for $1 \leqq \pi < \Omega$. On the other hand, the families $\mathscr{R}_{k,l}^*(S_\pi)$ and $\mathscr{R}_{k,l}^*(S_\pi^\infty)$ are both strictly increasing. An obvious corollary from this state of affairs is that Theorem I.8.3 fails for the systems S_π^∞ ($1 \leqq \pi < \Omega$) and hence that these systems do not satisfy condition (S).

III.3.9. *Systems* S_π^∞ *satisfy* (A) *and* (C) *for* $1 \leqq \pi < \Omega$.

Proof. (A) is obvious and (C) results from the evaluation of the predicate $\Phi \in Cn_\pi^\infty(X)$ which is analytic if X is a set of formulas whose Gödel numbers form an analytic set.

6 We assume for simplicity that $A \subseteqq R_{1,0}$.

III.4. **System** S_Ω^∞. The properties of this system are very similar to those of S_Ω.

III.4.1. *S_Ω^∞ is closed with respect to the rule ω.*

PROOF. If G_n is a graph of a proof of $\Phi(\delta_n)$ in S_{π_n}, where $\pi_n < \Omega$, then joining these graphs together we obtain a graph of a proof of $(x)\Phi(x)$ in S_π^∞, where $\pi = \sup \pi_n + 1$.

III.4.2. *$\mathscr{R}_{k,l}(S_\Omega^\infty)$ coincides with the family PCA.*

PROOF. Evaluation of the predicate $\varphi \in T_\Omega^\infty$ reveals that $T_\Omega^\infty \in PCA$. Hence $\mathscr{R}_{k,l}(S_\Omega^\infty) \subseteq PCA$. Now let $A \in PCA$ and assume for simplicity that $A \subseteq R_{1,0}$. It follows that there is an arithmetically definable relation Q such that $\varphi \in A \equiv (E\psi)(\vartheta)Q(\varphi, \psi, \vartheta)$. The right hand side of this equivalence can be transformed to $(E\psi)(f(\varphi, \psi) \in B_\Omega)$, where f is an arithmetically definable function. Hence $\varphi \in A \equiv (E\psi)[\text{Bord}\,(\gamma_{f(\varphi,\psi)}) \in Cn_\Omega^\infty(0)]$. If Φ strongly defines in S_Ω^∞ the set $\{(\varphi, \psi, \vartheta) : \vartheta = f(\varphi, \psi)\}$, then

$$\varphi \in A \equiv (E\xi, \zeta)[\Phi(\gamma_\varphi, \xi, \zeta) \,\&\, \text{Bord}\,(\zeta)] \in Cn_\Omega^\infty(0)$$

which proves that $A \in \mathscr{R}_{1,0}(S_\Omega^\infty)$.

II.4.3. *S_Ω^∞ satisfies conditions* (A) *and* (C).

This is an obvious corollary from III.4.2.

III.4.4. *S_Ω^∞ satisfies condition* (S).

The proof is obtained mutatis mutandis from the proof of II.5.6.

From III.4.3. and III.4.4. we can obtain various corollaries using the general theory of Chapter I. As an instance of such theorems we can quote the following: *two disjoint CPCA sets are separable by means of sets which are simultaneously PCA and CPCA sets.*

BIBLIOGRAPHY

1. A. Ehrenfeucht and S. Feferman, ***Representability of recursively enumerable sets in formal theories***, Arch. Math. Logik Grundlagenforsch. vol. 5 (1959) pp. 38–41.

2. K. Gödel, ***Über die Länge von Beweisen***, Ergebnisse eines mathematischen Kolloquiums vol. 7 (1936) pp. 23–24.

3. A. Grzegorczyk, A. Mostowski and Cz. Ryll-Nardzewski, ***The classical and the ω-complete arithmetic***, J. Symb. Logic vol. 23 (1958) pp. 188–206.

4. S. C. Kleene, ***Introduction to metamathematics***, Amsterdam, North-Holland Publishing Co.; Groningen, P. Noordhoff N.V.; 1952.

5. ———, ***Hierarchies of number-theoretic predicates***, Bull. Amer. Math. Soc. vol. 61 (1955) pp. 193–213.

6. S. C. Kleene and E. L. Post. ***The upper semi-lattice of degrees of recursive unsolvability***, Ann. of Math. vol. 59 (1954) pp. 379–407.

7. K. Kuratowski, ***Topologie***. I. Monografie Matematyczne XX, 2d ed. Warszawa-Wrocław, 1948.

8. G. Kreisel, ***Set theoretic problems suggested by the notions of potential totalities***, Infinitistic Methods (Proceedings of the Symposium on Foundations of Mathematics), Warszawa, 1961.

9. W. Markwald, ***Zur Theorie der konstruktiven Wohlordnungen***, Math. Ann. vol. 127 (1954) pp. 135–149.

10. A. Mostowski, ***Formal system of analysis based on an infinitistic rule of proof***, Infinitistic Methods (Proceedings of the Symposium on Foundations of Mathematics),

Warszawa, 1961.

11. ———, *A generalization of the incompleteness theorem,* Fund. Math. vol. 49 (1961) pp. 205–232.

12. S. Orey, *On ω-consistency and related properties,* J. Symb. Logic vol. 21 (1956) pp. 246–252.

13. H. Putnam and R. M. Smullyan, *Exact separation of recursively enumerable sets within theories,* Proc. Amer. Math. Soc. vol. 11 (1960) pp. 574–577.

14. J. B. Rosser, *Gödel theorems for nonconstructive logics,* J. Symb. Logic vol. 2 (1937) pp. 129–137.

15. J. C. Shepherdson, *Representabilily of recursively enumerable sets in formal theories,* Arch. Math. Logik Grundlagenforsch, 4,5 (1960) pp. 119–127.

16. C. Spector, *Inductively defined sets of natural numbers,* Infinitistic Methods (Proceedings of the Symposium on Foundations of Mathematics), Warszawa, 1961.

17. A. Tarski, *Fundamentale Begriffe der Methodologie der deduktiven Wissenschaften.* I, Monatshefte für Mathematik und Physik vol. 37 (1930) pp. 361–404.

UNIVERSITY OF WARSAW,
WARSAW, POLAND

Note to p. 33. The existence of the formula Π such that $Cn_\beta(\{\Pi\})$ is a consistent and complete system was claimed in [**10**, p. 166], Theorem 8.20. The proof of this theorem given in [**10**] was incorrect: see the review of [**10**] by H. B. Enderton which is to appear in the Journal of Symbolic Logic. As noted in this review the Theorem 8.20 of [**10**] while incorrectly proved is itself true: it follows directly from the existence (established by Gandy and Putnam) of a minimal β-model for the second order arithmetic.

HERBRAND-GÖDEL-STYLE RECURSIVE FUNCTIONALS OF FINITE TYPES

BY

S. C. KLEENE

In this paper we continue an investigation begun in § 8 of [RF] (cf. [RF, § 3.2]),[1] and continued in [**13**] (cf. the introduction there), [**14**; **15**].

Type-0 objects are natural numbers $0, 1, 2, \cdots$; *type*-$j+1$ objects are one-place (total) functions α^{j+1} from *type*-j objects β^j to natural numbers.

We shall consider an extension to (total and partial) functions with variables of types > 1 of the notion of general recursiveness originally given for total functions of type-0 variables by Gödel [**3**, pp. 26–27] building on a suggestion of Herbrand. We call this extension "Herbrand-Gödel-style recursiveness" to avoid confusion with the rather different notion for which we appropriated the terms "general" and "partial recursive" in [RF, §§ 3 ff.], and to forestall any misunderstanding in future references that Gödel is responsible for the present extension.[2]

In the cases of Turing-Post computability and Church-Kleene λ-definability, there seemed to be no doubt how the definition of the extended notion should be chosen, given that the independent variables are to range over all objects of their types (cf. [**13**; **15**]). In the case of Herbrand-Gödel-style recursiveness, it proves to be a somewhat delicate matter to choose an extension of the formalism of recursive functions of [IM, § 54] which is neither too weak nor too strong for our purpose. This is discussed in §§ 3.13–3.16, to which the reader may turn for summary and motivation.

1. Definition of Herbrand-Gödel-style recursiveness.

1.1. We introduce a formal language with symbols as follows (extending [IM, § 54]): $=$ (equals), 0 (zero), $'$ (successor), $\xi_1^j, \xi_2^j, \xi_3^j, \cdots$ (an infinite list of variables for type-j objects, $j = 0, 1, 2, \cdots$), $\mathrm{f}_1, \mathrm{f}_2, \mathrm{f}_3, \cdots$ (an infinite list of function letters, i.e. symbols for unspecified functions of various numbers of variables of types $0, 1, 2, \cdots$), λ (Church's lambda-operator), (,) (parentheses) and , (comma).

We shall make the type-j variables ξ^j for $j > 0$ serve also as type-j constants by assigning them various functions α^j as interpretations according to circumstances; otherwise we would need also $2^{\aleph_0}$ type-1 constants, $2^{2^{\aleph_0}}$

Received by the editor April 6, 1961.

[1] We cite [**11**] (§§ 1–8) and [**12**] (of which we need only §§ 9, 10) as "[RF]", and [**10**] as "[IM]".

[2] The extensions of the Herbrand-Gödel notion to partial functions and to type-1 variables are due to Kleene ([**7**] and [**9**], respectively). "General" and "partial recursive" in the literature, prior to [RF, § 3], refer to the Herbrand-Gödel notion (often in one of Kleene's versions [**6**; **8**; IM]), or to one of these two previous extensions of it.

type-2 constants, etc.

'Term' and 'type-j functor' ($j = 1, 2, 3, \cdots$) are defined inductively, thus. 1. 0 is a *term*. 2. If v is a *term*, so is $(\mathrm{v})'$. 3. A type-0 variable ξ^0 is a *term*. 4. For each $j > 0$, a type-j variable ξ^j is a *type-j functor*. 5. If f is a function letter, and each of $\mathrm{v}_1, \cdots, \mathrm{v}_m$ is a *term* or *functor* ($m \geqq 0$), then $\mathrm{f}(\mathrm{v}_1, \cdots, \mathrm{v}_m)$ is a *term*. 6. For each $j \geqq 0$, if t is a *term*, and ξ^j is a type-j variable, then $\lambda\xi^j(\mathrm{t})$ is a *type-$j+1$ functor*. 7. For $j = 0$ ($j > 0$), if s is a *type-$j+1$ functor*, and v is a *term* (*type-j functor*), then $(\mathrm{s})(\mathrm{v})$ is a *term*. (Also extremal clause.)

A formal expression $\mathrm{t}_1 = \mathrm{t}_2$ where t_1 and t_2 are terms is an *equation*. A finite sequence $\mathrm{e}_1, \cdots, \mathrm{e}_s$ ($s \geqq 0$) of equations is a *system of equations*.

A term, functor or equation is a *formula*.

In writing formulas, parentheses are omitted when no confusion will result.

Free and *bound* occurrences of variables, etc., are distinguished in the familiar way, now with λ-prefixes $\lambda\xi^j$ as the operators that bind (cf. [IM, § 18]). A part which is a term or functor of a formula is *free*, if it is not in the scope of any $\lambda\xi^j$ with ξ^j a free variable of the part (analogously to [IM, p. 410]). *Congruence* of two formulas is defined in the usual way (cf. [IM, p. 153]).

A formula containing no free occurrence of a number (i.e. type-0) variable is *closed*. In the situations in which we shall require that formulas be closed, each function (i.e. type-j, $j > 0$) variable ξ^j free in them will have been interpreted as expressing a function α^j.

A part of a formula of the form $(\lambda\xi^j\mathrm{t})(\mathrm{v})$ (ξ^j a type-j variable; t a term; v a term if $j = 0$, a type-j functor if $j > 0$) is a (λ-)*redex* (of *order* j). A formula containing no redex is (λ-)*normal*.

Numerals are defined and denoted as usual [IM, p. 195].

A closed normal functor having no free part which is a (closed normal) term but not a numeral is *regular*.

By $\mathrm{S}^{\xi^j}_{\mathrm{v}}\mathrm{t}\,|$ we mean the result of substituting v for (each free occurrence of) ξ^j in t.

1.2. We formulate rules of inference R1, R2, R3, R4.1, R4.2, R4.3, $\cdots$, R5, R6, R7 with $1, 2, 2, 0, \aleph_0, 2^{\aleph_0}, \cdots, 1, 1, 1$ premises, respectively.

R1: to pass from an equation e to the result d of substituting simultaneously for each of its free number variables ξ^0 (in all free occurrences) a respective numeral $\boldsymbol{u}$, and for each of its free type-j function variables ξ^j (in all free occurrences) a respective regular type-j functor u free for ξ^j in e. (Substitution.)

R2: to pass from a normal equation $\mathrm{p} = \mathrm{q}_\mathrm{r}$ where q_r contains a specified free occurrence of a term r (the *major* premise) and an equation $\mathrm{r} = \boldsymbol{x}$ where $\boldsymbol{x}$ is a numeral (the *minor* premise) to $\mathrm{p} = \mathrm{q}_{\boldsymbol{x}}$ where $\mathrm{q}_{\boldsymbol{x}}$ is the result of replacing that occurrence of r in q_r by $\boldsymbol{x}$. (Replacement on the right.)

R3: under the circumstances of R2, to pass from $\mathrm{q}_{\boldsymbol{x}} = \mathrm{p}$ (the *major* premise) and $\mathrm{r} = \boldsymbol{x}$ (the *minor* premise) to $\mathrm{q}_\mathrm{r} = \mathrm{p}$. (Replacement on the left.)

In the "type-j evaluation" rules R4.j ($j = 1, 2, 3, \cdots$), ξ^j is a type-j variable which is being interpreted by a type-j function α^j.

R4.1: (to pass from the null set of premises to) $\xi^1(\boldsymbol{u}) = \boldsymbol{z}$ where ($\boldsymbol{u}, \boldsymbol{z}$ are the numerals for u, z with) $z = \alpha^1(u)$.

R4.2: to pass from $\mathrm{u}(0) = \boldsymbol{z}_0, \mathrm{u}(1) = \boldsymbol{z}_1, \mathrm{u}(2) = \boldsymbol{z}_2, \cdots$ where u is a regular type-1 functor and $z_a = \alpha^1(a)$ $(a = 0, 1, 2, \cdots)$ to $\xi^2(\mathrm{u}) = \boldsymbol{z}$ where $z = \alpha^2(\alpha^1)$.

R4.j $(j = 3, 4, 5, \cdots)$: to pass to $\xi^j(\mathrm{u}) = \boldsymbol{z}$ where $z = \alpha^j(\alpha^{j-1})$ from a set of equations $\mathrm{u}(\xi^{j-2}) = \boldsymbol{z}_\alpha^{j-2}$ where u is a regular type-$j-1$ functor, ξ^{j-2} is a type-$j-2$ variable not occurring free in u, the interpretation α^{j-2} of ξ^{j-2} ranges over all type-$j-2$ functions (a different one for each premise), and $z_{\alpha^{j-2}} = \alpha^{j-1}(\alpha^{j-2})$.

R5: to pass from an equation e to an equation d congruent to e. (Change of bound variables; congruence transformation.)

R6: to pass from an equation e to the result d of replacing in e a redex $(\lambda\xi^j\mathrm{t})(\mathrm{v})$ where v is free for ξ^j in t by $\mathrm{S}^{\xi^j}_{\mathrm{v}}\mathrm{t}\,|$. (($\lambda$-)reduction, contracting the redex $(\lambda\xi^j\mathrm{t})(\mathrm{v})$.)

R7: inversely to R6, to pass from d to e. ((λ-)expansion.)

We may also use "(λ-)reduction" ["(λ-)expansion"] to describe an application of R6 [R7] preceded [followed] by one of R5.

1.3. A system E of equations $\mathrm{e}_1, \cdots, \mathrm{e}_s$ is *proper* if: (a) the left member of each equation e_i is of the form $\mathrm{f}(\mathrm{w}_1, \cdots, \mathrm{w}_m)$ where f is a function letter and $\mathrm{w}_1, \cdots, \mathrm{w}_m$ $(m \geqq 0)$ are only numerals, number variables with zero or more accents, and distinct function variables, (b) it is impossible by any substitutions of numerals for the number variables and of distinct function variables for the function variables to transform the left members of two equations e_i and e_j $(i \neq j)$ into the same term, (c) the right member of each equation e_i contains free only variables occurring (free) in the left member, and (d) (the right member of) each equation e_i is normal.

The function letter f occurring in the left member of the last equation e_s of E (or the first function letter f_1 if E is empty) we call the *principal function letter of* E.

1.4. We consider "deductions" in tree form, from proper systems E of equations, in which each branch headed by an equation e of E has at the top an application of R5 followed by an application of R1, which is not applied elsewhere.[3] That application of R5 may be the identical transformation; likewise the application of R1 if e contains no number variables. Furthermore, each deduction is constructed relative to an interpretation of each function variable of a list $\xi^1_1, \cdots, \xi^1_{n_1}, \cdots, \xi^r_1, \cdots, \xi^r_{n_r}$ including all which occur free in its conclusion or endequation.[4] The interpretation is extended to the

[3] Such a "global" restriction on the use of a rule may seem out of keeping with the notion of a "formal system" (here nonconstructive). But in effect we are simply employing the equations of E as axiom schemata, from each of which R1 (not otherwise used), with help from R5, extracts an infinite (or if no variables occur free in it, a unit) class of axioms.

[4] This, and the restriction on R1, could be avoided by using function constants as mentioned in § 1.1.

free function variables in all the equations not above (the conclusion of) an application of R1, as follows. Each free function variable of the (or either) premise of an inference (except by R1) which is free in the conclusion shall have the same interpretation as in the conclusion. The premises of an application of R4.j ($j \geqq 3$) correspond 1–1 to the type-$j-2$ objects, and in each premise the corresponding such object α^{j-2} is the interpretation of ξ^{j-2}. Should there be other cases (and applications of R2 and R6 are the only possibilities) in which a premise has a free function variable ξ^k not free in the conclusion, we may interpret it by α_0^k, where $\alpha_0^1, \alpha_0^2, \alpha_0^3, \cdots$ is a fixed list (but cf. § 1.8). This completes the definition of a *deduction* from E for a given *interpretation* (by functions) $\mathfrak{a}$ (corresponding 1-1 to a list of function variables including all which occur free in the conclusion), or briefiy an $\mathfrak{a}$-*deduction* from E. When such a deduction of d exists, we say d is $\mathfrak{a}$-*deducible* from E, and write: E $\mathfrak{a}\!\vdash$ d.

1.5. R5-R7 give (for equations) Church and Rosser's "third kind of conversion" [**2**, pp. 481–482], which differs from their first kind (using the rules I-III of Church [**1**] as modified in Kleene [**5**]) in that the x of a formula λxR need not occur free in the R. But the type restrictions in our present formation rules give us (a) and (b) of the following theorem in addition to their results. A normal formula into which a given formula A can be *converted*, i.e. transformed by R5-R7 (but applied to any formulas, not just to equations), we call a *normal form* of A. A *reduction* here is an application of R6 possibly preceded by one of R5 (cf. § 1.2). If F goes to G by a single reduction, we write: F imr G. If F is congruent to G, we write: F cong G.

THEOREM 1. (a) *Each formula* A *has a normal form* B. (b) *To each formula* A, *there is a number* m *such that any sequence of reductions starting from* A *will lead to a normal formula* B *in at most* m *reductions.* (c) *Any two normal forms* B_1 *and* B_2 *of a given formula* A *are congruent.*

PROOF. We prove (b) first (assuming (a)), because the proof illustrates a device which Church and Rosser gave (in an unpublished "Supplement to 'Some Properties of Conversion'") for adapting the proof of their Lemma 1 [**2**, p. 475] to the third kind of conversion p. 482.[5]

(b) Let each functor $\lambda\xi^j$t in A in which the t does not contain the ξ^j free be replaced by $\lambda\xi^j$f(t, ξ^j) where the additional symbols f(, ξ^j) are written in red, the original symbols in black; call the resulting formula A*. In each of a sequence of reductions starting from A*, the portion of the resulting formula that would be required if only the black portion of the given formula existed shall be written in black, the rest in red. To any sequence of n reductions A imr A_1 imr A_2 $\cdots$ imr A_n starting from A, there is a sequence of n reductions A* imr A_1^* imr A_2^* $\cdots$ imr A_n^* starting from A* such that the black portion of A_i^* is identical with A_i ($i = 1, \cdots, n$). Reductions starting from A* come under the first kind of conversion of Church-Rosser [**2**], for which

[5] Some expository remarks in [**15**, § 2] may also help the reader to take over what we use from Church-Rosser [**2**].

their Theorem 2, p. 479 holds. By (a), A* has a normal form. So there is a number m such that any sequence of reductions starting from A* terminates (in a normal formula B*) after at most m reductions. Therefore so does any sequence starting from A.

(a) We use induction on the maximum m of the orders of the redexes in A. First, we perform a "terminating sequence of contractions" on the redexes in A, with result C, as we can by Church-Rosser [**2**, Lemma 1, p. 475] for the third kind of conversion p. 482. But in a reduction from F to G by contracting an order-j redex $(\lambda\xi^j \mathrm{t})(\mathrm{v})$ to $\mathrm{S}^{\xi^j}_{\mathrm{v}}\mathrm{t}\,|$, the redexes in G not residuals of ones in F, if any, arise thus: $j > 0$, v is a type-j functor $\lambda\xi^{j-1}\mathrm{w}$, and t has one or more parts $\xi^j(\mathrm{u})$, which by the substitution become order-$j-1$ redexes $(\lambda\xi^{j-1}\mathrm{w})(\mathrm{u})$. Any such "new" redexes are thus of order one less than the redex contracted. So the maximum of the orders of the redexes in C is $\leqq m-1$. So by the hypothesis of the induction, C has a normal form B. Hence A has B as normal form.

(c) This is Church-Rosser [**2**, Corollary 2 Theorem 1, p. 479], which holds for the third kind of conversion p. 482. —

Using Theorem 1: *Each equation in a deduction under* § 1.4 *not above* (*the conclusion of*) *an application of* R1 (*is closed or*) *has a closed normal form.*

1.6. Consider a deduction as in § 1.4 of a normal equation d from a proper system E of equations. The role in such a deduction of R6 and R7 can be circumscribed. For, of the rules other than R5-R7, only R1 can have a conclusion not normal, and then by § 1.3 (a) only the right member. Using § 1.3 (d), each initial equation is normal and of those rules only R4.2, R4.3, $\cdots$ can have a premise not normal, and then only the left member. Consider any sequence of applications of R5-R7, beginning with c, an initial equation or the conclusion of an inference by another rule than R5-R7, and ending with d, the endequation or a premise of an inference by another rule than R5-R7. By Theorem 1 (b) and (c), we can effect the transformation of c into d by using (besides R5) a sequence of reductions (R6 with R5) to convert the right member to a normal term, non-empty exactly if c is a non-normal conclusion for R1, followed by an expansion (R7 with R5) to convert the left member to a non-normal term, exactly if d is a non-normal premise of R4.j ($j \geqq 2$).

Similarly, we can circumscribe the role of R6 and R7 in a deduction of an equation d not normal. If in a given deduction d comes by zero or more applications of R6 (with R5) from the conclusion of an application of R1, we alter nothing; otherwise, we substitute a deduction of a normal form d_{N} of d with R6 and R7 circumscribed as above, followed by a sequence of applications of R7 (with R5) to convert d_{N} to d.

Summarizing: *We need use* R6 (*with* R5) *only* (a) *applied to the right immediately following* R1; *and* R7 (*with* R5) *only* (b) *applied to the left immediately preceding* R4.j ($j \geqq 2$), *or* (c) *in converting a normal equation to a non-normal endequation.*

1.7. THEOREM 2. *In a deduction* (*as in* § 1.4) *from a proper system* E *of* *equations, no equation has a numeral or other accented term or a number variable as left member* (*or as normal form of the left member*).

PROOF. This is immediate for an equation by R4.1, and by § 1.3 (a) for an equation of E and thence for an equation resulting therefrom by R6 and R1. To deal with any other equation in the deduction, we can replace the subdeduction of it by a deduction of it with R5 and R7 circumscribed as in § 1.6. In this deduction, no inference with premise(s) not having an accented term or number variable as left member (or as normal form thereof) can lead to an equation with such a left member (R6 being excluded as applied to the left).

1.8. After the use of R6 and R7 has been circumscribed as in § 1.6, applications of R2 and R6 will not suppress any free function variable ξ^k (cf. § 1.4). For, apart from applications of R5, they can occur only in a sequence of applications of R6, R2 and R3 beginning with the result c of a substitution by R1 with R5 into an equation of E. Using § 1.3 (c), the left member of c will contain free all the variables free in the right. The applications will not decrease the set of variables free in the left, nor increase the set of those free in the right.

Now (with the need for the $\alpha_0^1, \alpha_0^2, \alpha_0^3, \cdots$ of § 1.4 removed), the interpretation of the free function variables can be managed conveniently thus. Say that in the endequation only function variables of the types $1, \cdots, r$ occur, and of a given such type only ones from among the first n_j variables $\xi_1^j, \cdots, \xi_{n_j}^j$ of that type. For the endequation, we can list functions $\alpha_1^1, \cdots, \alpha_{n_1}^1, \cdots, \alpha_1^r, \cdots, \alpha_{n_r}^r$, or briefly $\mathfrak{a}$, as the interpretation of those variables. At each application of a rule other than R4.j ($j \geqq 3$) or R1, the same list can be used to interpret the variables in the premise(s) as in the conclusion. At each application of R4.j ($j \geqq 3$), the ξ^{j-2} can (without loss of generality) be chosen to be the next type-$j-2$ variable after those for which an interpretation is in the list for the conclusion, and its interpretation α^{j-2} in each premise can be added to that list to obtain the list for that premise.

1.9. Consider a partial (perhaps total) function φ of say $n_0, \cdots, n_r$ variables of types $0, \cdots, r$ respectively (written in a given order). For given values $\mathfrak{b}$ of these variables, let D be respective numerals expressing the numbers, and distinct function variables interpreted by the functions $\mathfrak{a}$, in $\mathfrak{b}$ (e.g. if $\mathfrak{b}$ is $\alpha_1^1, a, b, \alpha_2^1, \alpha^2$, then D is $\xi_1^1, \boldsymbol{a}, \boldsymbol{b}, \xi_2^1, \xi^2$ where $\boldsymbol{a}, \boldsymbol{b}$ are the numerals for a, b and $\alpha_1^1, \alpha_2^1, \alpha^2$ is the interpretation $\mathfrak{a}$ of ξ_1^1, ξ_2^1, ξ^2). We say a system E of equations *defines* φ *Herbrand-Gödel-style recursively*, if (i) E is proper, and, where f is the principal function letter of E, (ii) for each $\mathfrak{b}$ for which $\varphi(\mathfrak{b})$ is defined, E $\mathfrak{a}$-$\vdash$ f(D) $= \boldsymbol{y}$ where $y = \varphi(\mathfrak{b})$, and (iii) for each $\mathfrak{b}$ for which $\varphi(\mathfrak{b})$ is undefined, E $\mathfrak{a}$-$\vdash$ f(D) $= \boldsymbol{y}$ for no numeral $\boldsymbol{y}$.

We say a partial (total) function φ is *Herbrand-Gödel-style partial* (*general*) *recursive*, if there is a system E of equations which defines φ Herbrand-Gödel-style recursively.

2. Recursiveness of Herbrand-Gödel-style recursive functions.

2.1. We shall need a result concerning sequences of replacements, starting from a term t, each of a free part r which is a term but not a numeral by a numeral $\boldsymbol{x}$. (For definiteness we stick to this situation.)

Consider a term t, and two distinct such parts r and s. As in [IM] by Lemmas 1–4, pp. 24 and 73, either (CASE 1) r and s are disjoint, or (CASE 2) r is a proper part of s, or (CASE 3) s is a proper part of r. In the first two cases, when the part r is replaced by a numeral $\boldsymbol{x}$, the part s "becomes" an identifiable part of the result t_1, which part we call the *residual* of s under the replacement. By identity and transitivity, we can thence identify the *residual* of s, when it exists, under any sequence of $k \geqq 0$ such replacements.

Inversely, given any sequence of k such replacements, the part replaced in each replacement can be identified as the residual of a part of the initial term t. We then say the sequence of replacements is *on* the parts $r_1, \cdots, r_k$ of t whose residuals are replaced.

THEOREM 3. *Let a term* t *be transformed to a term* u *by a sequence* S *of* k ($\geqq 1$) *replacements* (*by numerals*) *on distinct free term-parts* $r_1, \cdots, r_k$ (*not numerals*) *of* t. *Then there is a sequence* S' *of replacement on* $r_1, \cdots, r_k$ *starting from* t *in which, of any two of* $r_1, \cdots, r_k$ *which are disjoint, a residual of the one to the left is replaced first.* (*Of any two which are distinct but not disjoint, a residual of the one contained in the other must be replaced first.*)

PROOF, by induction on k. IND. STEP: $k > 1$. Let the leftmost of $r_1, \cdots, r_k$ which contains no other be r_i. If r_i is replaced in the first replacement, let S' be S. Otherwise, the part(s) of t altered by the replacements ρ preceding that of a residual of r_i must be disjoint from r_i. So the replacement of a residual of r_i can be advanced over ρ to produce a sequence S' of replacements with r_i replaced first. The hyp. ind. applies to what remains from S' after the initial replacement of r_i.

2.2. We call a term *ready* (for replacement in a 'standard deduction', below), if it is closed and normal but not a numeral, and it contains no free proper part which is a (closed normal) term but not a numeral.

A term is ready if and only if it is of one of the three forms $f(u_1, \cdots, u_m)$ where f is a function letter and $u_1, \cdots, u_m$ are numerals and regular functors, $\xi^1(\boldsymbol{u})$ where ξ^1 is a type-1 variable and $\boldsymbol{u}$ is a numeral, and $\xi^j(u)$ where ξ^j is a type-j variable ($j > 1$) and u is a regular type-$j-1$ functor.

By a *ready part* of a formula we mean a free part which is a ready term.

A *principal branch* of a deduction is one which at each application of R2 or R3 contains the major premise. Its top equation is its *principal equation*.

We call a deduction from (a proper system) E *standard*, if it meets the following conditions. (A1) The use in it of R6 and R7 is circumscribed as in § 1.6. (A2) Each application of R6 read downward [of R7 read upward] contracts the leftmost innermost redex. (A3) In the part of the deduction not above (the conclusion of) an application of R1, R5 is used exactly once preceding each application of R6 read downward [of R7 read upward], in such a way that the bound variables in the resulting equation are determined from those in the given equation by a suitable convention, and not elsewhere. Such a convention is given implicitly in [**15**, § 5]; but we now elect that the application of R5 be the identical one when possible. (A4) In each application of R1, given the equation of E and the regular functors to be substituted for its free function variables, the bound variables resulting from the preliminary application of R5 are determined by a suitable convention, the application of R5 being identical when possible. (B) In each application of R2 or R3 the r is the leftmost ready part of the q_r. (C) Each subdeduction with endequation not above an application of R1 and of the form $t = \boldsymbol{y}$ (t a normal term, $\boldsymbol{y}$ a numeral) has the following property. In it there is an equation, called the *midequation*, which either (CASE 1) is $f(u_1, \cdots, u_m) = w$ (f a function letter, $u_1, \cdots, u_m$ numerals and regular functors, w a closed normal term) and comes from an equation of E by using only an application of R5, an application of R1, and pairs of applications of R5 and R6, or (CASE 2) is $\xi^1(\boldsymbol{u}) = \boldsymbol{y}$ (ξ^1 a type-1 variable, $\boldsymbol{u}$ a numeral) and comes by R4.1, or (CASE 3) is $\xi^2(u) = \boldsymbol{y}$ (ξ^2 a type-2 variable, u a regular type-1 functor) and comes by R4.2, or (CASE 4) is $\xi^j(u) = \boldsymbol{y}$ ($j > 2$, ξ^j a type-j variable, u a regular type-$j-1$ functor) and comes by R4.j. Below the midequation there occur only applications of R3 and of R2 (only in Case 1 and above the applications of R3).

The left member of the midequation is ready.

THEOREM 4. *Let* E *be a proper system of equations,* t_0 *a term,* $\mathfrak{a}_0$ *an interpretation of certain function variables including all occurring free in* t_0*, and* $\boldsymbol{y}_0$ *a numeral; and suppose that* E $\mathfrak{a}_0$-⊢ $t_0 = \boldsymbol{y}_0$. *Then there is a standard* $\mathfrak{a}_0$*-deduction from* E *of* $t_0 = \boldsymbol{y}_0$.

PROOF. Given any $\mathfrak{a}_0$-deduction from E of $t_0 = \boldsymbol{y}_0$, we can find another in which the use of R6 and R7 conforms to § 1.6. We shall prove by (course-of-values) induction over the part of this deduction not above an application of R1 that: *To each equation* $t_1 = \boldsymbol{y}$ (t_1 *a term,* $\boldsymbol{y}$ *a numeral*) *in this part, to each term* t *congruent to* t_1*, there is a standard* $\mathfrak{a}$*-deduction from* E *of* $t = \boldsymbol{y}$ (*where* $\mathfrak{a}$ *is the interpretation in force at the position of the equation* $t_1 = \boldsymbol{y}$*, extending* $\mathfrak{a}_0$ *to the additional function variables if any introduced between* $t_0 = \boldsymbol{y}_0$ *and* $t_1 = \boldsymbol{y}$ *reading upward*). Applying this result to the endequation $t_0 = \boldsymbol{y}_0$ as the $t_1 = \boldsymbol{y}$, and to t_0 as the t, we shall have the required conclusion.

Using Theorem 2, the following cases are exhaustive (Cases 1–4 for normal t_1).

CASE 1: t_1 is $f(v_{11}, \cdots, v_{1m})$ where f is a function letter and $v_{11}, \cdots, v_{1m}$ are normal terms and normal functors. The principal equation of any principal branch of the subdeduction of $t_1 = \boldsymbol{y}$ is an equation of E with f as the initial symbol; for, from an equation $\xi^1(\boldsymbol{u}) = \boldsymbol{z}$ by R4.1 or an equation of E with a function letter g other than f as the initial symbol no sequence of passages from premise (the major one in the case of R2 or R3) to conclusion can introduce the function letter f in the initial position (not R3 because by Theorem 2 the whole left will never be a numeral, nor R6 because it is applied under § 1.6 only to the right). Now consider in a principal branch of the subdeduction of $t_1 = \boldsymbol{y}$ the equation which results from the last application of R6 following the application of R1. By § 1.3 (a) it is of the form $f(u_{11}, \cdots, u_{1m}) = w_1$ where $u_{11}, \cdots, u_{1m}$ are numerals and regular functors and w_1 is a normal term. In a principal branch from this equation down to $t_1 = \boldsymbol{y}$ there can occur only applications of R2, R3 and R5; for, an application of R4.j $(j > 1)$ would introduce a function variable ξ^j as initial symbol after which (as above with ξ^1 or g) f could not be reintroduced as initial symbol, and under § 1.6 there are no lower applications of R6 and there are applications of R7 only preparatory to R4.j $(j > 1)$ or in the final stage of deducing a non-normal equation. Since applications of R4.j $(j > 1)$ are excluded, there is only one principal branch. Since no equation in the subdeduction of $t_1 = \boldsymbol{y}$ has a numeral as left member, the applications of R3 replace proper numeral-parts. So, beginning with the equation $f(u_{11}, \cdots, u_{1m}) = w_1$ in question (which is uniquely determined, since the principal branch is unique), the replacements by R3 [by R2] with applications of R5 read downward take $u_{11}, \cdots, u_{1m}$ respectively [w_1] into $v_{11}, \cdots, v_{1m}$ [$\boldsymbol{y}$]. Now t, being congruent to t_1, is $f(v_1, \cdots, v_m)$ where $v_1, \cdots, v_m$ are congruent respectively to $v_{11}, \cdots, v_{1m}$. Consider the tranformation by applications of R3 and R5 upward of each v_{1i} into u_{1i}. Making corresponding applications of R3 (i.e. to correspondingly located parts as the r's, with the same respective $\boldsymbol{x}$'s) while omitting the applications of R5, we obtain a transformation upward of v_i into a u_i congruent to u_{1i}. For each of these applications the minor premise $r = \boldsymbol{x}$ is congruent to a minor premise $r_1 = \boldsymbol{x}$ for the transformation of v_{1i} into u_{1i}. Hence by the hyp. ind., there is a standard α-deduction from E of the $r = \boldsymbol{x}$. We can alter each of these applications, if r is not ready, so that, instead of replacing r by $\boldsymbol{x}$ outright, the same replacements by R3 are performed successively reading upward within the part r of the major premise $p_r = q$ as are performed by R3 below the midequation in the standard α-deduction of $r = \boldsymbol{x}$ provided by the hyp. ind., followed by a final replacement by $\boldsymbol{x}$. These replacements have as minor premises, instead of $r = \boldsymbol{x}$ if r is not ready, equations $s = \boldsymbol{z}$ with s ready deduced by standard α-deductions which occur as parts of the standard α-deduction of $r = \boldsymbol{x}$. We shall make these alterations; but first we reorder the applications to be altered, as we may do by Theorem 3, so that of two disjoint parts of $f(v_1, \cdots, v_m)$ residuals of which are replaced a residual of the one to the left is replaced first. Now making the described alterations of each of the applications of R3, we obtain a sequence of replacements by R3 read upward satisfying (B) (with no application of R5, so thus far (A3) is satisfied) which transforms $f(v_1, \cdots, v_m) = \boldsymbol{y}$ into $f(u_1, \cdots, u_m) = \boldsymbol{y}$. The equation $f(u_{11}, \cdots, u_{1m}) = w_1$ comes from an equation of E by an application of R5, followed by an application of R1, followed by pairs of applications of R5 and R6. Let us modify the initial application of R5 to conform to (A4) relative to the same equation of E and as the regular functors to be substituted those among $u_1, \cdots, u_m$ instead of the respective ones among $u_{11}, \cdots, u_{1m}$. Then the application of R1 leads to $f(u_1, \cdots, u_m) = v$ where v is congruent to the right side of the equation obtained by the former application of R1. By Theorem 1 (b) and (c) a sequence of zero or more pairs of applications of R5 and R6 satisfying (A3) and (A2) converts this equation to an equation $f(u_1, \cdots, u_m) = w$ where w cong w_1. We have now constructed a beginning down to $f(u_1, \cdots, u_m) = w$ of the required standard $\mathfrak{a}$-deduction of $t = \boldsymbol{y}$, and an end from $f(u_1, \cdots, u_m) = \boldsymbol{y}$ down to $t = \boldsymbol{y}$. To construct a middle to bridge the gap between $f(u_1, \cdots, u_m) = w$ and $f(u_1, \cdots, u_m) = \boldsymbol{y}$, consider the applications of R2 and R5 read downward in the subdeduction of $t_1 = \boldsymbol{y}$ by which w_1 is tranformed into $\boldsymbol{y}$. Treating these in the same manner as we treated the applications of R3 and R5 read upward by which each v_{1i} is transformed into u_{1i}, we obtain a sequence of applications of R2 (with none of R5) satisfying (B) which transforms $f(u_1, \cdots, u_m) = w$ downward into $f(u_1, \cdots, u_m) = y$ where y is congruent to $\boldsymbol{y}$ and therefore is $\boldsymbol{y}$. Thus we have bridged the gap. Now, when we include, for these applications of R2 in the middle, and for those of R3 in the end, the standard $\mathfrak{a}$-deductions of the minor premises $s = \boldsymbol{z}$, we have altogether a $\mathfrak{a}$-standard deduction from E of $t = \boldsymbol{y}$ as required.

CASE 2: t_1 is $\xi^1(v_1)$ where ξ^1 is a type-1 variable and v_1 is a normal term. If the principal equation of a principal branch were an equation of E or an equation by R4.1 with a type-1 variable other than ξ^1 as the initial symbol, ξ^1 could not be introduced as the initial symbol. Likewise R4.j ($j > 1$) is not used (so the principal branch is unique), nor then by § 1.6 R6 or R7. So the principal equation is an equation $\xi^1(\boldsymbol{u}) = \boldsymbol{z}$ by R4.1 with the same ξ^1, and below it only R2, R3 and R5 are used. But by Theorem 2, R2 is and remains inapplicable, so $\boldsymbol{z}$ is $\boldsymbol{y}$. Now t is $\xi^1(v)$ where v cong v_1. By treating the applications of R3 and R5 read upward by which v_1 is transformed into $\boldsymbol{u}$ in the same manner as those by which in Case 1 each v_{1i} was tranformed into u_{1i}, we are led to a sequence of applications of R3 (without R5) which conforms to (B) and transforms $\xi^1(v) = \boldsymbol{y}$ upward into $\xi^1(\boldsymbol{u}) = \boldsymbol{y}$. Using these downward, with the standard $\mathfrak{a}$-deductions of the minor premises $s = \boldsymbol{z}$, we have a standard $\mathfrak{a}$-deduction of $t = \boldsymbol{y}$.

CASE 3: t_1 is $\xi^2(v_1)$ where ξ^2 is a type-2 variable and v_1 is a normal type-1 functor. Similar to:

CASE 4: t_1 is $\xi^j(v_1)$ ($j > 2$) where ξ^j is a type-j variable and v_1 is a normal type-$j$$-$1 functor. A principal branch contains an application of R4.j which introduces $\xi^j(u_1) = \boldsymbol{z}$ with the same ξ^j followed by applications of R3 and R5 leading to $\xi^j(v_1) = \boldsymbol{y}$ (so $\boldsymbol{z}$ is $\boldsymbol{y}$). Now t is $\xi^j(v)$ where v cong v_1. Treating the applications of R3 and R5 read upward which transform v_1 into u_1 as we did those which in Case 1 transformed each v_{1i} into u_{1i}, we are led to ones of R3 (without R5) satisfying (B) which transform $\xi^j(v) = \boldsymbol{y}$ upward into $\xi^j(u) = \boldsymbol{y}$ for a u cong u_1. But $\xi^j(u) = \boldsymbol{y}$ will come by R4.j from premises $u(\xi^{j-2}) = \boldsymbol{z}_{\alpha^{j-2}}$ with (common) right member $u(\xi^{j-2})$ congruent to the right member $u_1(\xi^{j-2})$ of the premises $u_1(\xi^{j-2}) = \boldsymbol{z}_{\alpha^{j-2}}$ of the application in question of R4.j in the subdeduction of $t_1 = \boldsymbol{y}$. By hyp. ind. there is, for each α^{j-2}, a standard $\mathfrak{a},\alpha^{j-2}$-deduction of $u(\xi^{j-2}) = \boldsymbol{z}_{\alpha^{j-2}}$. These, followed by the new application of R4.j, followed by the applications of R3 read downward, with the concurrent standard $\mathfrak{a}$-deductions of the minor premises $s = \boldsymbol{z}$, constitute a standard $\mathfrak{a}$-deduction from E of $t = \boldsymbol{y}$.

CASE 5: t_1 is not normal. Then under § 1.6 $t_1 = \boldsymbol{y}$ comes from a normal equation $t_2 = \boldsymbol{y}$ by applications of R7 and R5 read downward (it cannot come by R5 and R6 from a conclusion for R1, because its left member is not normal). Applying Theorem 1 (b) and (c), by pairs of applications of R5 and R7 read upward satisfying (A3) and (A2) we can convert $t = \boldsymbol{y}$ into an equation $t_3 = \boldsymbol{y}$ cong $t_2 = \boldsymbol{y}$. By hyp. ind. there is a standard $\mathfrak{a}$-deduction from E of $t_3 = \boldsymbol{y}$. This followed by the applications just described of R7 and R5 read downward constitutes a standard $\mathfrak{a}$-deduction of $t = \boldsymbol{y}$.

2.3. Consider a proper system E of equations, and an interpretation $\mathfrak{a}$ of certain function variables as in § 1.8. We describe a process for computing formally (the numeral expressing) the 'value' relative to E and $\mathfrak{a}$ of any term t_0 containing free only function variables interpreted in $\mathfrak{a}$. The process consists in applying repeatedly a step described by the six mutually exclusive cases listed below, so long as one of those cases applies. No step is applicable to a numeral. If the resulting sequence of terms $t_0, t_1, t_2, \cdots$ terminates in a numeral, that numeral expresses the *value* of t_0 for the given E and $\mathfrak{a}$. If $t_0, t_1, t_2, \cdots$ terminates otherwise or does not terminate, the *value* is undefined (or t_0 does not have a *value*). The E and $\mathfrak{a}$ are fixed throughout the sequence of steps, and none of $t_1, t_2, \cdots$ have free function variables not free in t_0 (by § 1.3 (c) in Case 1′). However certain steps depend on the values of other terms computed by the same process with the same E and the same $\mathfrak{a}$ or one $(\mathfrak{a}, \alpha^{j-2})$ extending $\mathfrak{a}$ to interpret by α^{j-2} the next type-$j-2$ variable ξ^{j-2} after those interpreted in $\mathfrak{a}$, where $2 < j \leqq r$. The steps are analogs of standard reductions in [15, end § 4].

CASE 0′: t is normal and contains a proper ready part. Replace the leftmost such part s by the numeral $\boldsymbol{z}$ for the value z of s computed by the same process with the same E and $\mathfrak{a}$.

CASE 1′: t is $f(u_1, \cdots, u_m)$ where f is a function letter and $u_1, \cdots, u_m$ are numerals and regular functors. Pass to the right side v of the result of that substitution (if any) into an equation of E which produces t on the left side, after changes in bound variables on the right side made in accordance with a suitable convention under which the substitution will be free, the same as for (A4) in § 2.2. By § 1.3 (b) with (a) the equation of E if it exists is unique, and using also (c) so are the respective numerals and regular functors which are substituted for its free variables; so the v if it exists is determined uniquely by the t and E.

CASE 2′: t is $\xi^1(\boldsymbol{u})$ where ξ^1 is a type-1 variable and $\boldsymbol{u}$ is a numeral. Pass to the numeral for $\alpha^1(u)$ where α^1 is the function in $\mathfrak{a}$ interpreting ξ^1.

CASE 3′: t is $\xi^2(u)$ where ξ^2 is a type-2 variable and u is a regular type-1 functor such that u(0) (hence $u(\boldsymbol{a})$ for any a) does not have an accented term as normal form. Pass to the numeral for $\alpha^2(\alpha^1)$ where α^2 is the function in $\mathfrak{a}$ interpreting ξ^2 and, for each a, $\alpha^1(a)$ is the value of $u(\boldsymbol{a})$ computed by the same process with the same E and $\mathfrak{a}$.

CASE 4′: t is $\xi^j(\mathrm{u})$ $(j > 2)$ where ξ^j is a type-j variable and u is a regular type-$j-1$ functor such that $\mathrm{u}(\xi^{j-2})$ (ξ^{j-2} the next type-$j-2$ variable) does not have an accented term as normal form. Pass to the numeral for $\alpha^j(\alpha^{j-1})$ where α^j is the function in $\mathfrak{a}$ interpreting ξ^j and, for each α^{j-2}, $\alpha^{j-1}(\alpha^{j-2})$ is the value of $\mathrm{u}(\xi^{j-2})$ computed by the same process with the same E and with the interpretation $(\mathfrak{a}, \alpha^{j-2})$ extending $\mathfrak{a}$ to include α^{j-2} as interpretation of ξ^{j-2}.

CASE 5′: t is not normal. Perform a reduction in which the leftmost innermost redex in t is contracted, choosing the bound variables in the resulting term in accordance with a suitable convention, the same as for (A3) in § 2.2.

The computation process just described can be construed as an inductive definition of the predicate 't has the $\mathfrak{a}$-value y' with a fixed E. There are seven direct clauses, the first six corresponding to the cases 0′–5′; e.g.: 0′. If t is normal and has an occurrrence of s as leftmost proper ready part, s has the $\mathfrak{a}$-*value* z, u comes from t by replacing that part by $\boldsymbol{z}$, and u has the $\mathfrak{a}$-*value* y, then t has the $\mathfrak{a}$-*value* y. 1′. If t and v are as described in Case 1′, and v has the $\mathfrak{a}$-*value* y, then t has the $\mathfrak{a}$-*value* y. 4′. Under the conditions of Case 4′, if, for each α^{j-2}, $\mathrm{u}(\xi^{j-2})$ has the $\mathfrak{a}, \alpha^{j-2}$-*value* $z_{\alpha^{j-2}}$, then t has the $\mathfrak{a}$-*value* $\alpha^j(\lambda\alpha^{j-2}\, z_{\alpha^{j-2}})$. 6′. A numeral $\boldsymbol{y}$ has the $\mathfrak{a}$-*value* y. For convenience, we are assimilating an application of 6′ into the clauses rendering Cases 2′–4′. (For another representation of the computation process, cf. § 3.11 below.)

By an induction of the form corresponding to this inductive definition (similar to that proving the uniqueness of the w in [RF, end § 3.8]), the value given to a term t by this process for a given E and $\mathfrak{a}$ (the $\mathfrak{a}$-value of t for that E) when it exists is unique (which vindicates our use of the definite article in describing the steps). The result of each step is a term determined uniquely by the t, E and $\mathfrak{a}$.

2.4. THEOREM 5. *Let* E *be a proper system of equations,* t *a term,* $\mathfrak{a}$ *an interpretation of certain function variables as in* § 1.8 *including all which occur free in* t, *and* $\boldsymbol{y}$ *a numeral. Then*: (a) *If* E $\mathfrak{a}$-$\vdash$ t $= \boldsymbol{y}$, *the term* t *receives the value* y *under the computation process of* § 2.3 *with that* E *and* $\mathfrak{a}$; *and* (b) *conversely, provided that* t *does not have an accented term as normal form.*

PROOF. (a) To match the notation in the proof of Theorem 4, rewrite $\mathfrak{a}$, t, $\boldsymbol{y}$ as $\mathfrak{a}_0, \mathrm{t}_0, \boldsymbol{y}_0$. Suppose that E $\mathfrak{a}_0$-$\vdash$ $\mathrm{t}_0 = \boldsymbol{y}_0$. By Theorem 4, there is a standard $\mathfrak{a}_0$-deduction from E of $\mathrm{t}_0 = \boldsymbol{y}_0$. We prove by induction over the part of this deduction not above an application of R1 that: *To each standard* $\mathfrak{a}$-*subdeduction with endequation* t $= \boldsymbol{y}$ (t *a term,* $\boldsymbol{y}$ *a numeral*) *in this part, the computation process* (*with* E *and with the interpretation* $\mathfrak{a}$ *in force at the position of* t $= \boldsymbol{y}$) *gives* y *as value to* t.

CASE 1: t is $\mathrm{f}(\mathrm{v}_1, \cdots, \mathrm{v}_m)$ where f is a function letter and $\mathrm{v}_1, \cdots, \mathrm{v}_m$ are normal terms and normal functors. By hyp. ind., for each minor premise s $= \boldsymbol{z}$ for R3 or R2 in the principal branch the computation process (with the same E and $\mathfrak{a}$) gives s the value z. So a series of zero or more steps under Case 0′ (corresponding to the applications of R3 read upward), then one under Case 1′ (corresponding to the pair of applications of R5 and R1 to the principal equation), then zero or more under Case 5′ (corresponding to the pairs of applications of R5 and R6 leading downward to the midequation $f(u_1, \cdots, u_m) = w$), and finally zero or more under Case 0′ (corresponding to the applications of R2 read downward) leads from t to $\boldsymbol{y}$; i.e. the process gives t the value y.

CASE 4: t is $\xi^j(v)$ $(j > 2)$ where ξ^j is a type-j variable and v is a normal type-$j-1$ functor. By hyp. ind., for each minor premise $s = \boldsymbol{z}$ for an application of R3 below the midequation $\xi^j(u) = \boldsymbol{y}$ the process gives s the value z. So steps under Case 0′ take t into $\xi^j(u)$. Applying Theorem 2 to any premise $u(\xi^{j-2}) = \boldsymbol{z}_{\alpha^{j-2}}$ for the application of R4.j which gives the midequation, $u(\xi^{j-2})$ does not have an accented term as normal form. By hyp. ind., for each α^{j-2}, $u(\xi^{j-2})$ receives the value $z_{\alpha^{j-2}}$ under the process with the same E and with $\mathfrak{a}, \alpha^{j-2}$ as the interpretation. So a step under Case 4′ takes $\xi^j(u)$ into the numeral for $\alpha^j(\alpha^{j-1}) = \alpha^j(\lambda\alpha^{j-2}\, z_{\alpha^{j-2}})$, i.e. by R4.$j$ into $\boldsymbol{y}$.

(b) We show by an induction of the form corresponding to the inductive definition of '$\mathfrak{a}$-value' in § 2.3 that: *If* t *(not having an accented term as normal form) has the* $\mathfrak{a}$*-value* y*, then* $E\ \mathfrak{a}\text{-}\vdash t = \boldsymbol{y}$.

CASE 0′: t is normal and contains a proper ready part. By beginning 2.2, the leftmost proper ready part s is not an accented term. By hyp. ind., $E\ \mathfrak{a}\text{-}\vdash s = \boldsymbol{z}$ and $E\ \mathfrak{a}\text{-}\vdash u = \boldsymbol{y}$. By R3 with $u = \boldsymbol{y}$ as major premise and $s = \boldsymbol{z}$ as minor premise, $E\ \mathfrak{a}\text{-}\vdash t = \boldsymbol{y}$.

CASE 1′: t is $f(u_1, \cdots, u_m)$ where f is a function letter and $u_1, \cdots, u_m$ are numerals and regular functors. The first step(s) in the computation process by Case 1′ (Cases 1′ and 5′) lead from t to w, where w is v if v is normal (a certain normal form of v otherwise). Correspondingly, by an application of R5, an application of R1, and zero or more pairs of applications of R5 and R6, $E\ \mathfrak{a}\text{-}\vdash t = w$. SUBCASE 1: w is a numeral. Then w is $\boldsymbol{y}$, so we already have $E\ \mathfrak{a}\text{-}\vdash t = \boldsymbol{y}$. SUBCASE 2: w is not an accented term. By hyp. ind., $E\ \mathfrak{a}\text{-}\vdash w = \boldsymbol{y}$. By R2, $E\ \mathfrak{a}\text{-}\vdash t = \boldsymbol{y}$. SUBCASE 3: w is an accented term not a numeral. Then one or more applications of Case 0′ (Cases 1′–5′ are and remain inapplicable) must lead to $\boldsymbol{y}$. By corresponding applications of R2, with minor premises $s = \boldsymbol{z}$ which are available by hyp. ind., $E\ \mathfrak{a}\text{-}\vdash t = \boldsymbol{y}$.

CASE 4′: t is $\xi^j(u)$ $(j > 2)$ etc. with $u(\xi^{j-2})$ not having an accented term as normal form. By hyp. ind., for each α^{j-2}, $E\ \mathfrak{a}, \alpha^{j-2}\text{-}\vdash u(\xi^{j-2}) = \boldsymbol{z}_{\alpha^{j-2}}$. But $y = \alpha^j(\lambda\alpha^{j-2}\, z_{\alpha^{j-2}})$. So by R4.$j$, $E\ \mathfrak{a}\text{-}\vdash t = \boldsymbol{y}$.

THEOREM 6. (a) *For each proper system* E *of equations, each term* t, *and each interpretation* $\mathfrak{a}$ *of certain function variables as in* § 1.8 *including all which occur free in* t, $E\ \mathfrak{a}\text{-}\vdash t = \boldsymbol{y}$ *for at most one numeral* $\boldsymbol{y}$. (b) *For each list of (informal) variables, each proper system* E *of equations defines Herbrand-Gödel-style recursively a unique partial function of those variables.* (c) *In the notation of* § 1.9: *If* E *defines* $\varphi(\mathfrak{d})$ *Herbrand-Gödel-style recursively, then, for each* $\mathfrak{d}$, *the computation process of* § 2.3 *with* E *and* $\mathfrak{a}$, *applied to the term* f(D), *gives* $\varphi(\mathfrak{d})$ *as value (no value) if* $\varphi(\mathfrak{d})$ *is defined (undefined).*

PROOF. (a) By Theorem 5 (a) with the uniqueness of the $\mathfrak{a}$-value (end § 2.3).

(b) Thence with § 1.9. (c) By Theorem 5 (a) and (b), with § 1.9.

2.5. THEOREM 7. *Each Herbrand-Gödel-style partial (general) recursive function* $\varphi(\mathfrak{b})$ *is partial (general) recursive in the sense of* [RF, § 3].

METHOD OF PROOF. Using Theorem 6 (c), we can give a proof based on [RF, §§ 9, 10] similar to that of Theorem 7 in [**15**, § 5]. A computation by the steps described in 2.3 plays the role here of a sequence of standard reductions there. It is a straightforward exercise to supply the details, paralleling [**15**, § 5].

2.6. REMARK 1. If in § 1.3 we had omitted (b), but in § 1.9 had required that (iv) for each $\mathfrak{b}$ there is at most one numeral $\boldsymbol{y}$ for which $\mathrm{E}\ \mathfrak{a}\vdash \mathrm{f}(\mathrm{D}) = \boldsymbol{y}$, then Theorem 7 would fail: there is a function $\varphi(a, \alpha^2)$ which would be Herbrand-Gödel-style general recursive under the definition thus modified but is not general recursive in the sense of [RF, § 3].

For, consider, first in the informal symbolism of [IM], the equations

$$(\alpha) \qquad \chi_1(t, a) \simeq \mu x T_1(a, a, x) ,$$

$$(\beta) \qquad \chi_2(t, a) \simeq \mu x \overline{T}_1(a, a, t) ,$$

$$(\gamma) \qquad \varphi(a, \alpha^2) \simeq \alpha^2(\lambda t\, \chi_1(t, a)) ,$$

$$(\delta) \qquad \varphi(a, \alpha^2) \simeq 1 + \alpha^2(\lambda t\, \chi_2(t, a)) .$$

Consider any a, α^2. CASE 1: $(Ex)T_1(a, a, x)$. Then $\lambda t\, \chi_1(t, a)$ is completely defined and $\lambda t\, \chi_2(t, a)$ is not completely defined. So (γ) applies and gives $\varphi(a, \alpha^2) = \alpha^2(\lambda t\, \chi_1(t, a))$, while (δ) is inapplicable. CASE 2: otherwise, i.e. $(x)\overline{T}_1(a, a, x)$. Now $\lambda t\, \chi_1(t, a)$ is completely undefined, and $\lambda t\, \chi_2(t, a)$ is completely defined. So (γ) is inapplicable, while (δ) gives $\varphi(a, \alpha^2) = 1 + \alpha^2(\lambda t\, \chi_2(t, a))$.

This can be expressed syntactically. Let $\mathrm{E}_0, \mathrm{E}_1, \mathrm{E}_2$ be systems of equations defining recursively [IM, §§ 54, 63] $+, \chi_1, \chi_2$, constructed by the methods of [IM, § 54 and beginning § 57], with disjoint sets of function letters and say $\mathrm{h}_0, \mathrm{h}_1, \mathrm{h}_2$ as principal function letters. Let $\mathrm{E}_\gamma, \mathrm{E}_\delta, \mathrm{E}$ come from $\mathrm{E}_0\mathrm{E}_1\mathrm{E}_2$ by suffixing the translation(s) of (γ), of (δ), of (γ) and (δ), with $\simeq, +, \chi_1, \chi_2, \varphi$ expressed by $=, \mathrm{h}_0, \mathrm{h}_1, \mathrm{h}_2, \mathrm{f}$ where f does not occur in $\mathrm{E}_0\mathrm{E}_1\mathrm{E}_2$. Then E satisfies § 1.3 (a), (c) and (d), while E_γ and E_δ are proper. Theorem 4 extends to deductions from this E, since § 1.3 (b) is not used in its proof. Consider any a, α^2. We easily see that in a standard α^2-deduction from E of an equation $\mathrm{f}(\boldsymbol{a}, \xi^2) = \boldsymbol{y}$ where $\boldsymbol{y}$ is a numeral, (the translation of) (γ) or (δ) can occur only as the principal equation (or just below it by R5), since outside the principal branch the function letter f will not occur.

By Theorem 6 (a) applied to E_γ (E_δ), using (γ) $((\delta))$ as the principal equation $\mathrm{f}(\boldsymbol{a}, \xi^2) = \boldsymbol{y}$ is deducible for at most one y. But using the appropriate one of (γ) and (δ), $\mathrm{f}(\boldsymbol{a}, \xi^2) = \boldsymbol{y}$ is deducible when y is the value of $\varphi(a, \alpha^2)$ as given above by cases, while using the other we cannot deduce such an equation because the $\aleph_0$ premises required for the application of R4.2 to evaluate the other of $\xi^2(\lambda\xi^0 \mathrm{h}_1(\xi^0, \boldsymbol{a}))$ and $\xi^2(\lambda\xi^0 \mathrm{h}_2(\xi^0, \boldsymbol{a}))$ will not all be available.

Thus $\varphi(a, \alpha^2)$ is Herbrand-Gödel-style general recursive under the modified definition.

But $\varphi(a, \alpha^2)$ is not general recursive [RF, § 3]. For if it were, by [RF, XXIII] (using S2 to get $\theta(a, \alpha^1) = 0$) the number-theoretic function $\varphi(a) = \varphi(a, \lambda\alpha^1 0)$ would be general recursive. But $\varphi(a) = 0$ or $\varphi(a) = 1$ according as $(Ex)T_1(a, a, x)$ or $(x)\overline{T}_1(a, a, x)$, i.e. $\varphi(a)$ is the representing function of $(Ex)T_1(a, a, x)$, which is not general recursive ([IM, p. 283] with [RF, XXXI]).

REMARK 2. Similarly, omitting 1.3 (c) but adding (iv) in § 1.9. For, consider with (α) and (β)

$$(\gamma') \qquad \tau(0, a, \alpha^2) \simeq \alpha^2(\lambda t\, \chi_1(t, a)) ,$$

$$(\delta') \qquad \tau(1, a, \alpha^2) \simeq 1 + \alpha^2(\lambda t\, \chi_2(t, a)) ,$$

$$(\varepsilon') \qquad \varphi(a, \alpha^2) \simeq \tau(b, a, \alpha^2) .$$

REMARK 3. The class of terms allowed for $w_1, \cdots, w_m$ in § 1.3 (a), though quite narrow, suffices for our purpose of proving Theorem 12 below. We could allow somewhat more and still prove Theorem 7.

For example, speaking informally, we could allow φ to appear in the left members of exactly two equations with the respective left members $\varphi(\psi_1(a, \alpha^2), \alpha^2)$ and $\varphi(\psi_2(a, \alpha^2), \alpha^2)$, when ψ_1 and ψ_2 are general recursive functions such that $\psi_1(a_1, \alpha^2)$ and $\psi_2(a_2, \alpha^2)$ are not the same number for any function α^2 and pair of numbers a_1 and a_2. To utilize these two equations, speaking formally now, another replacement rule would be required, namely R2′: to pass from normal equations $f(g_i(\boldsymbol{a}, u), u) = p$ and $g_i(\boldsymbol{a}, u) = \boldsymbol{x}$, where $i = 1$ or 2, f and g_i are the function letters expressing φ and ψ_i, $\boldsymbol{a}$ and $\boldsymbol{x}$ are numerals, and u is a type-2 functor, to $f(\boldsymbol{x}, u) = p$ (cf. Gödel [3, p. 27 (2a)]).[6] In the modified Theorem 4, under (C) R2′ is to be used below the midequation exactly in Case 1 for f the function letter expressing φ, exactly once, below the applications of R2 and above those of R3. In the proof of this theorem, the (first) application of Theorem 3 is replaced by an argument using [**15**, § 2, Theorems 1 and 2] (after Church and Rosser [**3**]) for the case of α-redexes only; moreover, to satisfy the conditions for an evaluation [**15**, § 1], the induction for the modified Theorem 5 (a) (which entails Theorem 6 (a)) should be combined with that for the modified Theorem 4.

But if in this example ψ_1 and ψ_2 be only partial recursive, Theorem 7 would no longer hold in general (though Theorem 6 (a) would); for, consider with (α) and (β)

$$(\gamma'') \qquad \psi_1(a, \alpha^2) \simeq a + 0 \cdot \alpha^2(\lambda t\, \chi_1(t, a)) ,$$

$$(\delta'') \qquad \psi_2(a, \alpha^2) \simeq a + 0 \cdot \alpha^2(\lambda t\, \chi_2(t, a)) ,$$

$$(\varepsilon'') \qquad \varphi(\psi_1(a, \alpha^2), \alpha^2) \simeq 0 ,$$

$$(\zeta'') \qquad \varphi(\psi_2(a, \alpha^2), \alpha^2) \simeq 1 .$$

3. Herbrand-Gödel-style recursiveness of recursive functions.

3.1. We define 'simple functor' inductively, thus. 1. A function variable is *simple*. 2. $\lambda\xi^{j-2}h(\xi^j, \xi^{j-2}, B)$ $(j \geqq 2)$, where h is a function letter, ξ^j, ξ^{j-2} are

[6] We give this rule a very special form, to limit the extent to which the above analysis must be redone for this example.

variables of the indicated types, and B is a list of numerals and distinct function variables not including ξ^j, ξ^{j-2}, is *simple* (cf. S8 in [RF, § 1.3]). 3. $\lambda\xi^0$h(ξ^0, D), where h is a function letter, ξ^0 is a number variable, and D is a list of numerals and distinct function variables, is *simple* (cf. S4′.1 in [RF, § 9.8]). 4. $\lambda\xi^{j-1}$h($\xi^j_1(\xi^{j-1}), \cdots, \xi^j_{n_j}(\xi^{j-1})$, D) ($j > 1, n_j > 0$), where h is a function letter, $\xi^j_1, \cdots, \xi^j_{n_j}, \xi^{j-1}$ are distinct variables of the indicated types, and D is a list of numerals and distinct function variables including exactly $\xi^j_1, \cdots, \xi^j_{n_j}$ of type j and not including ξ^{j-1}, is *simple* (cf. S4′.j in [RF, § 9.8]). 5. A functor resulting by a free substitution of *simple* functors simultaneously for the free function variables of a *simple* functor, followed by a sequence of reductions to normal form, is *simple*. (Also extremal clause.)

(a) Each simple functor is closed and normal.

(b) Each free occurrence of a function variable ξ^j in a simple functor either is the functor, or is as an argument of a function letter, or is in a part $\xi^j(\xi^{j-1})$ ($j > 1$) within the scope of a λ-prefix $\lambda\xi^{j-1}$.

(c) In each part $\lambda\xi^j$t of a simple functor, where t is a term, the term t contains the variable ξ^j free.

(d) Each simple functor is regular (end § 1.1).

(e) In a simple functor, each bound occurrence of a variable not in a λ-prefix is as an argument either of a function letter or of a free function variable of type > 1.

(f) Each free part of a simple functor which is a functor is simple.

(g) In each simple λ-functor $\lambda\xi^j$t, the free occurrences of ξ^j in t lie outside each proper part which is a λ-functor.

(h) If u is a simple functor of type $j > 1$ [type 1], and ξ^{j-1} is a type-$j-1$ variable [$\boldsymbol{a}$ is a numeral], then in a normal form of u(ξ^{j-1}) [u($\boldsymbol{a}$)] each free part which is a functor is simple (using (f), (g)), no ′ occurs except as part of a numeral, and that normal form is not a numeral.

(i) If u is a simple functor, and v cong u, v is simple.

A deduction under 1.4 in which each functor required for R1, R4.2, R4.j ($j > 2$) to be regular is simple is *simple*. If there is a simple $\mathfrak{a}$-deduction of d from E, we write: E $\mathfrak{a}\text{-}\vdash_S$ d.

In the definitions in § 1.9, when (ii) holds with $\vdash$ strengthened to $\vdash_S$ (call it then (ii)$_S$), we suffix "S" as subscript to "*recursive(ly)*".

REMARK 4. The set of the regular functors does not have the closure property given by Clause 5 to the set of the simple functors. EXAMPLE 1. Substitute λy0 for ξ^1 in λxf(ξ^1(x)). EXAMPLE 2. Substitute $\lambda\xi^1\xi^1(0)$ for ξ^2 in λyξ^2(λxf(y, g(x))).

3.2. THEOREM 8. *For each $j > 1$ and each interpretation α^j, α^{j-1} of ξ^j, ξ^{j-1}: $\alpha^j, \alpha^{j-1}\text{-}\vdash_S \xi^j(\xi^{j-1}) = z$ where $z = \alpha^j(\alpha^{j-1})$.*

PROOF, by ind. on j. BASIS: $j = 2$. By R4.1, for each a: $\alpha^1\text{-}\vdash_S \xi^1(\boldsymbol{a}) = z_a$ where $z_a = \alpha^1(a)$. So by R4.2 with ξ^1 as the u: $\alpha^2, \alpha^1\text{-}\vdash_S \xi^2(\xi^1) = z$ where $z = \alpha^2(\alpha^1)$. IND. STEP: $j > 2$. By hyp. ind., for each α^{j-2}: $\alpha^{j-1}, \alpha^{j-2}\text{-}\vdash_S \xi^{j-1}(\xi^{j-2}) = z_{\alpha^{j-2}}$ where $z_{\alpha^{j-2}} = \alpha^{j-1}(\alpha^{j-2})$. So by R4.$j$: $\alpha^j, \alpha^{j-1}\text{-}\vdash_S \xi^j(\xi^{j-1}) = z$ where

$z = \alpha^j(\alpha^{j-1})$.

3.3. THEOREM 9. *Let* E *be a proper system of equations*; t *a term containing a specified set of free occurrences of a type-k variable* ξ^k $(k > 0)$; s *a simple type-k functor free at the specified occurrences of* ξ^k *in* t; t_s *the result of replacing simultaneously those occurrences of* ξ^k *in* t *by* s; $\mathfrak{a}$ *an interpretation of certain function variables including all which occur free in* t *and in* s, *with* α^k *as the interpretation of* ξ^k; *and* $\boldsymbol{y}$ *a numeral. For* $k = 1$, *suppose that, for each* a, E $\mathfrak{a}$-$\vdash_S$ $s(\boldsymbol{a}) = z_a$ *where* $z_a = \alpha^k(a)$. *For* $k > 1$, *let* ξ^{k-1} *be a type-$k-1$ variable not interpreted in* $\mathfrak{a}$, *and suppose that, for each interpretation* α^{k-1} of ξ^{k-1}, E $\mathfrak{a}, \alpha^{k-1}$-$\vdash_S$ $s(\xi^{k-1}) = z_{\alpha^{k-1}}$ *where* $z_{\alpha^{k-1}} = \alpha^k(\alpha^{k-1})$. *If* E $\mathfrak{a}$-$\vdash_S$ $t = \boldsymbol{y}$, *then* E $\mathfrak{a}$-$\vdash_S$ $t_s = \boldsymbol{y}$.

PROOF, by induction on k. We can choose a simple $\mathfrak{a}$-deduction of $t = \boldsymbol{y}$ from E in which R6 and R7 are circumscribed as in § 1.6 and the additional free function variables above the endequation are as in § 1.8 so that ξ^k does not serve as the ξ^{j-2} of an application of R4.j $(j > 1)$. Furthermore we can clearly rearrange the bound variables if necessary so that in the part of the deduction not above an application of R1 but above a possible final new application of R5 s is free at all free occurrences of ξ^k. Within the induction on k, we use induction over this part of this last deduction to show that: *To each equation* $v = w$ *in this part, to each set of* (*zero or more*) *free occurrences of* ξ^k *in* v, *there is a set of free occurrences of* ξ^k *in* w *such that* E $\mathfrak{a}_1$-$\vdash_S$ $v_s = w_s$ *where* v_s, w_s *result from* v, w *by replacing the specified occurrences of* ξ^k *simultaneously by* s *and* $\mathfrak{a}_1$ *is the interpretation in force at the position of* $v = w$.

CASE R1: $v = w$ is the conclusion $f(u_1, \cdots, u_m) = w$ of an application of R1, with the functors substituted simple. The specified occurrences of ξ^k in v are in (zero or more of) the $u_1, \cdots, u_m$ which are functors and entered through the substitution by R1, whence we identify corresponding free occurrences of ξ^k in w. We can alter the substitution by R1 so that, of each u_i which contains some of the specified occurrences of ξ^k, a normal form u_{isN} of u_{is} is substituted instead, since by § 3.1, Clause 5 u_{isN} is also simple, and by our preliminary arrangements s is free at all the free occurrences of ξ^k so the new substitution is also free. Then by applications of R7 with R5 we can convert the resulting occurrences of each u_{isN} to u_{is}. Thus E $\mathfrak{a}_1$-$\vdash_S$ $v_s = w_s$.

CASE R2: $v = w$ is the conclusion $p = q_{\boldsymbol{x}}$ of an application of R2 with premises $p = q_r$ and $r = \boldsymbol{x}$. By the hypothesis of the induction over the part in question of the deduction in question of $t = \boldsymbol{y}$, for the specified occurrences of ξ^k in p, there is a set of free occurrences of ξ^k in q_r such that E $\mathfrak{a}_1$-$\vdash_S$ $p_s = q_{rs}$. Applying the hyp. ind. also to $r = \boldsymbol{x}$ for those occurrences of ξ^k in r which in the specified part r of q_r are among those just mentioned in q_r, E $\mathfrak{a}_1$-$\vdash_S$ $r_s = \boldsymbol{x}$. By Theorem 1 (b), a sequence ρ of zero or more reductions converts r_s to a normal form r_{sN}; and we can convert q_{rs} to a normal form q_{rsN} by first converting in situ by ρ to r_{sN} the part r_s (free by the preliminary arrangements) resulting from the specified part r of q_r by the replacement by s of certain ξ^k's, and then performing zero or more further reductions σ, via which we can identify zero or more free parts r_{sN} in the normal form q_{rsN} of q_{rs}. Also a sequence of reductions τ converts p_s to a normal form p_{sN}. Thus by applications of R6 with R5 we can pass from $p_s = q_{rs}$ and $r_s = \boldsymbol{x}$ to $p_{sN} = q_{rsN}$ and $r_{sN} = \boldsymbol{x}$. Then by zero or more new applications of R2 we can replace the identified parts r_{sN} in $p_{sN} = q_{rsN}$ by $\boldsymbol{x}$; and from the resulting equation by (a sequence of) expansions (R7 with R5) τ^{-1} inverse to τ and expansions ε inverse to σ except with $\boldsymbol{x}$ in place of r_{sN} we can pass to $p_s = q_{\boldsymbol{x}s}$. Thus E $\mathfrak{a}_1 \text{-}\vdash_s v_s = w_s$.

Case R3: v = w is the conclusion $q_r = p$ of an application of R3 with premises $q_{\boldsymbol{x}} = p$ and $r = \boldsymbol{x}$. The specified occurrences of ξ^k in q_r correspond to ones in $q_{\boldsymbol{x}}$ and r. By hyp. ind., E $\mathfrak{a}_1 \text{-}\vdash_s q_{\boldsymbol{x}s} = p_s$ for a certain specification of occurrences of ξ^k in p, and E $\mathfrak{a}_1 \text{-}\vdash_s r_s = \boldsymbol{x}$. As in Case R2, by reductions ε^{-1} and τ on $q_{\boldsymbol{x}s} = p_s$ and ρ on $r_s = \boldsymbol{x}$, new applications of R3, and finally expansions σ^{-1}, ρ^{-1} and τ^{-1}, we can pass to $q_{rs} = p_s$. Thus E $\mathfrak{a}_1 \text{-}\vdash_s v_s = w_s$.

Case R4.1: v = w is $\xi^1(\boldsymbol{u}) = \boldsymbol{z}$ (where $z = \alpha^1(u)$) by R4.1. Then v = w is already $v_s = w_s$, unless $k = 1$ and ξ^1 is a specified occurrence of ξ^k. In that case, by the supposition for $k = 1$ in the theorem, E $\mathfrak{a}_1 \text{-}\vdash_s s(\boldsymbol{u}) = \boldsymbol{z}_u$ where $z_u = \alpha^k(u) = \alpha^1(u) = z$, i.e. E $\mathfrak{a}_1 \text{-}\vdash_s v_s = w_s$.

Case R4.2. v = w is $\xi^2(u) = \boldsymbol{z}$ by R4.2, with u simple. Similar to:

Case R4.j ($j > 2$): v = w is $\xi^j(u) = \boldsymbol{z}$ by R4.j, with u simple. For each α^{j-2}, there is a respective premise $u(\xi^{j-2}) = \boldsymbol{z}_{\alpha^{j-2}}$ where $z_{\alpha^{j-2}} = \alpha^{j-1}(\alpha^{j-2})$; and $z = \alpha^j(\alpha^{j-1})$. By hyp. ind., for each α^{j-2}, E $\mathfrak{a}_1, \alpha^{j-2} \text{-}\vdash_s u_s(\xi^{j-2}) = \boldsymbol{z}_{\alpha^{j-2}}$, whence using R6 with R5 E $\mathfrak{a}_1, \alpha^{j-2} \text{-}\vdash_s u_{sN}(\xi^{j-2}) = \boldsymbol{z}_{\alpha^{j-2}}$. But u_{sN} is also simple. Subcase 1: the initial occurrence of ξ^j in v is not one of the specified occurrences of ξ^k. By a new application of R4.j E $\mathfrak{a}_1 \text{-}\vdash_s \xi^j(u_{sN}) = \boldsymbol{z}$, whence by R7 with R5 E $\mathfrak{a}_1 \text{-}\vdash_s \xi^j(u_s) = \boldsymbol{z}$, i.e. E $\mathfrak{a}_1 \text{-}\vdash_s v_s = w_s$. Subcase 2: otherwise (then $k = j$). If the ξ^{k-1} in the supposition for $k > 1$ in the theorem occurs among the variables interpreted in $\mathfrak{a}_1$, we can change it to a variable not so occurring. By that supposition for α^{j-1} as the α^{k-1}, with the hypothesis of the induction on k for u_{sN} as the t, E $\mathfrak{a}_1, \alpha^{j-1} \text{-}\vdash_s s(u_{sN}) = \boldsymbol{z}_{\alpha^{j-1}}$ where $z_{\alpha^{j-1}} = \alpha^k(\alpha^{j-1}) = \alpha^j(\alpha^{j-1}) = z$. Then since ξ^{k-1} does not occur free in $s(u_{sN})$, E $\mathfrak{a}_1 \text{-}\vdash_s s(u_{sN}) = \boldsymbol{z}$. Using R7 with R5, E $\mathfrak{a}_1 \text{-}\vdash_s s(u_s) = \boldsymbol{z}$, i.e. E $\mathfrak{a}_1 \text{-}\vdash_s v_s = w_s$.

Cases R5–R7: v = w comes from $v_1 = w_1$ by R5, R6 or R7. From the specified free occurrences of ξ^k in v we identify ones in v_1. By hyp. ind., E $\mathfrak{a}_1 \text{-}\vdash_s v_{1s} = w_{1s}$ for certain free occurrences of ξ^k in w_1, whence we identify ones in w. By R5, R6 or R7, E $\mathfrak{a}_1 \text{-}\vdash_s v_s = w_s$, since in both v = w and $v_1 = w_1$ s is free at all free occurrences of ξ^k.

3.4. Theorem 10. *Let* E *be the (proper) system of equations which results by translating the equations for the schemata applications of a primitive recursive description* $\varphi_1, \cdots, \varphi_k$ *of* $\varphi\ (= \varphi_k)$ [RF, § 1.3] *into the formal symbolism, using distinct function letters* $f_1, \cdots, f_k$ *to express* $\varphi_1, \cdots, \varphi_k$. *For each tuple*

$\mathfrak{d}$ *of natural numbers and functions as arguments of* φ, $\mathrm{E}\ \mathfrak{a}\text{-}\vdash_{\mathrm{S}} \mathrm{f}_k(\mathrm{D}) = \boldsymbol{y}$ *where* D *are respective numerals expressing the numbers, and distinct function variables interpreted by the functions* $\mathfrak{a}$, *in* $\mathfrak{d}$ *and* $y = \varphi(\mathfrak{d})$.

PROOF, by induction on k, with cases according to the schema by which φ is introduced, and in Case S5 induction on the first argument a.

CASE S7: $\varphi(\mathfrak{d}) = \varphi(\alpha^1, a, \mathfrak{b}) = \alpha^1(a)$. Using (R5 and) R1 to substitute D (write it $\xi^1, \boldsymbol{a}, \mathrm{B}$) into the last equation of E, $\mathrm{E}\ \mathfrak{a}\text{-}\vdash_{\mathrm{S}} \mathrm{f}_k(\xi^1, \boldsymbol{a}, \mathrm{B}) = \xi^1(\boldsymbol{a})$. By R4.1, $\mathrm{E}\ \mathfrak{a}\text{-}\vdash_{\mathrm{S}} \xi^1(\boldsymbol{a}) = \boldsymbol{y}$. Apply R2.

CASE S8.j ($j > 2$). $\varphi(\alpha^j, \mathfrak{b}) = \alpha^j(\alpha^{j-1})$ where $\alpha^{j-1} = \lambda\alpha^{j-2}\, \chi(\alpha^j, \alpha^{j-2}, \mathfrak{b})$. Using R5 and R1, $\mathrm{E}\ \mathfrak{a}\text{-}\vdash_{\mathrm{S}} \mathrm{f}_k(\xi^j, \mathrm{B}) = \xi^j(\lambda\xi^{j-2}\mathrm{h}(\xi^j, \xi^{j-2}, \mathrm{B}))$ where h (which is f_l for some $l < k$) expresses χ. By hyp. ind., for each interpretation α^{j-2} of ξ^{j-2}, $\mathrm{E}\ \mathfrak{a}, \alpha^{j-2}\text{-}\vdash_{\mathrm{S}} \mathrm{h}(\xi^j, \xi^{j-2}, \mathrm{B}) = \boldsymbol{z}_{\alpha^{j-2}}$ where $z_{\alpha^{j-2}} = \chi(\alpha^j, \alpha^{j-2}, \mathfrak{b}) = \alpha^{j-1}(\alpha^{j-2})$, whence using R7 $\mathrm{E}\ \mathfrak{a}, \alpha^{j-2}\text{-}\vdash_{\mathrm{S}} (\lambda\xi^{j-2}\mathrm{h}(\xi^j, \xi^{j-2}, \mathrm{B}))(\xi^{j-2}) = \boldsymbol{z}_{\alpha^{j-2}}$, whence for α^{j-1} as interpretation of a variable ξ^{j-1} not interpreted in $\mathfrak{a}$ $\mathrm{E}\ \mathfrak{a}, \alpha^{j-1}, \alpha^{j-2}\text{-}\vdash_{\mathrm{S}} (\lambda\xi^{j-2}\mathrm{h}(\xi^j, \xi^{j-2}, \mathrm{B}))(\xi^{j-2}) = \boldsymbol{z}_{\alpha^{j-2}}$. By Theorem 8, $\alpha^j, \alpha^{j-1}\text{-}\vdash_{\mathrm{S}} \xi^j(\xi^{j-1}) = \boldsymbol{y}$, whence since $\mathfrak{a}$ includes α^j as interpretation of ξ^j $\mathrm{E}\ \mathfrak{a}, \alpha^{j-1}\text{-}\vdash_{\mathrm{S}} \xi^j(\xi^{j-1}) = \boldsymbol{y}$. The functor $\lambda\xi^{j-2}\mathrm{h}(\xi^j, \xi^{j-2}, \mathrm{B})$ is simple. By Theorem 9, $\mathrm{E}\ \mathfrak{a}, \alpha^{j-1}\text{-}\vdash_{\mathrm{S}} \xi^j(\lambda\xi^{j-2}\mathrm{h}(\xi^j, \xi^{j-2}, \mathrm{B})) = \boldsymbol{y}$, whence since B does not contain ξ^{j-1} $\mathrm{E}\ \mathfrak{a}\text{-}\vdash_{\mathrm{S}} \xi^j(\lambda\xi^{j-2}\mathrm{h}(\xi^j, \xi^{j-2}, \mathrm{B})) = \boldsymbol{y}$. Apply R2.

3.5. THEOREM 11. *Let* E *be as in Theorem* 10; t *a closed normal term, not a numeral, not containing* ′ *except as part of a numeral, containing only the function letters* $\mathrm{f}_1, \cdots, \mathrm{f}_k$, *of which each free part which is a functor is simple*; $\mathfrak{a}$ *an interpretation of certain function variables including all occurring free in* t; *and t the natural number expressed by* t *under the interpretation* $\mathfrak{a}$ *with the interpretation of* $\mathrm{f}_1, \cdots, \mathrm{f}_k$ *by* $\varphi_1, \cdots, \varphi_k$. *Then* $\mathrm{E}\ \mathfrak{a}\text{-}\vdash_{\mathrm{S}} \mathrm{t} = \boldsymbol{t}$.

PROOF. Let the number of occurrences of λ, function letters, and of function variables applied to (rather than as) arguments, in t be the *grade* g of t. We use induction on g. By beginning § 2.2 the following cases are exhaustive (for t not ready, Case 1; for t ready, Cases 2–4).

CASE 1: t contains as a free proper part a term r not a numeral; write t also as t_{r}. Say that under the interpretation $\mathfrak{a}$ the value of r is r. By hyp. ind., $\mathrm{E}\ \mathfrak{a}\text{-}\vdash_{\mathrm{S}} \mathrm{r} = \boldsymbol{r}$ and (using § 3.1 (d) in verifying that $\mathrm{t}_{\boldsymbol{r}}$ satisfies the hypotheses) $\mathrm{E}\ \mathfrak{a}\text{-}\vdash_{\mathrm{S}} \mathrm{t}_{\boldsymbol{r}} = \boldsymbol{t}$. By R3, $\mathrm{E}\ \mathfrak{a}\text{-}\vdash_{\mathrm{S}} \mathrm{t} = \boldsymbol{t}$.

CASE 2: t is $\mathrm{f}(\mathrm{u}_1, \cdots, \mathrm{u}_m)$ where $\mathrm{u}_1, \cdots, \mathrm{u}_m$ are numerals and simple functors. Let $\mathrm{v}_1, \cdots, \mathrm{v}_m$ come from $\mathrm{u}_1, \cdots, \mathrm{u}_m$ replacing each one u which is a λ-functor by a different function variable ξ^j of the same type not interpreted in $\mathfrak{a}$, and let $\mathfrak{b}$ come from $\mathfrak{a}$ by adding as the interpretation of each such ξ^j the function α^j which the u it replaces expresses under $\mathfrak{a}$. By Theorem 10, $\mathrm{E}\ \mathfrak{b}\text{-}\vdash_{\mathrm{S}} \mathrm{f}(\mathrm{v}_1, \cdots, \mathrm{v}_m) = \boldsymbol{t}$. For each such u of type $j > 1$, let ξ^{j-1} be a function variable not interpreted in $\mathfrak{b}$. By the hypothesis of the induction applied to a normal form of $\mathrm{u}(\xi^{j-1})$ (which by § 3.1 (h) satisfies the hypotheses) and R7 with R5, for each interpretation α^{j-1} of ξ^{j-1}, $\mathrm{E}\ \mathfrak{a}, \alpha^{j-1}\text{-}\vdash_{\mathrm{S}} \mathrm{u}(\xi^{j-1}) = \boldsymbol{z}_{\alpha^{j-1}}$ where $z_{\alpha^{j-1}} = \alpha^j(\alpha^{j-1})$. Similarly, for each such u of type $j = 1$, for each a, $\mathrm{E}\ \mathfrak{a}\text{-}\vdash_{\mathrm{S}} \mathrm{u}(\boldsymbol{a}) = \boldsymbol{z}_a$ where $z_a = \alpha^j(a)$. So using Theorem 9 to replace successively each one v of $v_1, \cdots, v_m$ which differs from the respective u by u, E $\mathfrak{a}\text{-}\vdash_S f(u_1, \cdots, u_m) = \boldsymbol{t}$, i.e. E $\mathfrak{a}\text{-}\vdash_S t = \boldsymbol{t}$.

CASE 3: t is $\xi^1(\boldsymbol{u})$. Immediate by R4.1.

CASE 4: t is $\xi^{j+1}(u)$ where u is a simple type-j functor ($j > 0$). Let ξ^j be u if u is a variable, otherwise a new type-j variable with the same interpretation α^j. By Theorem 8, $\alpha^{j+1}, \alpha^j\text{-}\vdash_S \xi^{j+1}(\xi^j) = \boldsymbol{t}$. If ξ^j is u, E $\mathfrak{a}\text{-}\vdash_S t = \boldsymbol{t}$ is immediate. Otherwise, as in Case 2, using the hyp. ind., R7 with R5, and Theorem 9 to replace ξ^j by u, E $\mathfrak{a}\text{-}\vdash_S \xi^{j+1}(u) = \boldsymbol{t}$, i.e. E $\mathfrak{a}\text{-}\vdash_S t = \boldsymbol{t}$.

3.6. THEOREM 12. *Each function $\varphi(\mathfrak{d})$ partial (general) recursive in the sense of* [RF, § 3] *is Herbrand-Gödel-style partial (general) recursive.*

The proof occupies §§ 3.7–3.12.

3.7. By [RF, § 10.8, LXVIII] each function $\varphi(\mathfrak{d})$ partial recursive in the sense of [RF, § 3] is definable by a finite succession of recursions (a*) [RF, § 10.4] with normal recursive $\mathsf{F}(\zeta, \Psi; \mathfrak{a})$ (as operations for the definition of a function φ_m from $l \geqq 0$ functions $\psi_1, \cdots, \psi_l$ or briefly Ψ) and of applications of S4$'.j$ ($j \geqq 0$) [RF, §§ 10.8 and 9.8]. Each recursion (a*) can be rendered here by the succession of applications of the schemata S0$'$, S0$'.i$ ($i = 1, \cdots, l$), S1–S8, S4$'.j$ ($j \geqq 1$), S5$'$ (cf. [RF, §§ 10.1, 1.3]) in a uniform derivation of $\varphi_m = \lambda\mathfrak{a}\ \mathsf{F}(\zeta, \Psi; \mathfrak{a})$ from ζ, Ψ, with ζ as introduced by S0$'$ identified with φ_m and each ψ_i ($i = 1, \cdots, l$) as introduced by S0$'.i$ identified with the previously defined function in question. Thus we are led to a finite sequence $\varphi_1, \cdots, \varphi_k$ of functions, each of which is either an initial function by S1–S3 or S7 or comes from other functions of the list (earlier ones except in the case of S0$'$) by S0$'$, S0$'.i$, S4–S6, S8, S4$'.j$ ($j \geqq 0$) or S5$'$, such that the last function $\varphi_k = \varphi$. The schema applications introducing $\varphi_1, \cdots, \varphi_k$ can be translated into the formal symbolism, using distinct function letters $f_1, \cdots, f_k$ to express the respective functions. The resulting system E of equations is proper (cf. §§ 1.3, 1.9 (i)).

We shall establish also § 1.9 (ii)$_S$ and (iii), so as to show that E defines φ Herbrand-Gödel-style recursively$_S$ (cf. end § 3.1).

3.8. In this proof we call a functor *special* if it belongs to the class defined by the inductive definition of 'simple functor' in § 3.1 when Clause 2 is omitted and in Clause 3 (Clause 4) the h is restricted to be a function letter translating the θ of one of the applications of S4$'.1$ (S4$'.j$, $j > 1$) in the above "description" $\varphi_1, \cdots, \varphi_k$ of φ.

A special functor is simple and hence regular (cf. § 3.1 (d)).

A special λ-functor of type $j > 1$ is of the form $\lambda\xi^{j-1}$r where r is a closed normal term with one or more proper ready parts, all of which are of the form $\xi^j(\xi^{j-1})$ (cf. § 2.2).

3.9. Consider, for a given set of arguments $\mathfrak{d}$, the tree for the computation of $\varphi(\mathfrak{d})$ via the description $\varphi_1, \cdots, \varphi_k$. This is to be constructed in the same manner as in [RF, § 10.2] (continuing from [RF, § 9.1]), except that now we are dealing with a succession of applications of (a*) [RF, § 10.4] and of R4$'.j$ ($j \geqq 0$) instead of with just one application of (a) [RF, § 10.1]. Thus we use indices $d_1, \cdots, d_k$ to specify the schemata introducting the functions $\varphi_1, \cdots, \varphi_k$, so that each position γ in the tree is occupied by a tuple $(d_\gamma, \mathfrak{d}_\gamma)$ where d_γ is an index of φ_γ and $\mathfrak{d}_\gamma$ is a sequence of arguments for φ_γ. The 0-position is occupied by $(d_k, \mathfrak{d})$. After a position γ occupied by a tuple $(d_\gamma, \mathfrak{b}_\gamma, \mathfrak{c}_\gamma)$ coming under S0′, there is a single next position occupied by $(d_m, \mathfrak{b}_\gamma)$ where d_m is the index of the function φ_m defined by the recursion (a*) in question for that application of S0′. After a position γ occupied by a tuple $(d_\gamma, \mathfrak{b}_\gamma, \mathfrak{c}_\gamma)$ coming under S0′.i ($i = 1, \cdots, l$, for the l of the recursion (a*) in question for that application of S0′.i), there is a single next position occupied by $(e_i, \mathfrak{b}_\gamma)$ where e_i is the index of ψ_i in the list $d_1, \cdots, d_k$. As in [RF, § 9.1], we terminate a branch when S1–S3 or S7 applies without making the application. Departing from [RF, § 9.8], we find it more convenient here not to consider as part of the tree the computations of the values of the first function argument $\lambda\tau^{j-1}\,\theta(\cdots)$ at applications of S4′.j ($j \geqq 1$), which we don't now class as "nodes".

As in [RF, § 9.1 LIII], there are two cases. In one, $\varphi(\mathfrak{d})$ is defined, at each node the upper next position exists, each tuple in the tree has a value, and each branch terminates at an application of S1–S3 or S7. In the other, $\varphi(\mathfrak{d})$ is undefined, there is an infinite branch, each infinite branch is uppermost and no (each) tuple on (below) it has a value.

To each position γ in the tree (occupied by $(d_\gamma, \mathfrak{d}_\gamma)$), beginning with the 0-position and proceeding from n-positions to $n+1$-positions, we correlate a term $\mathrm{f}_\gamma(\mathrm{D}_\gamma)$ (a translation of $\varphi_\gamma(\mathfrak{d}_\gamma)$ into the formal symbolism) and an interpretation $\mathfrak{a}_\gamma$ as in § 1.8, where f_γ is the function letter expressing φ_γ and D_γ is a sequence of numerals and special functors expressing $\mathfrak{d}_\gamma$ under the interpretation $\mathfrak{a}_\gamma$. Thus, to the 0-position we correlate $\mathrm{f}_k(\mathrm{D})$ and $\mathfrak{a}$, where D are respective numerals expressing the numbers, and distinct function variables interpreted by the functions $\mathfrak{a}$, in $\mathfrak{d}$. In an application from position γ of S8.j, a new function argument α^{j-2} enters, which for the various next positions δ ranges over all type-j−2 objects; this function argument we translate by the next type-$j-2$ variable ξ^{j-2} after those interpreted in $\mathfrak{a}_\gamma$, and we adjoin the respective type-$j-2$ object α^{j-2} as its interpretation to $\mathfrak{a}_\gamma$ to obtain $\mathfrak{a}_\delta$. In an application from γ of S4′.j ($j \geqq 1$), a new function argument enters at the next position δ which is not expressed simply by a function variable but by a special functor containing free exactly the function variables already occurring free. Further details are combined with the discussion in § 3.10.

3.10. We shall prove by induction over the appropriate portion of the computation tree for $\varphi(\mathfrak{d})$ that: *At each position γ at which $\varphi_\gamma(\mathfrak{d}_\gamma)$ is defined (so $(d_\gamma, \mathfrak{d}_\gamma)$ has a value),* $\mathrm{E}\ \mathfrak{a}_\gamma\text{+}_{\mathrm{S}}\ \mathrm{f}_\gamma(\mathrm{D}_\gamma) = \boldsymbol{y}_\gamma$ *where* $y_\gamma = \varphi_\gamma(\mathfrak{d}_\gamma)$. Applying this result to the 0-position when $\varphi(\mathfrak{d})$ is defined, we shall have (ii)$_{\mathrm{S}}$ for § 1.9.

Cases arise according to the schema S0′, S0′.i, S1–S8, S4′, S5′ which applies to φ_γ, and in the cases of S5 and S5′ according to whether the first or second equation of the schema applies. This determines which equation of E we take as the principal equation for the deduction of $\mathrm{f}_\gamma(\mathrm{D}_\gamma) = \boldsymbol{y}_\gamma$ (cf. § 2.2). In the applications of R5 preceding ones of R1 (of R6) used in the cases, the convention of (A4) (of (A3)) shall determine the bound variables in the result (cf. § 2.2).

CASE S0′: $\varphi_\gamma(\mathfrak{d}_\gamma) = \varphi_\gamma(\mathfrak{b}_\gamma, \mathfrak{c}_\gamma) = \varphi_m(\mathfrak{b}_\gamma)$ by S0′. The position γ is occupied by $(d_\gamma, \mathfrak{b}_\gamma, \mathfrak{c}_\gamma)$, the sole next position δ by $(d_m, \mathfrak{b})$. Substituting by R1 (after the identical application of R5) into the appropriate equation of E as the principal equation, E $\mathfrak{a}_\gamma$-$\vdash_S$ $f_\gamma(B_\gamma, C_\gamma) = f_m(B_\gamma)$. By hyp. ind. over the tree, E $\mathfrak{a}$-$\vdash_S$ $f_m(B_\gamma) = \boldsymbol{y}_\gamma$. By R2, E $\mathfrak{a}_\gamma$-$\vdash_S$ $f_\gamma(B_\gamma, C_\gamma) = \boldsymbol{y}_\gamma$.

CASE S0′.i ($i = 1, \cdots, l$): $\varphi_\gamma(\mathfrak{d}_\gamma) = \varphi_\gamma(\mathfrak{b}_\gamma, \mathfrak{c}_\gamma) = \psi_i(\mathfrak{b}_\gamma)$. The position γ is occupied by $(d_\gamma, \mathfrak{b}_\gamma, \mathfrak{c}_\gamma)$. Using R1 on the appropriate principal equation, E $\mathfrak{a}_\gamma$-$\vdash_S$ $f_\gamma(B_\gamma, C_\gamma) = g_i(B_\gamma)$. By hyp. ind., E $\mathfrak{a}_\gamma$-$\vdash_S$ $g_i(B_\gamma) = \boldsymbol{y}$. Use R2.

CASE S4 (AND S4′.0): $\varphi_\gamma(\mathfrak{d}_\gamma) = \psi(\chi(\mathfrak{d}_\gamma), \mathfrak{d}_\gamma)$ with the tuples for $\chi(\mathfrak{d}_\gamma)$ and $\psi(z, \mathfrak{d}_\gamma)$ for $z = \chi(\mathfrak{d}_\gamma)$ at the lower and upper next positions. Using R1 on the principal equation, E $\mathfrak{a}_\gamma$-$\vdash_S$ $f_\gamma(D_\gamma) = g(h(D_\gamma), D_\gamma)$. By hyp. ind., E $\mathfrak{a}_\gamma$-$\vdash_S$ $h(D_\gamma) = \boldsymbol{z}$ and E $\mathfrak{a}_\gamma$-$\vdash_S$ $g(\boldsymbol{z}, D_\gamma) = \boldsymbol{y}_\gamma$. Use R2 twice.

CASE S4′.1 is similar to:

CASE S4′.j ($j > 1$): $\varphi_\gamma(\mathfrak{d}_\gamma) = \psi(\lambda\tau^{j-1}\, \theta(\alpha_1^j(\tau^{j-1}), \cdots, \alpha_{n_j}^j(\tau^{j-1}), \mathfrak{d}_\gamma), \mathfrak{d}_\gamma)$. Using R1 after R5 (not necessarily the identical application), E $\mathfrak{a}_\gamma$-$\vdash_S$ $f_\gamma(D_\gamma) = g(\lambda\xi^{j-1}\, h(s_1(\xi^{j-1}), \cdots, s_{n_j}(\xi^{j-1}), D_\gamma), D_\gamma)$ where ξ^{j-1} is a type-$j-1$ variable (not necessarily the same one as in the principal equation) and $s_1, \cdots, s_{n_j}$ are the (special) type-j functors in D_γ. Thence by $\leqq n_j$ pairs of applications of R5 and R6 (cf. § 3.1 (a) and (e)), E $\mathfrak{a}_\gamma$-$\vdash_S$ $f_\gamma(D_\gamma) = g(u, D_\gamma)$ where u is a normal type-j functor. By Clauses 4 and 5 of the definition of 'special functor' (§ 3.8 with § 3.1), u is special. We take $g(u, D_\gamma)$ as the $f_\delta(D_\delta)$ correlated to the position δ next after γ. By hyp. ind., E $\mathfrak{a}_\gamma$-$\vdash_S$ $g(u, D_\gamma) = \boldsymbol{y}_\gamma$. Use R2.

CASE S7: $\varphi_\gamma(\mathfrak{d}_\gamma) = \varphi_\gamma(\alpha^1, a, \mathfrak{b}_\gamma) = \alpha^1(a)$. Write D_γ as s, $\boldsymbol{a}$, B_γ, where s is a special functor expressing α^1 under the interpretation $\mathfrak{a}_\gamma$. Using R1, E $\mathfrak{a}_\gamma$-$\vdash_S$ $f_\gamma(s, \boldsymbol{a}, B_\gamma) = s(\boldsymbol{a})$. Thence by R6, E $\mathfrak{a}_\gamma$-$\vdash_S$ $f_\gamma(s, \boldsymbol{a}, B_\gamma) = t$ where t is normal. Since s is special, Theorem 11 applies to this t (using § 3.1 (h)) and the subsequence of the present system E constituted by those equations which translate the primitive recursive descriptions of the functions θ for the applications of S4′.j ($j \geqq 1$) in the description $\varphi_1, \cdots, \varphi_k$ as the E; thus E $\mathfrak{a}_\gamma$-$\vdash_S$ $t = \boldsymbol{y}_\gamma$. Use R2.

CASE S8.2 is similar to:

CASE S8.j ($j > 2$): $\varphi_\gamma(\mathfrak{d}_\gamma) = \varphi_\gamma(\alpha^j, \mathfrak{b}_\gamma) = \alpha^j(\lambda\alpha^{j-2}\, \chi(\alpha^j, \alpha^{j-2}, \mathfrak{b}_\gamma))$. Write D_γ as s, B_γ, where s is a special functor expressing α^j under $\mathfrak{a}_\gamma$. Using R1 after R5, E $\mathfrak{a}_\gamma$-$\vdash_S$ $f_\gamma(s, B_\gamma) = s(u)$ where u is $\lambda\xi_1^{j-2} h(s, \xi_1^{j-2}, B_\gamma)$, which is simple by Clauses 2 and 5 in § 3.1. By hyp. ind., for each interpretation α^{j-2} of ξ^{j-2} (cf. § 3.9), E $\mathfrak{a}_\gamma, \alpha^{j-2}$-$\vdash_S$ $h(s, \xi^{j-2}, B_\gamma) = \boldsymbol{z}_{\alpha^{j-2}}$ where $z_{\alpha^{j-2}} = \chi(\alpha^j, \alpha^{j-2}, \mathfrak{b}_\gamma)$, whence by R7 E $\mathfrak{a}_\gamma, \alpha^{j-2}$-$\vdash_S$ $u(\xi^{j-2}) = \boldsymbol{z}_{\alpha^{j-2}}$. So by R4.$j$, E $\mathfrak{a}_\gamma, \alpha^j$-$\vdash_S$ $\xi^j(u) = \boldsymbol{y}_\gamma$ where ξ^j is a new function variable interpreted by α^j and $y_\gamma = \alpha^j(\lambda\alpha^{j-2}\, \chi(\alpha^j, \alpha^{j-2}, \mathfrak{b}_\gamma)) = \varphi_\gamma(\mathfrak{d}_\gamma)$. Let ξ^{j-1} be a function variable not interpreted in $\mathfrak{a}_\gamma$. Since s is special, Theorem 11 applies to a normal form of $s(\xi^{j-1})$ as the t (using § 3.1 (h)) with an E selected from the present E as in Case S7; thus using also R7 with R5, for each interpretation α^{j-1} of ξ^{j-1}, E $\mathfrak{a}_\gamma, \alpha^{j-1} \vdash_S$ s(ξ^{j-1}) $= z_{\alpha^{j-1}}$ where $z_{\alpha^{j-1}} = \alpha^j(\alpha^{j-1})$. So using Theorem 9 to replace ξ^j by s, E $\mathfrak{a}_\gamma \vdash_S$ s(u) $= y_\gamma$. Use R2.

3.11. The terms which enter in the computation process of 2.3 applied to a term t for a given E and $\mathfrak{a}$, including those which enter in the steps by the same process required for a step under one of Cases 0′, 3′ and 4′, can be arranged on a tree with t at the 0-position. Terms to which Case 0′, 3′ or 4′ applies occupy nodes, with 1, $\aleph_0$ or more lower next positions, respectively.

As before (§ 3.9, [RF, § 9.1, LIII]), there are two cases. When t has a value, at each node the upper next position exists, each term in the tree has a value, and each branch terminates at a numeral. When t has no value, there is a branch either infinite or terminating other than at a numeral, each such branch is uppermost and no (each) term on (below) it has a value.

3.12. Assuming $\varphi(\mathfrak{d})$ undefined (the second case in § 3.9), we shall show that: *Corresponding to a given infinite branch in the computation tree* T *for* $\varphi(\mathfrak{d})$, *there is an infinite branch in the tree* U *representing the attempted computation of the term* $f_k(D)$ *under the process of* § 2.4 *with the present system* E *and interpretation* $\mathfrak{a}$. To each n-position γ occupied by $(d_\gamma, \mathfrak{d}_\gamma)$ in the given infinite branch in T, there is an m-position with $m \geqq n$ occupied by $f_\gamma(D_\gamma)$ in the corresponding branch in U.

This result with 3.11 and Theorem 5 (a) contraposed will give us (iii) for § 1.9.

Suppose we have constructed the required branch in U as far as the m-position corresponding to the n-position γ in the given branch in T. We use the same cases and notation as in § 3.10.

Case S0′. The position in U corresponding to γ in T is occupied by $f_\gamma(B_\gamma, C_\gamma)$. That corresponding to the $n+1$-position δ is occupied by $f_m(B_\gamma)$, coming from $f_\gamma(B_\gamma, C_\gamma)$ by a step under Case 1′ of § 2.4.

Case S0′.i $(i = 1, \cdots, l)$. Similarly, with $g_i(B_\gamma)$ instead of $f_m(B_\gamma)$.

Case S4 (and S4′.0). The position in U corresponding to γ in T is occupied by $f_\gamma(D_\gamma)$. We first pass by Case 1′ to $g(h(D_\gamma), D_\gamma)$. Subcase 1: the given branch in T includes the lower next position δ to the node γ (the upper one being unfilled). Then we further pass to $h(D_\gamma)$ as the initial term toward the computation of the leftmost proper ready part as called for under Case 0′; this occupies the lower next position to the node in U occupied by $g(h(D_\gamma), D_\gamma)$. Subcase 2: the given branch in T includes the upper next position ε, the value at the lower one δ being $z = \chi(\mathfrak{d}_\gamma)$. By § 3.10 and Theorem 5 (a) the value of the term $h(D_\gamma)$ under the process is also z, so a step by Case 0′ takes us from $g(h(D_\gamma), D_\gamma)$ to $g(z, D_\gamma)$; this occupies the upper next position to the node in U occupied by $g(h(D_\gamma), D_\gamma)$.

Case S4′.j $(j > 1)$. We pass by Case 1′ to $g(\lambda\xi^{j-1}h(s_1(\xi^{j-1}), \cdots, s_{n_j}(\xi^{j-1}), D_\gamma), D_\gamma)$, and thence by $\leqq n_j$ applications of Case 5′ to the normal term $g(u, D_\gamma)$.

CASE S7. Excluded because the given branch in T is infinite.

CASE S8.j ($j > 2$). We pass by Case 1′ to s(u) where u is $\lambda\xi_1^{j-2}h(s, \xi_1^{j-2}, B_\gamma)$. SUBCASE 1: s is a λ-functor. Then we further pass to a normal term t by one application of Case 5′ (cf. § 3.1 (a), (e)). By end § 3.8 t comes under Case 0′ with the leftmost proper ready part of the form $\xi^j(u)$. From t we pass to $\xi^j(u)$ as the initial term of the subsidiary computation necessary for a step by Case 0′, at the lower next position to a node in U. Next we pass to $u(\xi^{j-2})$ for the interpretation α^{j-2} of ξ^{j-2} which obtains at the next position after γ in the given branch in T, as the initial term of one of the subsidiary computations necessary for a step by Case 4′, at a lower next position to a node in U. Finally, we pass by Case 5′ to $h(s, \xi^{j-2}, B_\gamma)$. SUBCASE 2: s is a function variable. Then s(u) itself is of the form $\xi^j(u)$. As before, we pass to $u(\xi^{j-2})$ and thence to $h(s, \xi^{j-2}, B_\gamma)$. —

Theorem 12 is now proved.

3.13. Summarizing the definition in § 1.9 with §§ 1.1–1.4: a function φ is *Herbrand-Gödel-style partial* (and if total, *general*) *recursive*, if the equations expressing its values can be deduced formally from a (proper) system E of equations by use of substitution R1, replacement R2 and R3, the obvious rules R4.1, R4.2, R4.3, $\cdots$ for evaluating the functions for arguments expressed in the formalism, and the rules R5–R7 for the λ-operator as the obvious device for mediating substitutions for function variables. To this extent the definition is in the spirit of the Herbrand-Gödel definition (Gödel [**3**]; Kleene [**6**; **8**; IM]).

Some restrictions on the formalism have been made only for convenience to limit the amount of analysis required in § 2.2 (e.g., in R2 and R3 to normal equations, (d) in § 1.3, and the global restriction on R1 in deductions in § 1.4).

3.14. The restrictions § 1.3 (b) and (c) with (a) on the E are not merely for convenience. Rather, they supplement, or depart from, what would be the most straightforward extension of the Herbrand-Gödel definition. In the Herbrand-Gödel definition it was merely postulated as a general condition on E that, for each set of arguments, only one equation expressing a value of the function should be deducible from E by the given rules. Now instead we impose special restrictions on the form of E, which do indeed entail that only one such equation can be deduced.

These restrictions are natural ones from the point of view that E should have the form of a definition (with the help of recursion and auxiliary functions); namely, at most one clause of the definition should be applicable in a given case (which one being effectively recognizable), and the free variables in the definiens (on the right) should all appear in the definiendum (on the left). Such conditions were satisfied before by our particular applications of the Herbrand-Gödel definition (as in [IM, § 54, Lemmas IIa-IIe]), but they were not demanded by the definition itself.

It is shown in § 2.6 that these restrictions (other things being the same) are necessary in order that the Herbrand-Gödel-style partial recursive functions include only ones which are recursive in the sense of [RF, § 3] (or Turing-machine computable [**13**], or λ-definable [**15**]).

How are we to regard the examples in § 2.6 of functionals which would come to be included among the Herbrand-Gödel-style general recursive functions upon dropping the restrictions? It seems these functionals should not be regarded as "effectively calculable", and thus that the straightforward extension of the Herbrand-Gödel definition without these restrictions (or others of like effect) encompasses too wide a class of functionals to conform to Church's thesis.

Under our conception of a calculation, there should be a deterministic procedure (at each stage, what is to be done next being determined) leading from arguments to value if defined, effective except in those steps which use the oracle for a type-1 or higher argument. To compute $\varphi(a, \alpha^2)$ by (α)-(δ) in § 2.6, we would not know in general whether to pursue a computation based on (γ) or one based on (δ), though non-effectively we have proved that, for each a, α^2, just one of those two procedures will succeed.

The deterministic feature is immediate in the formulation of Turing-machine computability [**13**], and in our new notion of partial recursiveness [RF, § 3], both at types 0 and 1 and above. It is not present directly in λ-definability, or in Herbrand-Gödel-style recursiveness (with or without the restrictions § 1.3 (a)-(c)). For λ-definability, we can introduce it by the Church-Rosser Theorem 2, as extended to types above 0 by [**15**, Theorem 4], by which without altering the outcome we can always choose say the first innermost redex as the one to be contracted next. For Herbrand-Gödel-style recursiveness at types 0 and 1 (with or without the restrictions), we can get it via the circumstance that all of the equations deducible from a given system E can be brought into an enumeration, through which we can then search deterministically for an equation of the form sought. This device does not extend to deductions using the evaluation rules R4.j for type $j > 1$ arguments. With the restrictions § 1.3 (a)-(c), however, we can get the deterministic feature instead via Theorem 6 (c) at all the types.

At types 0 and 1, reckonability in various other formal systems than the formalism of recursive functions is an equivalent of Herbrand-Gödel-style general recursiveness etc. (cf. Gödel [**4**, the added note], Rosser [**16**]; [IM, pp. 295, 298, 320–321]); and indeed for any (constructive) formal system the set of the reckonable (total) functions is either the set of the general recursive functions or a proper subset of them. This does not carry over generally to systems with the evaluation rules R4.j for types $j > 1$.

3.15. Compared to the Herbrand-Gödel definitions in Gödel [**3**] and Kleene [**6**; **8**; IM] (apart from [**6**, Definition 2b]), the present rule R3 for replacement on the left is new. It is necessary if we are to be able to deduce equations evaluating expressions not consisting of a function letter with "simple" arguments, e.g. to deduce f(g(0)) = 3 from g(0) = 1, f(1) = 3. Such evaluations were not needed before. But they enter at least in our method of proving (§§ 3.1–3.12) that all functions partial recursive in the sense of [RF, § 3] are Herbrand-Gödel-style partial recursive (Theorem 12). In any case, R3 is unobjectionable.

3.16. Consider the example of [RF, § 9.3, LVI] (in informal symbolism, with c empty):

(η) $$\chi(x, a) \simeq \begin{cases} 0 \text{ if } \bar{T}_1(a, a, x), \\ \text{undefined otherwise}, \end{cases}$$

(θ) $$\varphi(\sigma^2, a) \simeq \sigma^2(\lambda x\, \chi(x, a)),$$

(ι) $$\varphi(a) \simeq \varphi(\lambda \tau^1\, \theta(a, \tau^1), a),$$

where, say, θ is a given general recursive function. As was remarked there, when $\varphi(\sigma^2, a)$ is interpreted as a partial recursive function [RF, § 3], the specialization $\varphi(a)$ of $\varphi(\sigma^2, a)$ for $\sigma^2 = \lambda \tau^1\, \theta(a, \tau^1)$ is not partial recursive. However, a certain extension of $\varphi(a)$ is partial recursive, by [RF, § 4, XXII or XXIII].

When the equations (η)–(ι) are translated into the formal symbolism (using distinct function letters for the several functions, and supplying suitable systems E_1 and E_2 to define recursively the partial recursive number-theoretic function $\chi(x, a)$ of (η) and the given general recursive $\theta(a, \tau^1)$), the resulting system E is proper, and defines Herbrand-Gödel-style recursively a unique partial recursive function of a (Theorem 6 (b)). This function is the same as the extension of $\varphi(\alpha)$ just mentioned.

The evaluation process by which in Herbrand-Gödel-style recursiveness (θ) is used to help in evaluating the $\varphi(a)$ of (ι) does not consist in the use of R4.2 (or S8.2) to evaluate σ^2 applied to its argument. Rather, R4.2 is bypassed by substituting $\lambda \tau^1\, \theta(a, \tau^1)$ for σ^2 and performing a λ-reduction. Thereby, of the evaluations of $\chi(x, a)$ for $x = 0, 1, 2, \cdots$ that would be required as premises for R4.2, all but the ones necessary to the calculation of $\theta(a, \tau^1)$ for $\tau^1 = \lambda x\, \chi(x, a)$ are avoided.

In the definition of partial recursiveness in [RF, § 3], substitutions for function variables which would give rise to this divergence between two evaluation processes are avoided; thus, only evaluations by [RF, § 1.3] S8.j are used.

As just mentioned, and shown in [RF, § 4, XXIII] and in somewhat other terms in §§ 2.1–2.5 (Theorem 7) here, the evaluation process using substitutions and λ-reductions produces partial recursive functions (i.e. functions definable using only the evaluation process of S8.j), extensions of the functions obtained by direct evaluations by S8.j.

In [RF, § 10, LXVIII (ii) and (iii)] all partial recursive functions (of [RF, § 3]) are obtained without the feature of infinitely many indices in the computation of a given function which S9 introduces (cf. [RF, § 10.8, Discussion]); direct evaluations by [RF, § 9.8] S4'.j are used there. By §§ 3.1–3.12 (Theorem 12) here, those evaluations can be replaced by ones using substitutions and λ-reductions.

BIBLIOGRAPHY

1. Alonzo Church, *A set of postulates for the foundation of logic*, Ann. of Math. vol.

33 (1932) pp. 346-366.

2. Alonzo Church and J. B. Rosser, *Some properties of conversion*, Trans. Amer. Math. Soc. vol. 39 (1936) pp. 472-482.

3. Kurt Gödel, *On undecidable propositions of formal mathematical systems*, Notes by S. C. Kleene and Barkley Rosser on lectures at the Institute for Advanced Study, mimeographed, Princeton, N. J., 1934, 30 pp.

4. ———, *Über die Länge von Beweisen*, Ergebnisse eines mathematischen Kolloquiums, Heft 7 (for 1934-1935, pub. 1936, with note added in press), pp. 23-24.

5. S. C. Kleene, *Proof by cases in formal logic*, Ann. of Math. vol. 35 (1934) pp. 529-544.

6. ———, *General recursive functions of natural numbers*, Math. Ann. vol. 112 (1936) pp. 727-742.

7. ———, *On notation for ordinal numbers*, J. Symb. Logic vol. 3 (1938) pp. 150-155.

8. ———, *Recursive predicates and quantifiers*, Trans. Amer. Math. Soc. vol. 53 (1943) pp. 41-73.

9. ———, *Recursive functions and intuitionistic mathematics*, Proceedings of the International Congress of Mathematicians (Cambridge, Massachusetts, August 30-September 6, 1950), vol. 1, 1952, pp. 679-685.

10. ———, *Introduction to metamathematics*, Amsterdam, North-Holland; Groningen, Noordhoff; New York and Toronto, Van Nostrand; 1952, x + 550 pp.

11. ———, *Recursive functionals and quantifiers of finite types I*, Trans. Amer. Math. Soc. vol. 91 (1959) pp. 1-52.

12. ———, *Recursive functionals and quantifiers of finite types II*, Trans. Amer. Math. Soc. vol. 108 (1963), pp. 106–142.

13. ———, *Turing-machine computable functionals of finite types I*, Proceedings of the International Congress of Logic, Methodology and Philosophy of Science (Stanford, California, August 24–September 2, 1960), Stanford University Press, 1962, pp. 38–45.

14. ———, *Turing-machine computable functionals of finite types II*, Proc. London Math. Soc., vol. 12 (1962) pp. 245-258.

15. ———, *Lambda-definable functionals of finite types*, Fundamenta Mathematicae, vol. 50 (1962) pp. 281-303.

16. J. Barkley Rosser, Review of [**4**], J. Symb. Logic vol. 1 (1936) p. 116.

UNIVERSITY OF WISCONSIN,
MADISON, WISCONSIN

INFINITE SERIES OF ISOLS

BY

J. C. E. DEKKER[1]

1. Introduction. This paper deals with non-negative integers (*numbers*), collections of numbers (*sets*) and collections of sets (*classes*). We consider the algebraic system $[\varepsilon, +, \cdot]$ consisting of the set ε of all numbers and the binary operations of ordinary addition and multiplication. This system can be extended to the system $[\Omega, +, \cdot]$ of all recursive equivalence types. We shall briefly discuss how this extension can be obtained. The reader is referred to [2] for a detailed exposition.

A mapping $f(x_1, \cdots, x_n)$ from a subcollection of ε^n into ε is called a *function*; its domain and its range are denoted by δf and ρf, respectively. We write c for the cardinality of the continuum, o for the empty set, $\subset$ for inclusion and $\subset_+$ for proper inclusion. While $+$ and $\cdot$ stand for addition and multiplication when applied to pairs of numbers, they denote union and intersection when applied to pairs of sets. Let

$$V = \text{class of all sets},$$
$$Q = \text{class of all finite sets},$$
$$I = \text{class of all isolated sets}.$$

We call a set *isolated* if it has no infinite r.e. (i.e., recursively enumerable) subset. Clearly, $Q \subset I$. This inclusion is proper and the class $I - Q$ has cardinality c; its members are the *immune* sets. If there exists a one-to-one correspondence between α and β we write $\alpha \sim \beta$. The set α is *recursively equivalent* to β (written: $\alpha \simeq \beta$), if there exists a partial recursive one-to-one function $p(x)$ such that $\alpha \subset \delta p$ and $p(\alpha) = \beta$. The $\simeq$ relation is reflexive, symmetric and transitive. If α and β can be separated by r.e. sets, we say that α is *separable* from β (written: $\alpha \mid \beta$). We need the primitive recursive functions j, j_3, k, l defined by

$$j(x, y) = x + (x + y)(x + y + 1)/2,$$
$$j_3(x, y, z) = j(x, j(y, z)),$$
$$j(k(n), l(n)) = n.$$

The function j maps ε^2 one-to-one onto ε and the function j_3 maps ε^3 one-to-one onto ε. The following propositions are basic.

(a) *Let α and β be finite. Then*

(i) $\alpha \simeq \beta \Longleftrightarrow \alpha \sim \beta$,

(ii) $\alpha \mid \beta \Longleftrightarrow \alpha \cdot \beta = o$.

Received by the editor May 12, 1961.

[1] Research on this paper was supported by a grant from the Rutgers Research Council.

(b) *A set is isolated if and only if it is not recursively equivalent to a proper subset.*

The $\simeq$ relation defines a partition of the class V of all sets.

NOTATIONS.

$$\text{Req}(\alpha) = \{\sigma \in V \mid \sigma \simeq \alpha\},$$
$$\text{Req}(\alpha) + \text{Req}(\beta) = \text{Req}(\alpha + \beta), \text{ provided } \alpha \mid \beta,$$
$$\text{Req}(\alpha) \cdot \text{Req}(\beta) = \text{Req}\, j(\alpha \times \beta),$$
$$\Omega = \{\text{Req}(\alpha) \mid \alpha \in V\},$$
$$\Lambda = \{\text{Req}(\alpha) \mid \alpha \in I\}.$$

The members of Ω are *recursive equivalence types* (RETs) and those of Λ are *isols*. We identify Req (o) with 0 and Req $(1, \cdots, n)$ with n. We call a RET *infinite* if it consists of infinite sets, and *finite* if it consists of finite sets, i.e., if it is identified with a number. The operations of addition and multiplication of RETs are well-defined; both Ω and Λ are closed under these operations. They are associative and commutative; moreover, multiplication is distributive over addition. The essential difference between Λ and $\Omega - \Lambda$ is expressed by

(c) *for* $X \in \Omega, \quad X \in \Lambda \Longleftrightarrow X \neq X + 1$.

The next paper in this volume is devoted to $\Omega - \Lambda$; in the present paper we shall in general restrict our attention to $[\Lambda, +, \cdot]$ or, in short, Λ. In this system we have the cancellation laws

(d) $X + Z = Y + Z \Rightarrow X = Y$,

(e) $XZ = YZ \ \& \ Z \neq 0 \Rightarrow X = Y$.

The first of these laws naturally leads to a definition of subtraction.

NOTATIONS.

$A \leqq B$ means $(\exists X)[A + X = B]$,

$A < B$ means $(\exists X)[A + X = B \ \& \ X \neq 0]$,

$C \geqq D$ means $D \leqq C$,

$C > D$ means $D < C$.

If $A \leqq B$, $B - A$ is the unique solution of $A + X = B$. Further algebraic properties of Λ are

(f) $\leqq$ *is a partial, but not a total ordering relation,*

(g) $A \in \Lambda - \varepsilon \Longleftrightarrow A > A - 1 > A - 2 \cdots$ *ad inf.*,

(h) $A + C \leqq B + C \Longleftrightarrow A \leqq B$,

(i) $A \leqq B \Rightarrow AC \leqq BC$,

(j) $AC \leqq BC \ \& \ C \neq 0$ *does not imply* $A \leqq B$.

Proofs of the statements (a)–(j) can be found in [2].

2. Heuristic remarks. In exploring the possibility of defining $\sum_0^\infty A_n$ for some sequences $\{A_n\}$ of isols we were led to a definition of the type,

$$(\alpha) \qquad \sum_0^\infty A_n = \begin{cases} \text{Req } \sum_0^\infty \alpha_n, \text{ where } \{\alpha_n\} \text{ is a sequence of mutually} \\ \text{disjoint sets such that } \alpha_n \in A_n \text{ for every } n, \end{cases}$$

satisfying the following conditions.

(I) Given any number $x \in \sum_0^\infty \alpha_n$, one should be able to determine effectively which of the sets $\alpha_0, \alpha_1, \cdots$ has contributed x. More precisely, if

$$(\beta) \qquad \text{for } x \in \sum_0^\infty \alpha_n, \qquad n_x = \text{the number } n \text{ such that } x \in \alpha_n,$$

then n_x should have a partial recursive extension.

(II) $\sum_0^\infty A_n$ should be an isol, i.e., $\sum_0^\infty \alpha_n$ should be an isolated set.

(III) A procedure for selecting representatives $\alpha_n \in A_n$ should be specified; let $\{\alpha_n\}$ be called coherent if $\{\alpha_n\}$ is chosen according to this procedure. We require that

$$\left.\begin{matrix} \{\alpha_n\} \text{ coherent } \& \ (\forall n)[\alpha_n \in A_n] \\ \{\beta_n\} \text{ coherent } \& \ (\forall n)[\beta_n \in A_n] \end{matrix}\right\} \Rightarrow \sum_0^\infty \alpha_n \simeq \sum_0^\infty \beta_n .$$

Condition (I) is suggested by the importance of separability in the definition of $A + B$. Condition (III) is natural, since one cannot expect Req $(\sum_0^\infty \alpha_n)$ to be independent of the sequence $\{\alpha_n\}$, if $(\forall n)[\alpha_n \in A_n]$ is the only restriction imposed on $\{\alpha_n\}$. It is harder to defend (II). For why should, e.g.,

$$1 + 1 + 1 + \cdots,$$

if definable at all, belong to Λ rather than to $\Omega - \Lambda$? One would rather expect this sum to depend *on the way in which* 1, 1, 1, $\cdots$ *are added.*

Let us now consider the specific infinite series $\sum_0^\infty A_n$, where $A_n = n$, i.e.,

$$0 + 1 + 2 + \cdots .$$

NOTATIONS.

For $n \in \varepsilon$, $\nu_n = \{x \mid x < n\}$.

For $\sigma \in V$, $j(x, \sigma) = \{j(x, y) \mid y \in \sigma\}$.

Thus $\nu_n \in A_n$ for every n. However, $\nu_0, \nu_1, \cdots$ are not even disjoint. We therefore put

$$\alpha_n = j(n, \nu_n) .$$

Clearly, $\alpha_n \in A_n$ for every n. Moreover, the function n_x related to $\{\alpha_n\}$ by (β) has a partial recursive extension, namely $k(x)$. This takes care of (I). On the other hand,

$$\delta = (j(1, 0), j(2, 0), j(3, 0), \cdots)$$

is an infinite r.e. subset of $\sum_0^\infty \alpha_n$, contrary to (II). To remedy this defect we *immunize* the representatives by defining

$$\beta_n = j(t_n, \nu_n) ,$$

where t_n is any one-to-one function which ranges over an immune set.

Again, $\beta_n \in A_n$ for every n, but, now it is readily verified that $\sum_0^\infty \beta_n$ is an isolated set. However, while the number $t_n = k(x)$ can be effectively found from $x = j(t_n, y) \in \sum_0^\infty \beta_n$, this will not give us (at least in general) the value of n. We therefore define

$$\gamma_n = j_3(n, t_n, \nu_n) .$$

It is not hard to see that the sequence $\gamma_0, \gamma_1, \cdots$ of representatives of A_0, $A_1, \cdots$ respectively, complies with both (I) and (II). Let

$$(\gamma) \qquad S = \text{Req} \sum_0^\infty j_3(n, t_n, \nu_n) .$$

Put $\tau = \rho t$ and $T = \text{Req}(\tau)$. It would still be unsatisfactory to define $\sum_0^\infty A_n$ as S, not because S depends on the isol T, but because S also depends on the function t_n. For, while subjected to the restriction $\rho t \in T$, the function t_n can still be chosen in c ways.

Let us call a one-to-one function t_n from ε into ε *regressive*, if there exists an effective procedure which when applied to t_{n+1} yields t_n. It can be shown that there exist c regressive functions with immune ranges. Moreover, if t_n and t_n^* are regressive functions such that $\rho t \simeq \rho t^*$,

$$\sum_0^\infty j_3(n, t_n, \nu_n) \simeq \sum_0^\infty j_3(n, t_n^*, \nu_n) .$$

Thus, if in (γ) we take for t_n any regressive function with an immune range, the isol S will depend on $T = \text{Req}(\rho t)$, but not on the function t_n.

In the present paper only infinite series of finite isols will be considered. The sums of such infinite series will be defined in § 4. The properties of regressive functions relevant to this definition are proved in § 3. In § 5 we prove a theorem which enables us to evaluate the sums of many infinite series of isols. Some properties of regressive sets which are not needed in §§ 4 and 5 are discussed in § 6.

3. Regressive functions. We shall here define the notion of a regressive function only for one-to-one functions of one variable which are everywhere defined.

DEFINITIONS. A one-to-one function t_n from ε into ε is *regressive*, if there exists a partial recursive function $p(x)$ such that

$$(1) \qquad \rho t \subset \delta p ,$$

$$(2) \qquad p(t_0) = t_0 \ \&\ (\forall n)[p(t_{n+1}) = t_n] .$$

A set is *regressive*, if it is finite or the range of a regressive function. A set is *retraceable*, if it is finite or the range of a strictly increasing regressive function.

Let for every partial recursive function $f(x)$,

$$f^0(x) = x,\ f^{n+1}(x) = f(f^n(x)) .$$

It is readily seen that for every regressive function t_n there also exists a partial recursive function $p(x)$ which satisfies, besides (1) and (2) the conditions

$$\rho p \subset \delta p \,, \tag{3}$$

$$(\forall x)[x \in \delta p \Rightarrow (\exists n)[p^{n+1}(x) = p^n(x)]] \,. \tag{4}$$

DEFINITION. Let t_n be a regressive function. Every partial recursive function satisfying (1), (2), (3), (4) is a function which *regresses* t_n or a *regressing* function of t_n.

NOTATION. Let $p(x)$ be any partial recursive function satisfying (3) and (4). Then

$$p^*(x) = (\mu y)[p^{y+1}(x) = p^y(x)] \,, \qquad \text{for} \quad x \in \delta p \,.$$

DEFINITION. Let t_n be a regressive function and τ its range. For every $x \in \tau$, the unique number n such that $x = t_n$ is the *t-rank* of x (written: $r_t(x)$).

Let $p(x)$ be a regressing function of the regressive function t_n with τ as range. Note that $p^*(x)$ is partial recursive with δp as its domain; also, $r_t(x)$ is the restriction of $p^*(x)$ to τ. Thus, while $r_t(x)$ is uniquely determined by t_n [in fact, $r_t(x) = t_x^{-1}$], $p^*(x)$ is not uniquely determined by t_n and need not be one-to-one. The distinction is important. For it will be shown that τ is in general immune. Hence, since δp^* is r.e. (being the domain of a partial recursive function), the set $\delta p^* - \tau$ is in general infinite.

NOTATIONS.

$$\rho_0 = o \,,$$

$$\rho_{x+1} = \begin{cases} (a(1), \cdots, a(k)), \text{ where } a(1), \cdots, a(k) \text{ are the distinct} \\ \text{numbers such that } x + 1 = 2^{a(1)} + \cdots + 2^{a(k)} \,, \end{cases}$$

$$r(x) = \text{the cardinality of } \rho_x \,.$$

The effective enumeration $\{\rho_x\}$ of the class Q of all finite sets is without repetitions and has the following properties

(5) given any number x we can effectively find both the members and the cardinality of ρ_x,

(6) given any finite set σ (i.e., by its members and cardinality) we can effectively find the unique number i such that $\sigma = \rho_i$ (the so-called *canonical index* of σ).

The function $r(x)$ is clearly recursive.

NOTATION. Let $p(x)$ be any partial recursive function satisfying (3) and (4). Then $\bar{p}(x)$ is the partial recursive function such that

$$\delta \bar{p} = \delta p \,, \tag{7}$$

$$x \in \delta \bar{p} \Rightarrow \rho_{\bar{p}(x)} = (x, p(x), \cdots, p^n(x)), \text{ where } n = p^*(x) \,. \tag{8}$$

Thus, if $p(x)$ is a regressing function of the regressive function t_n with range τ, we have

$$\tau \subset \delta \bar{p} \;\&\; (\forall n)[\rho_{\bar{p}(t_n)} = (t_n, t_{n-1}, \cdots, t_0)] \,.$$

PROPOSITION 1. *Let the sets α and β and the partial recursive functions $f(x)$ and $g(x)$ be related by the three conditions*

(a) $\alpha \subset \delta f$ & $f(\alpha) = \beta$ & f *is one-to-one on* α,

(b) $\beta \subset \delta g$ & $g(\beta) = \alpha$ & g *is one-to-one on* β,

(c) $f(x) = g^{-1}(x)$, *for* $x \in \alpha$.

Then there exists a partial recursive one-to-one function $h(x)$ such that

(d) $\alpha \subset \delta h$ & $h(\alpha) = \beta$,

(e) $h(x) = f(x)$, *for* $x \in \alpha$.

PROOF. Under the hypotheses

$$\sigma = \{x \in \delta f \mid gf(x) = x\}$$

is a r.e. set. Hence, if $h(x)$ is the restriction of $f(x)$ to σ, $h(x)$ satisfies the requirements.

PROPOSITION 2. *Every one-to-one recursive function is regressive. However, there exist exactly c other regressive functions and their ranges are immune.*

PROOF. The first part is immediate. Let $\{a_n\}$ be any infinite sequence whose elements are chosen from $(0, \cdots, 9)$, but such that $a_0 \neq 0$. Put

$$b_0 = a_0, \quad b_{n+1} = 10b_n + a_{n+1}.$$

Then b_n is a strictly increasing regressive function, different choices of $\{a_n\}$ yield different functions b_n and $\{a_n\}$ can be chosen in c distinct ways. Hence there exist at least c, and therefore exactly c regressive functions which are not recursive. Let t_n be such a regressive function and $\tau = \rho t$. Suppose there were a one-to-one recursive function ranging over a subset of τ, say $a(n)$. For any given n, the number t_n could now be computed as follows: determine the t-ranks of $a(0)$, $a(1)$, $\cdots$ until the first element is found whose t-rank is $\geqq n$, say $a(m) = t_k$. From t_k we can compute $t_0, \cdots, t_k$, hence in particular t_n. Thus t_n would be a recursive function of n, contrary to our hypothesis. We conclude that the infinite set τ has no infinite r.e. subset, i.e., that τ is immune.

DEFINITION. The one-to-one functions t_n and t_n^* from ε into ε are *recursively equivalent* (written: $t_n \simeq t_n^*$), if there exists a partial recursive one-to-one function $f(x)$ such that

$$\rho t \subset \delta f, \tag{9}$$

$$(\forall n)[f(t_n) = t_n^*]. \tag{10}$$

PROPOSITION 3. *Let $\tau = \rho t$ and $\tau^* = \rho t^*$, where t_n and t_n^* are regressive functions. Then*

$$\tau \simeq \tau^* \Longleftrightarrow t_n \simeq t_n^*.$$

PROOF. The conditional from the right to the left is trivial. Assume $\tau \simeq \tau^*$; let $f(x)$ be a partial recursive one-to-one function for which $\tau \subset \delta f$ and $f(\tau) = \tau^*$. Suppose that $p(x)$ and $q(y)$ are regressing functions of t_n and t_n^*

respectively. We may assume without loss of generality that $\delta p = \delta f$ and $\delta q = \rho f$. We proceed to define a function $g(x)$. Let $x \in \delta p$. Compute

$$x, p(x), \cdots, p^n(x), \quad \text{where} \quad n = p^*(x)$$

and their images under f, say

$$b_0, b_1, \cdots, b_n$$

respectively. Compute

$$q^*(b_0), q^*(b_1), \cdots, q^*(b_n)$$

and let $k = q^*(b_i)$ be their maximum. If either k is assumed more than once by $q^*(b_m)$ for $0 \leqq m \leqq n$, or $k < n$, we don't define $g(x)$ at all; otherwise we put $y = q^*(b_i)$, compute

$$c_k = y, c_{k-1} = q(y), \cdots, c_0 = q^k(y),$$

and define $g(x) = c_n$. The function $g(x)$ is partial recursive and $\tau \subset \delta g \subset \delta f$. Moreover, if $x \in \tau$,

$$x = t_n, p(x) = t_{n-1}, \cdots, p^n(x) = t_0,$$
$$b_0, b_1, \cdots, b_n \text{ are distinct elements of } \tau^*,$$
$$k \geqq n \text{ and } g(x) = t_n^*.$$

While the function $g(x)$ need not be one-to-one (i.e., on its domain), it is one-to-one on τ. Similarly we prove the existence of a partial recursive function $h(y)$ such that

$$\tau^* \subset \delta h \subset \rho f \qquad \& \qquad (\forall n)[h(t_n^*) = t_n].$$

Thus $g(x) = h^{-1}(x)$ for $x \in \tau$, and it follows by Proposition 1 that $t_n \simeq t_n^*$.

PROPOSITION 4. (a) *Let* $t_n \simeq t_n^*$. *Then* t_n *is a regressive function if and only if* t_n^* *is a regressive function.*

(b) *Let* $\tau \simeq \tau^*$. *Then* τ *is a regressive set if and only if* τ^* *is a regressive set.*

PROOF. (a) Let $f(x)$ be a partial recursive one-to-one function such that $\rho t \subset \delta f$ and for every n, $f(t_n) = t_n^*$. Let $p(x)$ be a regressing function of t_n. Put

$$\sigma = \{y \in \rho f \mid f^{-1}(y) \in \delta p\},$$
$$q(y) = fpf^{-1}(y), \quad \text{for} \quad y \in \sigma.$$

Then σ is a r.e. set and $q(y)$ a regressing function of t_n^*.

(b) Let τ be an infinite regressive set, $f(x)$ partial recursive and one-to-one, $\tau \subset \delta f$ and $f(\tau) = \tau^*$. Put $t_n^* = f(t_n)$, where t_n is any regressive function ranging over τ. Then t_n^* is a regressive function by (a), hence $\tau^* = \rho t^*$ is an infinite regressive set. If τ is finite the statement is trivial.

PROPOSITION 5. (a) *Every regressive function is recursively equivalent to a strictly increasing regressive function.*

(b) *Every regressive set is recursively equivalent to a retraceable set.*

PROOF. It is readily seen that (b) follows from (a). For let τ be regressive. If τ is finite, τ itself is retraceable. If τ is infinite, τ is recursively equivalent to a retraceable set by (a). To prove (a), assume t_n is a regressive function, $p(x)$ is a regressing function of t_n, and $\tau = \rho t$. Put

$$s(n) = 2^{t_0} + 2^{t_1} + \cdots + 2^{t_n}\,, \qquad \sigma = \rho s\,.$$

Let for $x \in \delta p$,

$$f(x) = 2^x + 2^{p(x)} + \cdots + 2^m\,, \quad \text{where} \quad m = p^n(x), \quad n = p^*(x)\,,$$

then $f(x)$ is a partial recursive function and for every n,

$$f(t_n) = 2^{t_n} + 2^{t_{n-1}} + \cdots + 2^{t_0} = s(n)\,.$$

Thus $f(x)$ is one-to-one on τ. Let α denote the set of all numbers y such that

(11) $$y > 0 \;\&\; \rho_y \subset \delta p\,,$$

(12) $$(\forall u)(\forall v)[u \in \rho_y \;\&\; v \in \rho_y \Rightarrow p^*(u) \neq p^*(v)]\,,$$

then α is r.e. We define for $y \in \alpha$

$$g(y) = \text{number } u \in \rho_y \text{ for which } p^*(u) \text{ is maximal}\,.$$

Thus $g(y)$ is a partial recursive function. It follows from the definitions of $s(n)$ and ρ_x that

$$\rho_{s(n)} = (t_0, \cdots, t_n)\,,$$
$$\sigma \subset \delta g \;\&\; (\forall n)[gs(n) = t_n]\,.$$

Hence $f(x) = g^{-1}(x)$ for $x \in \tau$, and $s_n \simeq t_n$ by Proposition 1.

DEFINITION. A RET is *regressive* if it consists of regressive sets.

NOTATION. Λ_R = the collection of all regressive isols.

A RET is therefore regressive if and only if it contains at least one regressive set (or, equivalently, at least one retraceable set). There is only one regressive RET which is not an isol, namely Req (ε), i.e., the class of all infinite r.e. sets. On the other hand, Λ_R has cardinality c. For there exist exactly c sets which are both isolated and regressive, while every non-zero isol contains exactly denumerably many sets.

4. The definition. We recall that $\nu_0 = o$ and $\nu_n = (0, \cdots, n-1)$ for $n > 0$. Also, $j(p, \sigma) = \{j(p, y) \mid y \in \sigma\}$.

DEFINITION. Let $\{a_n\}$ be any infinite sequence of numbers and T any regressive isol. If T is finite, say $T = k$,

$$\textstyle\sum_T a_n = \sum_{n<k} a_n \qquad (0 \text{ for } k = 0)\,.$$

If T is infinite,

$$\textstyle\sum_T a_n = \text{Req} \sum_0^\infty j(t_n, \nu(a_n))\,,$$

where t_n is any regressive function ranging over any set in T.

REMARK. We could have defined

$$\textstyle\sum_T a_n = \text{Req} \sum_0^\infty j_3(n, t_n, \nu(a_n))$$

as was suggested in § 2. This is, however, not necessary, since n can be effectively computed from t_n; for if $p(x)$ is a regressing function of t_n, we have $n = r_t(t_n) = p^*(t_n)$.

THEOREM 1. *For every infinite sequence* $\{a_n\}$ *of numbers* $\sum_T a_n$ *is a function from* Λ_R *into* Λ.

PROOF. We disregard the trivial case that T is finite. Assume T is an infinite regressive isol and $\{a_n\}$ any sequence of numbers. Let t_n and t_n^* be any two regressive functions ranging over sets in T. We have to prove:

$$\sum_0^\infty j(t_n, \nu(a_n)) \text{ is an isolated set }, \tag{13}$$

$$\sum_0^\infty j(t_n, \nu(a_n)) \simeq \sum_0^\infty j(t_n^*, \nu(a_n)) . \tag{14}$$

To prove (13), assume

$$\delta \subset \sum_0^\infty j(t_n, \nu(a_n)) ,$$

where δ is r.e. Then $k(\delta)$ is a r.e. subset of the immune set $\tau = \rho t$, hence $k(\delta)$ is finite, say $k(\delta) = (i(1), \cdots, i(r))$. Then

$$\delta \subset \sum_{s=1}^r j(t_{i(s)}, \nu(a_{i(s)})) \in Q ,$$

and δ is finite. To prove (14), denote the set on the left by λ and the one on the right by ρ. Let $f(x)$ be a partial recursive one-to-one function such that

$$\tau \subset \delta f \ \& \ (\forall n)[f(t_n) = t_n^*] .$$

Such a function $f(x)$ exists by Proposition 3. Put

$$g(z) = j(fk(z), l(z))$$

then $g(z)$ is a partial recursive one-to-one function which maps λ onto ρ.

5. The main result. Every sequence $\{f(n)\}$ of integers uniquely determines a sequence $\{c_i\}$ of integers such that

$$(\forall n)\left[f(n) = \sum_{i=0}^n c_i C_{n,i} \right] . \tag{15}$$

A function $f(n)$ from ε into ε is *combinatorial*, if the sequence $\{c_i\}$ associated with $\{f(n)\}$ by (15) consists of non-negative integers. It is readily seen that the combinatorial function $f(n)$ is recursive if and only if the corresponding function c_i is recursive. Combinatorial functions were introduced and studied by Myhill [**3**]. He calls the mapping

$$\Phi(\alpha) = \{j(x, y) \mid \rho_x \subset \alpha \ \& \ y < c_{r(x)}\}$$

from V into V the *normal* mapping corresponding to the combinatorial function $f(n) = \sum_{i=0}^{n} c_i C_{n,i}$. It has the properties

(i) $\alpha \in Q \Rightarrow \Phi(\alpha) \in Q$,

(ii) $\alpha \in I \Rightarrow \Phi(\alpha) \in I$,

(iii) $\alpha, \beta \in Q \ \& \ \alpha \sim \beta \Rightarrow \Phi(\alpha) \sim \Phi(\beta)$,

(iv) $\alpha, \beta \in I \ \& \ \alpha \simeq \beta \Rightarrow \Phi(\alpha) \simeq \Phi(\beta)$.

Note that (iv) implies (iii). The mapping

$$F(X) = \text{Req}\, \Phi(\xi), \qquad \xi \in X \in \Lambda,$$

is therefore well-defined, maps Λ into Λ and is an extension of $f(n)$. It is called the *canonical* extension of $f(n)$.

Definition. Let a_n be any function from ε into ε. The function $s(n)$ such that

$$s(0) = 0, \ \ s(n) = \sum_{i=0}^{n-1} a_i, \quad \text{for} \quad n > 0,$$

is the *partial sum* function of a_n.

Proposition 6. *If the function a_n is recursive and combinatorial, so is its partial sum function $s(n)$. In fact, if $a_n = \sum_{i=0}^{n} c_i C_{n,i}$,*

$$s(n) = \sum_{i=0}^{n} c_{i-1} C_{n,i}, \quad \textit{where} \quad c_{-1} = 0.$$

Proof. Left to the reader.

Theorem 2. *Let a_n be a recursive combinatorial function and $s(n)$ its partial sum function. Let $S(X)$ be the canonical extension of $s(n)$ to Λ. Then, for every regressive isol. T,*

$$\textstyle\sum_T a_n = S(T).$$

Proof. If T is finite the statement is trivial. Assume T is infinite. Let $\tau \in T$ and let t_n be any regressive function ranging over τ. Suppose c_i is the function from ε into ε such that $a_n = \sum_{i=0}^{n} c_i C_{n,i}$. Define

$$\alpha_n = \{x \mid x < a_n\}, \ \ \sigma = \sum_{n=0}^{\infty} j(t_n, \alpha_n),$$

$$\sigma^* = \{j(x, y) \mid x \neq 0 \ \& \ \rho_x \subset \tau \ \& \ y < c_{r(x)-1}\}.$$

We now have

$$\sigma \in \textstyle\sum_T a_n, \ \ \sigma^* \in S(T).$$

It therefore suffices to prove that $\sigma \simeq \sigma^*$. We note that

$$\begin{aligned}
a_0 &= c_0 , \\
a_1 &= c_0 + c_1 , \\
a_2 &= c_0 + 2c_1 + c_2 , \\
\alpha_3 &= c_0 + 3c_1 + 3c_2 + c_3 , \\
&\vdots \\
a_n &= c_0 + C_{n,1}c_1 + \cdots + C_{n,i}c_i + \cdots + c_n , \\
&\vdots
\end{aligned}$$

For every number n we decompose α_n into $n + 1$ mutually disjoint sets

$$\gamma_{n0} , \gamma_{n1} , \cdots, \gamma_{ni} , \cdots, \gamma_{nn}$$

such that γ_{ni} has $C_{n,i} \cdot c_i$ members. We shall write p for $C_{n,i}$, keeping in mind that p depends on n and i. Let

$$\begin{aligned}
\gamma_{n0} &= (e_{n0}, \cdots, e_{n1} - 1) , \quad \text{where } e_{n0} = 0 , \\
\gamma_{n1} &= (e_{n1} , \cdots, e_{n2} - 1) , \quad \text{where } e_{n1} = c_0 + C_{n,1}c_1 , \\
&\vdots \\
\gamma_{ni} &= (e_{ni} , \cdots, e_{n,i+1} - 1) , \text{ where } e_{ni} = \sum_{k=0}^{i} c_k C_{n,k} , \\
&\vdots \\
\gamma_{nn} &= (e_{nn} , \cdots, e_{n,n+1} - 1) , \text{ where } e_{n,n+1} = a_n .
\end{aligned}$$

The set $\gamma_{ni} = o$ in case $c_i = 0$. If $c_i \neq 0$ we decompose γ_{ni} into $p = C_{n,i}$ mutually disjoint sets, each of cardinality c_i in the following manner

γ_{ni1} consists of the first c_i elements of γ_{ni} ,

$\vdots$

γ_{nik} consists of the kth c_i elements of γ_{ni} ,

$\vdots$

γ_{nip} consists of the last c_i elements of γ_{ni} .

Thus, if we tabulate the members of these p sets, we get

$$\begin{aligned}
\gamma_{ni1} &= (e_{ni} , e_{ni} + 1, \cdots, e_{ni} + c_i - 1) , \\
&\vdots \\
\gamma_{nik} &= (e_{ni} + (k - 1)c_i , e_{ni} + (k - 1)c_i + 1, \cdots, e_{ni} + kc_i - 1) , \\
&\vdots \\
\gamma_{nip} &= (e_{ni} + (p - 1)c_i , e_{ni} + (p - 1)c_i + 1, \cdots, e_{n,i+1} - 1) .
\end{aligned}$$

Let $p(x)$ be a regressing function of t_n. We proceed to define a function $f(z)$ with σ as its domain. Let $z = j(x, y) \in \sigma$. From x and y we can compute

(i) the number n such that $x = t_n$, namely $p^*(x)$,

(ii) the number i such that $y \in \gamma_{ni}$, where $0 \leqq i \leqq n$,

(iii) the number $\bar{y}$ such that $y = e_{ni} + \bar{y}$, where $0 \leqq \bar{y} < e_{n,i+1}$,

(iv) the number s such that $y \in \gamma_{nis}$, where $0 \leqq s < p$,

(v) the number y^* such $\bar{y} = (s-1)c_i + y^*$, where $0 \leqq y^* < c_i$.

Thus, given the number $z = j(x, y) \in \sigma$, the numbers $n, i, \bar{y}, s, y^*$ such that

$$z = j(x, y) = j(t_n, e_{ni} + (s-1)c_i + y^*)$$

can be effectively computed. We now define $f(z) = j(u, v)$, where u and v are determined as follows. Consider all p subsets of τ of cardinality $i+1$ which are of the form

$$(t_n) + \text{some subset of } (t_0, \cdots, t_{n-1}).$$

Let their canonical indices in $\{\rho_x\}$ be

$$a_1, \cdots, a_p,$$

when arranged according to size. Then

$$u = a_s,$$

$$v = y^*, \text{ i.e., } v = y - [e_{ni} + (s-1)c_i].$$

It is readily seen that $f(z) = j(u, v) \in \sigma^*$. For first of all

$$t_n \in \rho_u \subset (t_0, \cdots, t_n),$$

so that u is the canonical index of a non-empty subset of τ, and secondly

$$v = y^* < c_i, \text{ i.e., } v < c_{r(u)-1}.$$

Thus, for every $z \in \sigma$ the number $f(z)$ can be effectively computed and belongs to σ^*. We now define a function $g(w)$ with σ^* as its domain. Let $w = j(u, v) \in \sigma^*$, then ρ_u is a non-empty subset of τ, and $v < c_i$, where $i = r(u) - 1$. Let t_n be the element of highest t-rank in ρ_u, then t_n, n and i can be effectively computed from $w = j(u, v)$. We define

$$g(w) = j(x, y),$$

where $x = t_n$, and y is determined as follows. Let $p = C_{n,i}$. Put

$$\alpha = \{x \mid t_n \in \rho_x \subset (t_0, \cdots, t_n) \ \& \ r(x) = i + 1\},$$

and let the elements of α be $a_1, \cdots, a_p$, when arranged according to size. Then $u \in \alpha$, say $u = a_s$ and s can be effectively computed from u. Then we define y by

$$y = e_{ni} + (s-1)c_i + v.$$

It now follows that

$$j(x, y) \in \sigma \ \& \ fj(x, y) = j(u, v).$$

Hence f maps σ one-to-one onto σ^*, g maps σ^* one-to-one onto σ and $f(z) = g^{-1}(z)$ on σ. The functions $f(z)$ and $g(w)$ have partial recursive extensions, though these need not be one-to-one (i.e., on their domains). It follows by Proposition 1 that $\sigma \simeq \sigma^*$.

Let a recursive combinatorial function a_n be given. The evaluation of $\sum_T a_n$ for some regressive isol T is now reduced to the evaluation of the partial sum function $s(n)$ of a_n. Proposition 6 and the formula

$$c_i = [\Delta^i a(n)]_{n=0}$$

supply us with an effective procedure for evaluating $s(n)$. In many cases the function $s(n)$ can be found from a_n by the finite integration procedure of the calculus of finite differences, without first computing the c_i's associated with a_n. In some cases a formula for $s(n)$ is known from elementary algebra. Let us write

$$\underbrace{a_0 + a_1 + a_2 + \cdots}_{T} \qquad \text{for } \textstyle\sum_T a_n .$$

In the following simple applications of Theorem 2, T denotes any regressive isol.

$$\underbrace{1 + 1 + 1 + \cdots}_{T} = T,$$

$$\underbrace{1 + 2 + 3 + \cdots}_{T} = \frac{T(T+1)}{2},$$

$$\underbrace{1^2 + 2^2 + 3^2 + \cdots}_{T} = \frac{T(T+1)(2T+1)}{6},$$

$$\underbrace{1 + 3 + 5 + \cdots}_{T} = T^2,$$

$$\underbrace{1 + r + r^2 + \cdots}_{T} = \frac{r^T - 1}{r - 1}, \qquad \text{for } r > 1 .$$

The last formula involves exponentiation with a possibly infinite exponent. This operation in Λ is defined in [2].

REMARK. Let us denote Req (ε) by R. Define for any infinite sequence $\{a_n\}$ of numbers

$$\textstyle\sum_R a_n = \text{Req} \sum_0^\infty j(f(n), \nu(a_n)) ,$$

where $f(n)$ is any one-to-one recursive function. There is clearly no loss in generality if one takes $f(n)$ as the identity function. It is readily seen that $\sum_R a_n = R$ for every recursive function a_n which is positive for infinitely many values of n.

6. Regressive sets. Let us compare the four statements:

(i) a set is recursive if and only if it is finite or the range of a strictly increasing recursive function,

(ii) a set is r.e. if and only if it is finite or the range of a one-to-one recursive function,

(iii) a set is r.e. if and only if it is recursively equivalent to a recursive set,

(iv) every recursive set is r.e., but not conversely,

with the four statements:

(i*) a set is retraceable if and only if it is finite or the range of a strictly increasing regressive function,

(ii*) a set is regressive if and only if it is finite or the range of a (one-to-one) regressive function,

(iii*) a set is regressive if and only if it is recursively equivalent to a retraceable set,

(iv*) every retraceable set is regressive, but not conversely.

The first four statements are well known, (i*) and (ii*) constitute our definitions of retraceable and regressive sets, and (iii*) follows from Proposition 5. It is proved in [**1**] that every retraceable set is recursive or immune. Combining this with (iii*) we see that every regressive set is r.e. or immune; this also follows from Proposition 2. Every recursive set is retraceable, but there are c retraceable sets which are immune. The first part of (iv*) is immediate. The simplest examples of regressive sets which are not retraceable are the r.e. sets which are not recursive. We now give an example of an immune set which is regressive, but not retraceable. It is known [**1**, Proof of T5] that there exist sets α and β which are immune and retraceable such that

$$\alpha \simeq \beta \;\&\; \alpha \mid \beta \;\&\; \alpha + \beta \text{ is not retraceable .}$$

It is readily seen [cf. the remark at the end of this paper] that the union of two recursively equivalent, separable regressive sets is again regressive. Hence $\alpha + \beta$ is regressive, but not retraceable; moreover, $\alpha + \beta$ is immune.

We do not know how far the analogy between recursive and r.e. sets on the one hand, and retraceable and regressive sets on the other hand extends. The following two propositions show that this analogy is not exhausted by (i) — (iv) and (i*) — (iv*).

PROPOSITION 7. *Every infinite regressive set α has an infinite retraceable subset β such that $\beta \mid \alpha - \beta$.*

PROOF. Let $a(n)$ be a regressive function, $p(x)$ a regressing function of $a(n)$ and $\alpha = \rho a$. Put

$$i_0 = 0\,, \qquad i_{n+1} = (\mu y)[a(y) > a(i_n)]\,,$$
$$b(n) = a(i_n)\,, \qquad \beta = \rho b\,.$$

The function $b(n)$ is strictly increasing and β is an infinite subset of α. Also, for every n,

$$b(n) = \max\,(a(0), \cdots, a(i_{n+1} - 1))\,.$$

We recall that $\bar{p}(x)$ is a partial recursive function with the same domain as $p(x)$ and such that for all x

$$x \in \alpha \Rightarrow \rho_{\bar{p}(x)} = (a(0), \cdots, a(n))\,, \quad \text{where} \quad a(n) = x\,.$$

Let for $x \in \delta p$,

$$q(x) = \begin{cases} a(0)\,, & \text{if } (x) = \rho_{\bar{p}(x)}\,, \\ \max(\rho_{\bar{p}(x)} - (x))\,, & \text{if } (x) \subset_+ \rho_{\bar{p}(x)}\,, \end{cases}$$

then $q(x)$ is partial recursive and

$$qb(0) = qa(i_0) = qa(0) = a(0) = b(0)\,,$$
$$qb(n+1) = qa(i_{n+1}) = \max(a(0), \cdots, a(i_n))$$
$$= a(i_n) = b(n)\,.$$

Hence $q(x)$ is a regressing function of $b(n)$ and β is an infinite retraceable subset of α. For $x \in \alpha$

$$x \in \beta \Longleftrightarrow x = \max \rho_{\bar{p}(x)}\,.$$

Since the right side of this biconditional can be effectively decided for $x \in \delta p$, it follows that $\beta \mid \alpha - \beta$.

If one of two r.e. sets α and β is finite, the set $\alpha \cdot \beta$ is r.e., because it is finite. If, however, α and β are infinite r.e. sets, the set $\alpha \cdot \beta$ is also r.e. Note that in this case α and β are recursively equivalent.

PROPOSITION 8. *The intersection of any two regressive sets which are recursively equivalent is a regressive set.*

PROOF. Let α and β be regressive sets. If $\alpha \cdot \beta$ is finite we are through. Now assume that $\alpha \cdot \beta$ is infinite and that $\alpha \simeq \beta$. Let $f(x)$ be a partial recursive one-to-one function such that $\alpha \subset \delta f$ and $f(\alpha) = \beta$. Let $a(n)$ and $b(n)$ be regressive functions ranging over α and β respectively. We may assume without loss of generality that $a(0) = b(0)$. Let $p(x)$ and $q(x)$ be regressing functions of $a(n)$ and $b(n)$ respectively. We now define a function c_n which is recursive in $a(n)$ and $b(n)$ together and ranges over $\alpha \cdot \beta$. Put $c_0 = a(0)$. For $i = 1, 2, \cdots$, we compare the finite sequences $\{a(0), \cdots, a(i)\}$ and $\{b(0), \cdots, b(i)\}$. Assume that $c_0, \cdots, c_t$ have been recognized as elements of $\alpha \cdot \beta$ after comparing $\{a(0), \cdots, a(n)\}$ with $\{b(0), \cdots, b(n)\}$. Now compare $\{a(0), \cdots, a(n+1)\}$ with $\{b(0), \cdots, b(n+1)\}$ and arrange all elements (if any) which occur in both of these finite sequences, but not in $\{c_0, \cdots, c_t\}$ according to size, and write them down in this order after c_t in

$$c_0, \cdots, c_t\,.$$

We then compare $\{a(0), \cdots, a(n+2)\}$ with $\{b(0), \cdots, b(n+2)\}$ and continue the procedure. Then c_{t+1} is the element written down after c_t in $c_0, \cdots, c_t$. Obviously, c_n is a one-to-one function which ranges over $\alpha \cdot \beta$. We claim that c_n is a regressive function. For let $x = c_{t+1}$ be given. We can compute the numbers i and k such that $x = a(i) = b(k)$, since $i = p^*(x)$ and $k = q^*(x)$. We can also compute

$$a(0), \cdots, a(i) \quad \text{and} \quad b(0), \cdots, b(k)\,.$$

If $i < k$ we can find $k+1$ distinct elements of α, namely $f^{-1}b(0), \cdots, f^{-1}b(k)$, and if $k < i$ we can find $i+1$ distinct elements of β, namely $fa(0), \cdots, fa(i)$.

Thus, if $\max(i, k) = s$ we can find at least $s + 1$ elements of each of the two sets α and β, hence the two finite sequences

$$\Gamma_a = \{a(0), \cdots, a(s)\},$$
$$\Gamma_b = \{b(0), \cdots, b(s)\}.$$

The number x occurs in both Γ_a and Γ_b. From the information supplied by Γ_a and Γ_b we can now compute $\{c_0, \cdots, c_{t+1}\}$. Hence c_t is the immediate predecessor of c_{t+1} in this finite sequence.

A set α is *decomposable* if there exist infinite sets β and γ such that

$$\alpha = \beta + \gamma \quad \text{and} \quad \beta \mid \gamma .$$

An isol A is *decomposable* if there exist infinite isols B and C such that $A = B + C$. A set or an isol is *indecomposable* if it is not decomposable. An isol is clearly decomposable (indecomposable) if and only if all its members are decomposable (respectively, indecomposable). Every finite isol is trivially indecomposable. There exist c decomposable and c indecomposable isols [**2**, Theorem 43]. Let X be any infinite regressive isol. Then X contains an immune, retraceable set. Hence X is decomposable [**2**, Theorem 49*]. Every infinite indecomposable isol is therefore not regressive. This implies that Λ_R is properly included in Λ and that $\Lambda - \Lambda_R$ has cardinality c.

PROPOSITION 9. (a) *Let $\beta \subset \alpha$, where α is regressive and $\beta \mid \alpha - \beta$. Then β is regressive.*

(b) $B \leqq A$ & $A \in \Lambda_R \Rightarrow B \in \Lambda_R$.

PROOF. If $B \leqq A$ there exists for every set $\alpha \in A$ a set $\beta \in B$ such that $\beta \subset \alpha$ and $\beta \mid \alpha - \beta$. Hence (a) implies (b). To prove (a) assume its hypothesis. Let α_1 be a retraceable set and $p(x)$ a partial recursive one-to-one function such that $\alpha_1 \subset \delta p$ and $p(\alpha_1) = \alpha$. Put $\beta_1 = p^{-1}(\beta)$. Then $\beta_1 \subset \alpha_1$ and $\beta_1 \mid \alpha_1 - \beta_1$. It is readily seen that β_1 is retraceable (cf. [**2**, top of p. 113]). Hence $\beta = p(\beta_1)$ is regressive.

A set σ is *cofinite, cosimple, cohypersimple*, if σ' is finite, simple, hypersimple respectively. A set α is *isomorphic* with β (written: $\alpha \cong \beta$), if there exists a recursive permutation which maps α onto β. In §3 we proved that every regressive set is recursively equivalent to some retraceable set. This raises the question whether this is also true in the class of all sets with a r.e. complement.

PROPOSITION 10. *Every regressive set with a r.e. complement is isomorphic with a retraceable set with a r.e. complement.*

PROOF. Let τ be a regressive set with a r.e. complement. If τ or τ' is finite, τ is recursive, hence retraceable; then there is nothing to prove. Now assume that both τ and τ' are infinite; let t_n be a regressive function ranging over τ and $a(n)$ a recursive function ranging over τ'. As in the proof of Proposition 5(a) we define

$$s(n) = 2^{t_0} + 2^{t_1} + \cdots + 2^{t_n}, \qquad \sigma = \rho s,$$
$$f(x) = 2^x + 2^{p(x)} + \cdots + 2^m,$$

where

$$x \in \delta p, \quad m = p^n(x), \quad n = p^*(x) .$$

Then $\tau \simeq \sigma$, where σ is a retraceable set. Both τ' and σ' are infinite. Hence to prove that $\tau \cong \sigma$ it suffices to show that σ' is r.e. [2, Theorem 6]. We recall that $\bar{p}(z)$ is a partial recursive function such that $\tau \subset \delta\bar{p} = \delta p$ and for all n,

$$\rho_{\bar{p}(t_n)} = (t_n, t_{n-1}, \cdots, t_0) .$$

Let for every number z such that $\rho_z \subset \delta p$,

$$q(z) = \begin{cases} \text{unique element } x \in \rho_z \text{ such that} \\ ((\forall y)[y \in \rho_z \ \& \ y \neq x \Rightarrow p^*(x) > p^*(y)] , \end{cases}$$

provided such a number x exists. Thus $q(z)$ is a partial recursive function such that $\rho_z \subset \tau$ implies $z \in \delta q$. Since for every n

$$\rho_{s(n)} = (t_0, \cdots, t_n) ,$$

we see that a number z belongs to σ if and only if it satisfies the two conditions

(i) $\rho_z \neq o$ (i.e., $z \neq 0$), and $\rho_z \subset \tau$,

(ii) $\rho_z = (t_0, \cdots, t_n)$, where t_n is the element of highest t-rank in ρ_z .

Assuming (i) is satisfied, (ii) is equivalent to

(iii) $\rho_z = \rho_{\bar{p}q}(z)$.

It follows that $z \in \sigma'$ if and only if z satisfies at least one of the three r.e. conditions

$$z = 0, \quad (\exists n)[a(n) \in \rho_z], \quad z \in \delta q \ \& \ \rho_z \neq \rho_{\bar{p}q(z)} .$$

Hence σ' is r.e. and $\tau \cong \sigma$.

COROLLARY. *Every regressive set with a r.e. complement is recursive or cohypersimple.*

PROOF. Use [1, Proposition P5].

NOTATIONS.

$$Z_0 = \text{the class of all hypersimple sets} ,$$
$$Z_1 = \text{the class of all simple sets} ,$$
$$\Lambda_0 = \{\text{Req}(\alpha) \mid \alpha \in Q \quad \text{or} \quad \alpha' \in Z_0\} ,$$
$$\Lambda_1 = \{\text{Req}(\alpha) \mid \alpha \in Q \quad \text{or} \quad \alpha' \in Z_1\} .$$

It is known that Λ_0 and Λ_1 are closed under addition and multiplication.

PROPOSITION 11. *Let $F(X)$ be the canonical extension to Λ of a recursive combinatorial function $f(n)$ from ε into ε. Then*

$$X \in \Lambda_1 \Rightarrow F(X) \in \Lambda_1$$

PROOF. Let $f(n) = \sum_{i=0}^{n} c_i C_{n,i}$ and

$$\Phi(\xi) = \{j(x, y) \mid \rho_x \subset \xi \ \& \ y < c_{r(x)}\} .$$

Let ξ be finite or cosimple, i.e., let ξ be an isolated set with a r.e. complement. Then

$$j(x, y) \notin \Phi(\xi) \iff \rho_x \cdot \xi' \neq o \quad \text{or} \quad y \geqq c_{r(x)} ,$$
$$z \notin \Phi(\xi) \iff \rho_{k(z)} \cdot \xi' \neq o \quad \text{or} \quad l(z) \geqq c_{rk(z)} .$$

Since ξ' is a r.e. set and c_i a recursive function it follows that $\Phi(\xi)$ has a r.e. complement.

REMARK. In a similar fashion one proves that if $F(X_1, \cdots, X_k)$ is the canonical extension to Λ^k of any recursive combinatorial function from ε^k into ε,

$$X_1, \cdots, X_k \in \Lambda_1 \Rightarrow F(X_1, \cdots, X_k) \in \Lambda_1 .$$

THEOREM 3. *Let $F(X)$ be the canonical extension to Λ of a recursive combinatorial function $f(n)$ from ε into ε. Then*

(a) $T \in \Lambda_R \Rightarrow F(T) \in \Lambda_R$,

(b) $T \in \Lambda_R \cdot \Lambda_0 \Rightarrow F(T) \in \Lambda_R \cdot \Lambda_0$.

PROOF. In view of Proposition 11, the fact that Λ_0 is properly included in Λ_1 and the corollary of Proposition 10, it is clear that (a) implies (b). We therefore restrict our attention to (a). It T is finite, so is $F(T)$. Assume that T is infinite. Let τ be a retraceable set in T and let t_n be the strictly increasing function ranging over τ. Suppose $p(x)$ is a regressing function of t_n. Let $u(n)$ be the strictly increasing function which ranges over the set

$$2^\tau = \{x \mid \rho_x \subset \tau\} .$$

We first prove that $u(n)$ is a regressive function. The enumeration $\{\rho_x\}$ of the class Q of all finite sets defined in §3 is such that $\rho_0 = o$ and

$$\rho_{k+1} = \{n \mid 0 \leqq n \leqq p \ \& \ c_n = 1\} ,$$

where

$$p = \max\{y \mid 2^y \leqq k + 1\} ,$$
$$k + 1 = c_p \cdot 2^p + c_{p-1} \cdot 2^{p-1} + \cdots + c_1 \cdot 2^1 + c_0 ,$$
$$c_i \in (0, 1) , \quad \text{for} \quad 0 \leqq i \leqq p .$$

This implies

(a) $k \geqq 1 \Rightarrow \max(\rho_k) \leqq \max(\rho_{k+1})$,

(b) $\rho_{u(0)} = o, \ \rho_{u(k+1)} = \{t_n \mid n \in \rho_{k+1}\} = t(\rho_{k+1})$,

(c) $\rho_{k+1} = \{p^*(y) \mid y \in \rho_{u(k+1)}\} = p^*(\rho_{u(k+1)})$.

Thus, given $u(k + 1)$, the number $k + 1$ can be effectively computed by (c). Also,

$$\rho_{u(k+1)} = (t_{i(0)}, \cdots, t_{i(p)}) \Rightarrow \rho_{k+1} = (i(0), \cdots, i(p)) .$$

If $k + 1 = 1$, $u(0) = 0$. Now suppose $k + 1 > 1$. We may assume without loss of generality that $i(0) < i(1) < \cdots < i(p)$. From the number $k + 1$ we can compute the numbers $j(0), \cdots, j(q)$ such that

$$\rho_k = (j(0), \cdots, j(q)), \quad j(0) < j(1) < \cdots < j(q) .$$

Note that (a) guarantees that $j(q) \leqq i(p)$. Hence from ρ_k we can effectively find the number $u(k)$ such that

$$\rho_{u(k)} = (t_{j(0)}, \cdots, t_{j(q)}) .$$

We conclude that $u(n)$ is a regressive function and 2^τ an infinite retraceable set. Let c_i be the recursive function such that $f(n) = \sum_{i=0}^{n} c_i C_{n,i}$. Put

$$\Phi(\tau) = \{j(x, y) \mid \rho_x \subset \tau \ \&\ y < c_{r(x)}\} .$$

We wish to prove that $\Phi(\tau)$ is a retraceable set. We may assume without loss of generality that $c_i > 0$ for every value of i. For if this is not the case we consider

$$\Psi(\tau) = \{j(x, y) \mid \rho_x \subset \tau \ \&\ y < d_{r(x)}\} ,$$

where $d_i = \max(c_i, 1)$. We then have

$$\Phi(\tau) \subset_+ \Psi(\tau) \quad \text{and} \quad \Phi(\tau) \mid \Psi(\tau) - \Phi(\tau)$$

hence

$$\Psi(\tau) \text{ retraceable} \Rightarrow \Phi(\tau) \text{ retraceable} .$$

Define $e_n = c_{ru(n)}$. We therefore assume that $e_n > 0$ for every n. The elements of $\Phi(\tau)$ are listed without repetitions in the array

$$\begin{array}{lcl} j(u(0), 0), & \cdots, & j(u(0), e_0 - 1) , \\ j(u(1), 0), & \cdots, & j(u(1), e_1 - 1) , \\ \vdots & & \vdots \end{array}$$

In fact,

$$j(u(0), 0), \cdots, j(u(0), e_0 - 1), j(u(1), 0), \cdots$$

is an enumeration of $\Phi(\tau)$ according to size. Let the function $f(x)$ be defined on $j(2^\tau \times \varepsilon)$ as follows

$$\begin{aligned} fj(u(0), 0) &= j(u(0), 0) , \\ fj(u(n + 1), y + 1) &= j(u(n + 1), y) , \\ fj(u(n + 1), 0) &= j(u(n), e_n - 1) . \end{aligned}$$

If $q(x)$ is a regressive function of $u(n)$

$$\begin{aligned} &2^\tau \subset \delta q, \quad qu(0) = u(0), \quad qu(n + 1) = u(n) , \\ &q^* u(n + 1) = n + 1 . \end{aligned}$$

It readily follows that $f(x)$ has a partial recursive extension. The strictly increasing function which ranges over $\Phi(\tau)$ is therefore regressive. Hence $\Phi(\tau)$ is an infinite retraceable set and $F(T)$ an infinite regressive isol.

COROLLARY. *Let a_n be a recursive combinatorial function from ε into ε. Then*

(a) $T \in \Lambda_R \Rightarrow \sum_T a_n \in \Lambda_R$,

(b) $T \in \Lambda_R \cdot \Lambda_0 \Rightarrow \sum_T a_n \in \Lambda_R \cdot \Lambda_0$.

PROOF. The sum function $s(n)$ of the function a_n is recursive and combinatorial. If $S(X)$ is the canonical extension to Λ of $s(n)$ and T is a regressive isol, then $\sum_T a_n = S(T)$.

REMARK. In view of Theorem 3,

$$X \in \Lambda_R \Rightarrow 2X \in \Lambda_R \ \&\ X^2 \in \Lambda_R .$$

Thus, if α and β are any two regressive sets, *which are recursively equivalent*, it follows that $j(\alpha \times \beta)$ is regressive and

$$\alpha | \beta \Rightarrow \alpha + \beta$$

is regressive.

REFERENCES

1. J. C. E. Dekker and J. Myhill, *Retraceable sets*, Canad. J. Math. vol. 10 (1958) pp. 357–373.

2. ———, *Recursive equivalence types*, Univ. California Publ. Math. (N. S.) vol. 3 (1960) pp. 67–214.

3. J. Myhill, *Recursive equivalence types and combinatorial functions*, Bull. Amer. Math. Soc. vol. 64 (1958) pp. 373–376.

4. A. Nerode, *Extensions to isols*, Ann. of Math. vol. 73 (1961) pp. 362–403.

RUTGERS, THE STATE UNIVERSITY,
NEW BRUNSWICK, NEW JERSEY

$\Omega - \Lambda$

BY

JOHN MYHILL[1]

Ω denote the collection of all recursive equivalence types, Λ the collection of all isols. Therefore this paper deals with recursive equivalence types which are not isols. Let us recall the definitions.

A *partial isomorphism* is a one-one partial recursive function. Two sets (of non-negative integers) α and β are called *recursively equivalent* ($\alpha \simeq \beta$) if there is a partial isomorphism defined at least on α, and mapping it on β. It is easily verified that this is indeed an equivalence relation; the equivalence classes into which it partitions the class V of all sets of non-negative integers are called *recursive equivalence types* (R.E.T.'s). Ω is the collection of all R.E.T.'s.

Addition and multiplication of recursive equivalence types are defined as follows: if α and β are sets having the recursive equivalence types A and B respectively, and if they are separated by disjoint recursively enumerable (r.e.) sets, then $\alpha + \beta$ has the recursive equivalence type $A + B$. (The condition that α and β are separated by disjoint r.e. sets is necessary in order to guarantee uniqueness of the sum so defined.) Again, if α and β have recursive equivalence types A and B respectively, the product AB is defined to be the recursive equivalence type of the set

$$\alpha \times \beta = \{2^m \cdot 3^n \mid m \in \alpha \;\&\; n \in \beta\} .$$

There is one other definition which is necessary for the understanding of this paper: for any set α with R.E.T. A, we define 2^α as the set of all indices of finite subsets of α in some fixed effective enumeration without repetitions of the class of all finite sets (cf. [1, p. 87, Theorem 26]), and 2^A as the recursive equivalence type of 2^α. This is a special case of a general definition of exponentiation which we shall not need today.

Hitherto attention has been largely concentrated on a subcollection of the R.E.T.'s, namely the collection Λ of *isols*, which may be defined as those R.E.T.'s X which satisfy the condition $X \neq X + 1$, or equivalently as R.E.T.'s of sets which contain no infinite recursively enumerable subsets. They are the recursive analogs of cardinal numbers which are finite in the sense of Dedekind; they contain the natural numbers as a proper subcollection, where we identify the natural number n with the R.E.T. of a set (of natural numbers) containing just n elements.

The reasons for the concentration on Λ have been twofold; firstly, the fact that the arithmetic of isols is sufficiently close to that of the natural numbers

---

Received by the editor April 6, 1961.

[1] This is the third in a series of expository papers collectively entitled *Recursive equivalence types and combinatorial functions*. The first two parts are items [**4**] and [**5**] in the Bibliography.

to suggest plausible conjectures about them, and secondly the development of a tool, namely the notion of a *combinatorial function*, which permits theorems about Λ to be established *en masse* as special cases of general metatheorems to the effect that all theorems of certainsyntactical forms which hold in the natural numbers hold likewise in the isols. A simple example of such a metatheorem is the following which is due to Nerode and myself: Every universally quantified Horn sentence built up from equations between recursive combinatorial functions holds in Λ provided that it holds in the natural numbers (cf. [**6**]).

In order to motivate what follows, let us consider more closely the meaning of this last sentence. If a sentence is built up by truth-functional composition and quantification from sentences of the form

$$\mathfrak{f}(\mathfrak{t}_1, \cdots, \mathfrak{t}_n) = \mathfrak{s} ,$$

where $\mathfrak{f}$ denotes a specific number-theoretic function and $\mathfrak{t}_1, \cdots, \mathfrak{t}_n$, $\mathfrak{s}$ are either particular natural numbers or variables, then to say that this sentence holds in Λ is to say that it is true when its numerical variables are reconstrued as ranging over Λ and its functional constants are reconstrued as denoting appropriate functions on Λ. Evidently such a theorem as the one just cited does not make sense unless we specify in what way the functional constants are to be reinterpreted. The notion of combinatorial function is an attempt to specify a canonical reinterpretation for certain function-symbols. We now give the definition of combinatorial functions, and the method whereby they are extended from natural numbers arguments to isol arguments.

Every number-theoretic function $f(n)$ can be expressed as

$$f(n) = \sum_i c_i C_{n,i} ; \tag{1}$$

f is called combinatorial if all c_i are non-negative. Thus a set of $f(n)$ elements, where f is defined by (1), can be regarded as composed of c_i replicas of each i-element subset of a set with n elements. This suggests immediately the following definition of the canonical extension to Λ (in fact to Ω) of a combinatorial function f. Let α have the R.E.T. A; then $F(A)$, the image of A under the canonical extension F of f, is defined as the R.E.T. of the set

$$\{2^m \cdot 3^n \mid \rho_m \subset \alpha \;\&\; n < c_{r(m)}\}$$

where $\{\rho_i\}$ is a fixed effective enumeration of finite sets and where $r(m)$ is the number of elements in the set ρ_m. The justification of this particular method of extension lies both in the beauty of the resulting theory and in the fact that in all cases (e.g., addition, multiplication, exponentiation, the factorial function) in which a 'natural' extension of some number-theoretic function to isols (or R.E.T.'s generally) had been given before the invention of combinatorial function theory, this extension turned out to coincide with the one required by the definition. (To keep the exposition simple, we have stated the definition only for functions of one variable; the generalization to functions of more than one variable presents no difficulties.)

In order to study $\Omega - \Lambda$ (recursive equivalence types other than isols), we first ask what is known of the extendibility of sentences true in the non-negative integers to the whole of Ω rather than just Λ.

The simplest statements about combinatorial functions, i.e., universally quantified equations, *do* extend to the whole of Ω. (I first stated this in [4, Theorem 3*(b)]; the proof is practically trivial.) We have for example

(2) $$(A + B) + C = A + (B + C); (AB)C = A(BC)\,,$$

(3) $$A + B = B + A;\ AB = BA\,,$$

(4) $$A(B + C) = AB + AC\,,$$

(5) $$A + 0 = A = A \cdot 1$$

which formulas however are very easily proved directly. It ought to be noted that something more than the mere extendibility of equations from the natural numbers to Ω is needed in order to merely rewrite an equation like (4), holding in the natural numbers, in capital letters with the assurance that it will automatically hold for all R.E.T.'s A, B, C; namely we require the knowledge that the canonical extension of the composition of two or more functions coincides with the composition of their canonical extensions. In [4] I overlooked the necessity for a proof of this fact, and later I found that it only holds on the additional assumption that the functions involved are recursive as well as combinatorial. However, this condition is satisfied in all cases in which we are interested.

Universally quantified Horn sentences more complex than equations, however, do not in general go over from the natural numbers to Ω. In fact, the sentence

$$A \neq A + 1$$

holds for natural numbers, but fails (by definition) for the whole of $\Omega - \Lambda$. On the other hand, one very important implication does go over; for we have

THEOREM 1 (Friedberg; cf. [5, Theorem 11]).

$$nA = nB \rightarrow A = B \qquad (n = 1, 2, 3, \cdots)\,.$$

The proof of this cancellation law is not easy [2]; it results from a combination of the Friedberg priority method (cf. [7]) with the Sierpinski-Tarski proof [8; 9] of the corresponding theorem for cardinals without the axiom of choice.

In the letter in which Friedberg informed me of this result and outlined its proof, he raised two questions concerning possible generalizations:

Q1. Does there exist for every recursive combinatorial function F, not a constant, a number n_F such that

$$F(A) = F(B) \rightarrow A = B\,? \qquad (A, B \in \Omega,\ A, B \geqq n_F)\,?$$

(This is a natural conjecture to make since the corresponding statement is true if the variables A, B are restricted to Λ rather than Ω; cf. [4, Theorem 1].)

Q2. Is it the case that for any nonzero isol X we have

$$XA = XB \rightarrow A = B? \quad (A, B \in \Omega).$$

(This too is a natural conjecture since in general isols behave more like natural numbers than do other recursive equivalence types, and since in particular a R.E.T. X satisfies

$$X + A = X + B \rightarrow A = B$$

for all $A, B \in \Omega$, precisely if it is an isol.)

I shall return to Friedberg's Q1 later. Meanwhile I shall show that the answer to his Q2 is negative. For we can prove by a direct construction of a mapping the identity

$$R \cdot 2^X = R \cdot 2^{2X} = R \cdot 2^X \cdot 2^X$$

where R is the R.E.T. of the set of all natural numbers; from this we can infer, where X is any isol which is not a natural number, that 2^X is an isol and

$$2^X \cdot R = 2^X \cdot (2^X \cdot R)$$

while yet

$$R \neq 2^X \cdot R,$$

so that the supposed generalization of the Friedberg cancellation law is false. That such an easy counterexample was missed by Friedberg and myself for a couple of years probably reflects the intensity of our conviction that no counterexample existed. (This result is due to my student Erik Ellentuck, who used the same method to get a counterexample to the corresponding law for cardinal numbers.)

In attempting to develop a theory of combinatorial functions of non-isols, the first thing one notices is the extent to which such functions degenerate. For we have rather easily the

THEOREM 2 (cf. [**5**, § 9]). *Every recursive combinatorial function F of one variable reduces on $\Omega - \Lambda$ to either* (a) *a constant,* (b) *a function of the form $mA^n/n!$ or* (c) *the function 2^A.* (The quotient $mA^n/n!$ exists by [**1**, Theorems 109 (b) and 113]; it is unique by the Friedberg cancellation law.)

The essential facts required for proving Theorem 2 are the *absorption laws* [**3**; **1**, pp. 77, 138]

$$A + B = A \longleftrightarrow RB \leqq A,$$
$$AB = A \longleftrightarrow B^R \text{ divides } A.$$

Here $C \leqq A$ means $(\exists D)(C + D = A)$; for the definition of B^R we refer the reader to [**1**, pp. 136, 183].

The combinatorial functions of a fixed $A_0 \in \Omega - \Lambda$ can be arranged in a well-ordered sequence of order type $\omega^2 + 1$ as follows:

$$
(6)\qquad
\begin{array}{ll}
0 < 1 < 2 < & \cdots < \\
A_0 \leqq 2A_0 \leqq 3A_0 \leqq & \cdots \leqq \\
\frac{1}{2} A_0^2 \leqq A_0^2 \leqq \frac{3}{2} A_0^2 \leqq & \cdots \leqq \\
\frac{1}{6} A_0^3 \leqq \frac{1}{3} A_0^3 \leqq \frac{1}{2} A_0^3 \leqq & \cdots \leqq \\
\cdots\cdots & \cdots \leqq \\
\frac{1}{n!} A_0^n \leqq \frac{2}{n!} A_0^n \leqq \frac{3}{n!} A_0^n \leqq & \cdots \leqq \\
\cdots\cdots & \cdots \leqq 2^{A_0}
\end{array}
$$

in which each term is less than or equal to all of its successors. Thus a first step in the study of combinatorial functions of recursive equivalence types which are not isols consists in classifying such R.E.T.'s according to which of the terms of the sequence (6) are distinct. This classification is simplified by the following chain of theorems:

Theorem 3. *For any recursive equivalence type A, either all its existent rational multiplies are identical or all are distinct.* (*In the former case A is called idemmultiple.*)

Theorem 4. *For any recursive equivalence type A, either all powers of A are distinct, all are equal, or there is a number $m \geqq 2$ for which*

$$A < A^2 < \cdots < A^m = A^{m+1} = A^{m+2} = \cdots .$$

Theorem 5. *If any power of A is idemmultiple, so are all higher powers.*

Theorem 6. *If $A^n = A^{n+1}$, then A^n is idemmultiple.*

Theorem 7. *If all powers of A are distinct, 2^A exceeds all powers of A. If $A^m = A^{m+1}$ for some $m \geqq 1$, then $A^m = 2^A$.*

Of these theorems, Theorem 3 is a corollary of the Friedberg cancellation law and the rest admit of very easy direct proofs. A special case of Theorem 6 is: Every idempotent R.E.T. (i.e., every R.E.T. A satisfying $A = A^2$) is idemmultiple. Dekker asked whether the converse of this is true; we shall return to this point later.

Theorems 3–7, combined, yield two numerical invariants which characterize completely the combinatorial functions of one variable. For denote by $\mathfrak{Comb}(A_0)$ the sequence (6); and call $\mathfrak{Comb}(A_0)$ and $\mathfrak{Comb}(A_1)$ *isomorphic* if there is a one-one mapping of $\mathfrak{Comb}(A_0)$ onto $\mathfrak{Comb}(A_1)$ which sends A_0 into A_1 and preserves sums, products, and inequalities. We easily establish (cf. [5, § 9]) the following

Theorem 8. *Let $A_0, A_1 \in \Omega - \Lambda$. A necessary and sufficient condition that $\mathfrak{Comb}(A_0)$ and $\mathfrak{Comb}(A_1)$ are isomorphic is that $m_0 = m_1$ and $n_0 = n_1$, where for $i = 0, 1$ we set*

$$m_i = (\mu x)(A_i^x = 2A_i^x) \quad (\textit{or } \infty \textit{ if there is no such } x)\,,$$
$$n_i = (\mu x)(A_i^x = A_i^{x+1}) \quad (\textit{or } \infty \textit{ if there is no such } x)\,.$$

m_i indicates the point at which the powers of A_i begin to absorb themselves additively, n_i the point at which they begin to absorb themselves multiplicatively. By Theorem 6, the first of these forms of degeneracy occurs no later than the second, i.e.,

$$1 \leqq m_i \leqq n_i \leqq \infty\,.$$

We call the pair (m_0, n_0) the *character* of A_0. To illustrate, A will have the character (1,1) if

$$\underbrace{A = 2A = 3A = \cdots = A^2 = A^3 = A^4}_{\text{multiplicative degeneracy}} = \cdots = 2^A\,;$$

the character (1, 2) if

$$\underbrace{A = 2A}_{\text{add. deg.}} < \underbrace{A^2 = A^3}_{\text{mult. deg.}} = \cdots = 2^A\,;$$

the character (1, ∞) if

$$\underbrace{A = 2A}_{\text{add. deg.}} < A^2 < A^3 < A^4 < \cdots < 2^A\,;$$

no multiplicative degeneracy

the character (3,3) if

$$A < 2A < 3A < \cdots < A^2 < 2A^2 < 3A^2 < \cdots < \underbrace{A^3 = A^4}_{\text{mult. deg.}} = \cdots = 2^A\,;$$

the character (3,4) if

$$A < 2A < 3A < \cdots < A^2 < 2A^2 < 3A^2 < \cdots < \underbrace{A^3 = 2A^3}_{\text{add. deg.}} < \underbrace{A^4 = A^5}_{\text{mult. deg.}} = \cdots = 2^A\,;$$

the character (3, ∞) if

$$A < 2A < 3A < \cdots < A^2 < 2A^2 < 3A^2 < \cdots$$
$$< \underbrace{A^3 = 2A^3}_{\text{add. deg.}} < A^4 < A^5 < \cdots < 2^A\,;$$

no multiplicative degeneracy

and the character (∞, ∞) if

$$A < 2A < 3A < \cdots < A^2 < 2A^2 < 3A^2 < \cdots$$
$$< A^3 < 2A^3 < \cdots < A^4 < \cdots < 2^A\,.$$

neither additive nor multiplicative degeneracy .

The principal theorem of this paper is

THEOREM 9. *Let* $1 \leqq m \leqq n \leqq \infty$. *Then there exists* $A \in \Omega - \Lambda$ *with character* (m, n).

I will not even attempt to outline the proof of this theorem; it is a quite nasty mixture of algebraic computations and category arguments. I will not publish it until I have it in a form that yields more insight.

Nonetheless the theorem already gives us some idea of what to expect in $\Omega - \Lambda$. It leads us to suppose that practically anything can happen except a violation of the Friedberg cancellation law. For a counterexample to almost any other supposed law beyond the trivial level of (2)–(5) can be concocted by an appropriate choice of characters. I conclude with two examples of this.

Friedberg's Problem $Q1$ (cancellation laws for arbitrary combinatorial functions of one variable). The answer is negative. For observe that for any R.E.T. A of character (1, 2) we have

$$A = 2A < A^2 = A^3 = 2^A$$

and so despite the fact that $A \neq A^2$, we have

$$F(A) = F(A^2) = A^2$$

for any nonlinear combinatorial function F. Thus we have a *uniform* counterexample to all cancellation laws other than Friedberg's. (Another such example was obtained by Erik Ellentuck before I proved Theorem 9; cf. [**5**, Theorem 12].)

Does idemmultiplicity imply idempotency? Again the answer is negative since any R.E.T. A of character (1, 2) satisfies

$$A = 2A < A^2$$

and is consequently idemmultiple without being idempotent.

The ease with which such counterexamples can be constructed using Theorem 9 will make credible the following pleasant

THEOREM 10. *Given any formula built up from equations between polynomials in the single variable A by means of truth-functional connectives, we can effectively tell whether it holds identically in* $\Omega - \Lambda$, *fails identically in* $\Omega - \Lambda$, *or neither*.

Of course such formulas are very simple, but not entirely trivial since they contain, for example, the formula

$$A = 2A \rightarrow A = A^2$$

which was just disproved. I am rash enough to hope that an extension of the methods of this paper may give a decision-procedure for the entire arithmetic of polynomials in $\Omega - \Lambda$. (A decision-procedure for the arithmetic of polynomials in Ω is rather obviously out of the question.)

BIBLIOGRAPHY

1. J. C. E. Dekker and J. Myhill, *Recursive equivalence types*, monograph, Univ.

California Publ. Math. (N.S.) vol. 3 (1960) pp. 67-214.

2. R. Friedberg, *The uniqueness of finite division for recursive equivalence types*, Math. Z. vol. 75 (1961) pp. 3-7.

3. J. Myhill, *Absorption laws in the arithmetic of recursive equivalence types*, Abstract 559-150, Notices Amer. Math. Soc. vol. 6 (1959) p. 526.

4. ———, *Recursive equivalence types and combinatorial functions* (Part I), Bull. Amer. Math. Soc. vol. 64 (1958) pp. 373-376.

5. ———, *Recursive equivalence types and combinatorial functions* (Part II), to appear in Proceedings of the International Congress in Logic and Methodology of Science held at Stanford in August 1960.

6. A. Nerode, *Extensions to isols* II, Abstract 564-271, Notices Amer. Math. Soc. vol. 7 (1960) p. 74.

7. G. E. Sacks, *On the priority method of Friedberg and Muchnik*, Abstract 60 T-15, Notices Amer. Math. Soc. vol. 7 (1960) p. 997.

8. W. Sierpinski, *Sur l'égalité* $2m = 2n$ *pour les nombres cardinaux*, Fund. Math. vol. 3 (1922) pp. 1-16.

9. A. Tarski, *Cancellation laws in the arithmetic of cardinals*, Fund. Math. vol. 36 (1949) pp. 77-92.

UNIVERSITY OF CALIFORNIA,
BERKELEY, CALIFORNIA

ARITHMETICALLY ISOLATED SETS AND NONSTANDARD MODELS

BY

A. NERODE[1]

1. Introduction. The ordinary theory of isolic integers begins with the following definitions. Let R be the set of general recursive functions, let $E=\{0,1,2,\cdots\}$, let $\mathscr{P}(E)$ be the collection of all subsets of E. Call α, $\beta\in\mathscr{P}(E)$ recursively equivalent if there is a 1-1 partial recursive function p such that α is a subset of the domain of p and $p(\alpha)=\beta$. Call α recursively isolated if α has no infinite recursively enumerable subset. The equivalence class $\langle\alpha\rangle$ of a recursively isolated set α under recursive equivalence is called a recursive isol, and $\Lambda(R)$ is the set of all recursive isols. Functions $+,\cdot:\Lambda(R)\times\Lambda(R)\rightarrow\Lambda(R)$ are defined by $\langle\alpha\rangle+\langle\beta\rangle=\langle[2x\mid x\in\alpha]\cup[2x+1\mid x\in\beta]\rangle$, $\langle\alpha\rangle\cdot\langle\beta\rangle=\langle[2^x\cdot 3^y\mid x\in\alpha \text{ and } y\in\beta]\rangle$. Then $\Lambda(R)$ is extended to a commutative ring with unit $\Lambda^*(R)$ by taking differences, the ring of isolic integers.

The principal object of the present paper is to modify this definition by letting the set $\tilde{A}$ of arithmetic functions replace the set R of general recursive functions everywhere. Namely, use arithmetically isolated sets (sets with no infinite arithmetic subset) instead of recursively isolated sets and 1-1 arithmetic partial functions instead of 1-1 partial recursive functions, and construct a corresponding ring $\Lambda^*(\tilde{A})$.

In § 3 the elementary (i.e. first order) theory of $\Lambda^*(\tilde{A})$ is determined. In fact, the elementary theory of $\Lambda^*(\tilde{A})$ coincides with that of the reduced product Q obtained by reducing the ring $E^{*\omega}$ of sequences of rational integers $E^*=\{0,\pm1,\pm2,\cdots\}$ under pointwise operations modulo the ideal I of sequences which vanish except at a finite number of coordinates. Equivalently, this is the elementary theory of the quotient Q' of the ring of (1-place) arithmetic functions on E^* reduced modulo the ideal I' of functions which vanish except at a finite number of arguments. Unfortunately, the analogues of these results fail for $\Lambda^*(R)$, a principal result of [**6**].

However, it must be noted that for fragments of arithmetic in which only recursive functions are definable, the equivalence result does hold for $\Lambda^*(R)$. In particular, for $\Lambda^*(R)$ as an Abelian group the elementary theory is the same as that of Q. This yields immediately the characterization of Myhill [**3**] of the elementary theory of the Abelian group $\Lambda^*(R)$.

Some facets of arithmetic which are not at all obvious for $\Lambda^*(\tilde{A})$ are entirely obvious for Q; for example, the nonexistence of primes. Hence much information about $\Lambda^*(\tilde{A})$ can be obtained easily from these results.

Received by the editor, April 28, 1961.

[1] Research for this paper was supported in part under National Science Foundation Contract Number 13263.

In § 4 it is shown that $\Lambda^*(\tilde{A})$ modulo a minimal prime ideal is a nonstandard model for the elementary theory of E^*. This is a strict analogue of the corresponding fact for Q (where Łoś ultrapowers are obtained) or Q' (where essentially Skolem's original nonstandard models are obtained). The corresponding question for $\Lambda^*(R)$ remains open.

There are quite a large number of ways to obtain these results. That chosen has the advantage that it works easily for Q and Q' as well and also parallels as closely as possible the elimination of quantifiers technique of Feferman-Vaught [**2**] for generalized powers. In fact E^* acts as if it were a base and the Boolean algebra $B(\tilde{A})$ of idempotents of $\Lambda^*(\tilde{A})$ acts as if it were an exponent. The principal difference is, of course, that $B(\tilde{A})$ is not given as an algebra of subsets of an index set and this has to be taken into account.

Finally, the exposition has been made self-contained by giving rather simple proofs in § 2 of the required results which could as well be obtained by the more complicated techniques of [**5**; **6**] (which yield more set-theoretic information).

2. Summary of required results. A set A of number-theoretic functions is called *recursively closed* if any function general recursive in a finite number of functions in A is also in A. The smallest such A is the set of general recursive functions, the largest is the set of all number-theoretic functions. Intermediate examples include the set of all functions expressible in both k-number quantifier forms for fixed k, the set of arithmetic functions, the set of hyperarithmetic functions, and the set of analytic functions. Functions in A will be called A-functions. A k-ary relation among natural numbers is A-*enumerable* if it corresponds (under one of the usual 1-1 onto fully effective mappings of n-tuples of natural number to natural numbers) to the range of an A-function. Such a relation is an A-relation if both it and its complement are A-enumerable. An n-ary partial function is a function on a set of n-tuples of natural numbers to the natural numbers. Such a partial function is a partial A-function if it is A-enumerable as a relation. An A-*equivalence* is a 1-1 1-ary partial A-function. Let $E = \{0, 1, 2, \cdots\}$ be the natural numbers, let $\mathscr{P}(E)$ be the set of all subsets of E. An $\alpha \in \mathscr{P}(E)$ is A-*isolated* if α has no infinite A-enumerable subset; or equivalently if there is no A-equivalence p with α a subset of the domain of p and $p(\alpha)$ a proper subset of α. Sets $\alpha, \beta \in \mathscr{P}(E)$ are A-*equivalent* if there is an A-equivalence p with α a subset of the domain of p and $p(\alpha) = \beta$. The equivalence class $\langle\alpha\rangle$ of a set α under A-equivalence is called an A-*equivalence type*, and $\Omega(A)$ denotes the set of A-equivalence types. The A-equivalence type $\langle\alpha\rangle$ of an A-isolated set α consists entirely of A-isolated sets and is called an A-*isol*. Then $\Lambda(A)$ denotes the set of A-isols. Each natural number n determines an A-isol consisting of all n-element sets in $\mathscr{P}(E)$. Identify n with this corresponding A-isol so that $E \subseteq \Lambda(A) \subseteq \Omega(A)$. Every function $f\colon X^kE \to E$ has a unique expansion $f(x_1, \cdots, x_k) = \sum c_{i_1 \cdots i_k} C_{x_1, i_1} \cdots C_{x_k, i_k}$, where the summation extends over all $(i_1, \cdots, i_k) \in X^kE$, each $c_{i_1 \cdots i_k}$ being a rational integer, $C_{x,i} = x!/i!(x-i)!$. The $c_{i_1 \cdots i_k}$ are called the Stirling numbers of f. Moreover, f is called *combinatorial* if all Stirling numbers of f are non-negative; addition and multiplication are combinatorial. If $\alpha \in X^k \mathscr{P}(E)$, let α_i be the ith coordinate of α. If $\alpha, \beta \in X^k \mathscr{P}(E)$, write $\alpha \leqq \beta$ if $\alpha_i \subseteqq \beta_i$, for $i = 1, \cdots, k$. Let $\langle \alpha \rangle = (\langle \alpha_1 \rangle, \cdots, \langle \alpha_k \rangle)$. Let $\mathscr{P}_{\text{fin}}(E)$ consist of all finite subsets of E. A map $\phi: X^k \mathscr{P}(E) \to \mathscr{P}(E)$ is a *combinatorial operator* if

(2.1) For $\alpha, \beta \in X^k \mathscr{P}_{\text{fin}}(E)$, with $\langle \alpha \rangle = \langle \beta \rangle$, we have $\phi(\alpha)$, $\phi(\beta) \in \mathscr{P}_{\text{fin}}(E)$ and $\langle \phi(\alpha) \rangle = \langle \phi(\beta) \rangle$.

(2.2) For any $x \in \bigcup_{\alpha \in X^k \mathscr{P}(E)} \phi(\alpha)$, there exists a unique $x_\phi \in X^k \mathscr{P}_{\text{fin}}(E)$ such that for any $\beta \in X^k \mathscr{P}(E)$, $x \in \phi(\beta)$ if and only if $x_\phi \leqq \beta$.

Suppose given for each k a fully effective 1-1 onto Gödel numbering $\mathfrak{G}: X^k \mathscr{P}_{\text{fin}}(E) \to E$. Call ϕ an A-combinatorial operator if ϕ is combinatorial and in addition

(2.3) The map on E to E given by $\mathfrak{G}(\alpha) \to \mathfrak{G}(\phi(\alpha))$ for $\alpha \in X^k \mathscr{P}_{\text{fin}}(E)$ is an A-function.

(2.4) (Myhill's Normal Form [**4**; **5**]). An A-combinatorial operator $\phi: X^k \mathscr{P}(E) \to \mathscr{P}(E)$ induces a combinatorial A-function $f: X^k E \to E$ given by $f(\langle \alpha \rangle) = \langle \phi(\alpha) \rangle$, for $\alpha \in X^k \mathscr{P}_{\text{fin}}(E)$. Every combinatorial A-function is so induced.

To prove (2.4), suppose ϕ is a combinatorial A-operator. Define $\phi^e(\alpha) = \phi(\alpha) - \bigcup_{\beta < \alpha} \phi(\beta)$ and observe that $\phi(\alpha)$ is a disjoint union of all $\phi^e(\beta)$ with $\beta \leqq \alpha$ by (2.2). Use (2.1) and induction to prove $c_{i_1 \cdots i_k}$ for f is $\langle \phi^e(\alpha) \rangle$ if $\langle \alpha_1 \rangle = i_1, \cdots, \langle \alpha_k \rangle = i_k$. Conversely, given a combinatorial A-function f, let $j: X^k \mathscr{P}_{\text{fin}}(E) \times E \to E$ be a 1-1 onto fully effective function and specify an A-combinatorial operator ϕ inducing f by requiring that $\phi^e(\alpha) = [j(\alpha, n) \mid 0 \leqq n < c_{\langle \alpha_1 \rangle \cdots \langle \alpha_k \rangle}]$, for $\alpha \in X^k \mathscr{P}_{\text{fin}}(E)$.

(2.5) (Myhill's Extension [**4**; **5**]). Every combinatorial A-function $f: X^k E \to E$ can be extended to a function $f_\Omega: X^k \Omega(A) \to \Omega(A)$ by the requirement that $f_\Omega(x) = \langle \phi(\alpha) \rangle$, if $x = \langle \alpha \rangle$ and ϕ is an A-combinatorial operator inducing ϕ.

To prove (2.5) suppose that ϕ, ψ are A-combinatorial operators inducing f. Suppose $\alpha, \beta \in X^k \mathscr{P}(E)$. Suppose for $i = 1, \cdots, k$, p_i is an A-equivalence such that α_i is a subset of the domain of p_i and $p_i(\alpha_i) = \beta_i$. Let D consist of all finite α' such that α'_i is a subset of the domain of p_i for $i = 1, \cdots, k$. Map $\bigcup_{\alpha' \in D} \phi^e(\alpha')$ 1-1 onto $\bigcup_{\alpha' \in D} \psi^e(p_1(\alpha'_1), \cdots, p_k(\alpha'_k))$ by mapping $\phi^e(\alpha')$ 1-1 onto $\psi^e(p_1(\alpha'_1), \cdots, p_k(\alpha'_k))$ in order of magnitude. This is an A-equivalence mapping $\phi(\alpha)$ 1-1 onto $\psi(\beta)$.

Without difficulty, composition commutes with extension, projections extend to projections, constant functions extend to constant functions with the same value; and hence identities involving A-functions true in E are true in $\Omega(A)$. Defining f_Λ as f_Ω restricted to $X^k \Lambda(A)$, without difficulty (2.2) and (2.3) imply $f_\Lambda: X^k \Lambda(A) \to \Lambda(A)$: for if β is an infinite A-enumerable subset of $\phi(\alpha)$, then one of $\bigcup_{x \in \beta} (x_\phi)_1 \subseteqq \alpha_1, \cdots, \bigcup_{x \in \beta} (x_\phi)_k \subseteqq \alpha_k$ is an infinite A-enumerable set.

A *Horn sentence* is a quantified conjunction of disjunctions of equations and their negations in which there is at most one occurrence of an unnegated equation in each conjunct. Suppose function symbols denote combinatorial

A-functions with E as domain of individuals, and corresponding extensions to $\Lambda(A)$ with $\Lambda(A)$ as domain of individuals.

THEOREM 2.1. *Suppose that A is recursively closed. Then a universal Horn sentence true in E is also true in* $\Lambda(A)$. (*See* [5, § 11].)

Distributing universal quantifiers over conjunctions and applying the fact that composition commutes with extension reduces this without difficulty to

(2.6) Suppose that $f_0^1, \cdots, f_0^k, f_1^1, \cdots, f_1^k \colon X^n E \to E$ are combinatorial A-functions. Suppose that for $x \in X^n E$, $f_0^k(x) = f_1^i(x)$ for $i = 1, \cdots, k$, implies $g_0(x) = g_1(x)$. Then for $y \in X^n \Lambda(A)$, $f_{0_\Lambda}^i(y) = f_{1_\Lambda}^i(y)$ for $i = 1, \cdots, k$, implies $g_{0_\Lambda}(y) = g_{1_\Lambda}(y)$.

PROOF. Suppose $y_i = \langle \alpha_i \rangle$, $i = 1, \cdots, n$, suppose $\phi_0^1, \cdots, \phi_0^k, \phi_1^1, \cdots, \phi_1^k, \phi_0, \phi_1$ are A-combinatorial operators inducing respectively $f_0^1, \cdots, f_0^k, f_1^1, \cdots, f_1^k, g_0, g_1$. By (2.5) there are A-equivalences p_i such that (2.7) below is satisfied by $\gamma = \alpha$.

(2.7) For $i = 1, \cdots, k$, $\phi_0^i(\gamma)$ is a subset of the domain of p_i and $p_i(\phi_0^i(\gamma)) = \phi_1^i(\gamma)$.

Let F consist of all finite β such that $\beta = \gamma$ satisfies (2.7). For $\beta \in F$, certainly $f_0^i(\langle\beta\rangle) = f_1^i(\langle\beta\rangle)$, so by assumption $g_0(\langle\beta\rangle) = g_1(\langle\beta\rangle)$. Thus $\phi_0(\beta)$ and $\phi_1(\beta)$ have the same finite number of elements. For $\beta \in F$, define $\phi_j^e(\beta) = \phi_j(\beta) - \bigcup_{\beta' < \beta, \beta' \in F} \phi_j(\beta')$. Using (2.2), $\phi_j(\beta)$ is a disjoint union of all $\phi_j^e(\beta')$ with $\beta' \leqq \beta$ and β' in F. Starting with the smallest member of F an easy induction shows that $\phi_0^e(\beta)$ and $\phi_1^e(\beta)$ have the same number of elements. Since for $j = 0, 1$, $\bigcup_{\beta \in F} \phi_j^e(\beta)$ is a disjoint union, there is a well-defined partial function p on $\bigcup_{\beta \in F} \phi_0^e(\beta)$ to $\bigcup_{\beta \in F} \phi_1^e(\beta)$ obtained by mapping $\phi_0^e(\beta)$ 1-1 onto $\phi_1^e(\beta)$ in order of magnitude. To prove that $g_{0_\Lambda}(y) = g_{1_\Lambda}(y)$ it suffices to show that $\phi_0(\alpha)$ and $\phi_1(\alpha)$ are A-equivalent; thus it suffices to show that p is a partial A-function and that $\phi_j(\alpha)$ is a union of all $\phi_j^e(\beta)$ for $\beta \in F, \beta \leqq \alpha$. Let $p_0^i = p_i$, $p_1^i = (p_i)^{-1}$. For any $\beta \in X^n \mathscr{P}(E)$, define $c(\beta)$ to be the smallest $\beta' \geqq \beta$ such that for $j = 0, 1$, (2.8) below holds.

(2.8) Suppose $1 \leqq i \leqq k$. Suppose $x \in E$. Suppose $x_{\phi_j^i}$ is defined. Suppose $x_{\phi_j^i} \leqq \beta'$. Suppose $p_j^i(x)$ is defined. Suppose $(p_j^i(x))_{\phi_{1-j}^i}$ is defined. Then $(p_j^i(x))_{\phi_{1-j}^i} \leqq \beta'$.

Obviously, there is a $\beta' \geqq \beta$ such that $\beta' = \gamma$ satisfies (2.7) if and only if $c(\beta) = \gamma$ satisfies (2.7); and then $c(\beta) \leqq \beta'$. But (2.8) is of such a character that if β is finite, $c(\beta)$ is A-enumerable. Combining these two remarks, if $\beta \leqq \alpha$ is finite, $c(\beta) \leqq \alpha$ and $c(\beta)$ is A-enumerable. But α_i is A-isolated for all i, so $c(\beta)$ is finite. Hence $c(\beta)$ is in F and α is the union of all $\beta \leqq \alpha$ with $\beta \in F$. That p is a partial A-function follows via (2.8) from the existence of a uniform effective procedure (relative to A-functions) which, applied to (a Gödel number of) a finite β such that $c(\beta) = \gamma$ is finite and satisfies (2.7), yields (a Gödel number of) $c(\beta)$.

Let $+$, $\cdot$ denote both the usual functions on E and their extensions to $\Lambda(A)$. Since the statement expressing the axioms for a commutative cancellation semigroup is Horn, $(\Lambda(A), +)$ is such and can thus be extended by forming differences to an Abelian group $(\Lambda^*(A), +)$. Since E is imbedded in $\Lambda(A)$, this imbeds $E^* = \{0, \pm 1, \pm 2, \cdots\}$ in $\Lambda^*(A)$. Extend A-notions (such as A-function, A-relation) to E^* via a fully effective 1-1 onto map from E^* to E.

Let $f: X^k E^* \to E^*$ be an A-function. Define an extension $f_{\Lambda^*}: X^k \Lambda^*(A) \to \Lambda^*(A)$ for $y_1, \cdots, y_k \in \Lambda^*(A)$ by requiring that $f_{\Lambda^*}(y_1, \cdots, y_k) = f_\Lambda^+(x_1, \cdots, x_{2k}) - f_\Lambda^-(x_1, \cdots, x_{2k})$ if $y_1 = x_1 - x_2, \cdots, y_k = x_{2k-1} - x_{2k}, x_1, \cdots, x_{2k} \in \Lambda(A)$; and f^+, $f^-: X^{2k}E \to E$ are combinatorial A-functions such that for all $a_1, \cdots, a_{2k}$ in E, $f(a_1 - a_2, \cdots, a_{2k-1} - a_{2k}) = f^+(a_1, \cdots, a_{2k}) - f^-(a_1, \cdots, a_{2k})$.

Appropriate universal Horn sentences falling under Theorem 2.1 show that this definition is independent of the choice of $x_1, \cdots, x_{2k}$ and f^+, f^-. That at least one such pair f^+, f^- exists is shown by representing the function $f(a_1 - a_2, \cdots, a_{2k-1} - a_{2k})$ of $2k$ natural number variables as $\sum c_{i_1 \cdots i_{2k}} C_{a_1, i_1} \cdots C_{a_{2k}, i_{2k}}$ and letting f^+ be $\sum \max(0, c_{i_1 \cdots i_{2k}}) C_{a_1, i_1} \cdots C_{a_{2k}, i_{2k}}$ and f^- be

$$\sum \max(0, -c_{i_1 \cdots i_{2k}}) C_{a_1, i_1} \cdots C_{a_{2k}, i_{2k}} .$$

Via Horn sentences again, the extension of addition, multiplication in E^* are respectively the usual addition, multiplication in $\Lambda^*(A)$; and thus Horn sentences assure that $(\Lambda^*(A), +, \cdot)$ is a commutative ring with unit since $(E^*, +, \cdot)$ is.

Corollary 2.2. *Let S be a Horn sentence with equations as atomic formulas. Suppose with E^* as domain of individuals function symbols denote A-functions, with $\Lambda^*(A)$ as domain of individuals function symbols denote corresponding extensions to $\Lambda^*(A)$. Suppose S is true in E^* with A-Skolem functions. Then S is true in $\Lambda^*(A)$ with corresponding extensions as Skolem functions.* (See [**6**].)

For without difficulty, extension to $\Lambda^*(A)$ commutes with composition (via Horn sentences falling under Theorem 2.1), so only universal sentences S need be considered. If S is universal, it is easily seen that the truth of S in E^* can be expressed (using f^+, f^- in place of f) by the truth of a corresponding universal Horn sentence $S^\frown$ in E. By Theorem 2.1, $S^\frown$ will be true in $\Lambda(A)$; and the truth of $S^\frown$ in $\Lambda(A)$ implies the truth of S in $\Lambda^*(A)$.

We remark that if A is arithmetically closed (i.e., contains any function arithmetic in a finite number of functions in A), then A-Skolem functions automatically exist.

Corollary 2.3. *Suppose A is arithmetically closed. Then any Horn sentence true in E^* is true in $\Lambda^*(A)$.*

This is false if A is merely recursively closed [**6**, § 5].

In any commutative ring with unit the set B of idempotents forms a Boolean algebra where $a \vee b = a + b - ab$, $a \wedge b = a \cdot b$, $a' = 1 - a$. For A recursively closed, the fact that these are identities and Corollary 2.2 imply that $\vee, \wedge, '$ for $\Lambda^*(A)$ are the extensions of the corresponding operations in E^*. Let $B(A)$ be the Boolean algebra of idempotents of $\Lambda^*(A)$.

Theorem 2.4. *If A is countable and recursively closed, then the Boolean* *algebra $B(A)$ is atomless.* (*See* [**6**].)

PROOF. Let $f: X^2E^* \to E^*$, $g, h: E^* \to E^*$ be the functions such that $f(x, 0) = 0$, $f(x, y) = x$ for $y \neq 0$, $g(0) = 0$, $g(x) = 1$ if $x \neq 0$, $h(0) = 1$, $h(x) = 0$ if $x \neq 0$. Due to Corollary 2.2 and a Horn sentence true in E^*, for x in $\Lambda^*(A)$: $f_{\Lambda^*}(x, g_{\Lambda^*}(x)) = 0$ if and only if $x = 0$, $f_{\Lambda^*}(x, h_{\Lambda^*}(x)) = x$ if and only if $x = 0$. Now suppose x is a nonzero idempotent. Then $f_{\Lambda^*}(x, g_{\Lambda^*}(x)) \neq 0$ and $f_{\Lambda^*}(x, h_{\Lambda^*}(x)) \neq x$. By (2.11) below, there is a y such that $f_{\Lambda^*}(x, y) \neq 0$ and $f_{\Lambda^*}(x, y) \neq x$. Due to a corresponding Horn sentence true in E^*, for x, y in $\Lambda^*(A)$ we have $x^2 = x$ implies $(f_{\Lambda^*}(x, y))^2 = f_{\Lambda^*}(x, y)$; and $x^2 = x$ implies $f_{\Lambda^*}(x, y) \cdot x = f_{\Lambda^*}(x, y)$. Thus $f_{\Lambda^*}(x, y) \in B(A)$ and $0 < f_{\Lambda^*}(x, y) < x$. Therefore B is atomless. For the remainder of the proof, topologize $X^2\mathscr{P}(E)$ as follows. For finite $\delta_1, \delta_2, \delta_1', \delta_2'$ with $\delta_1 \cap \delta_1'$ and $\delta_2 \cap \delta_2'$ null, a typical basic open set $U^{(\delta_1, \delta_2)(\delta_1', \delta_2')}$ consists of all $(\beta_1, \beta_2) \in X^2\mathscr{P}(E)$ such that $\delta_1 \subseteq \beta_1$ and $\delta_2 \subseteq \beta_2$ and $\delta_1' \cap \beta_1$ null and $\delta_2' \cap \beta_2$ null. It is easily seen that Baire's category theorem applies to $X^2\mathscr{P}(E)$ (a homeomorph of the Cantor set).

(2.9) Let $y \in \Lambda^*(A)$. Then the set of all $(\beta_1, \beta_2) \in X^2\mathscr{P}(E)$ such that β_1, β_2 are A-isolated and $y = \langle\beta_1\rangle - \langle\beta_2\rangle$ is dense.

PROOF. Let $U = U^{(\delta_1, \delta_2), (\delta_1', \delta_2')}$ be a nonempty neighborhood. Let $x = \langle\gamma_1\rangle - \langle\gamma_2\rangle$ with γ_1, γ_2 infinite. Let n be the largest integer in any of $\delta_1, \delta_2, \delta_1', \delta_2'$. Let $\phi: \mathscr{P}(E) \to \mathscr{P}(E)$ be defined by $\phi(\alpha) = [x + n + 1 \mid x \in \alpha]$. If $\eta_i \subseteq \phi(\gamma_i)$, $\langle\eta_i\rangle = \langle\delta_i\rangle$, then $((\phi(\gamma_1) - \eta_1) \cup \delta_1, (\phi(\gamma_2) - \eta_2) \cup \delta_2) \in U$ and $x = \langle(\phi(\gamma_1) - \eta_1) \cup \delta_1\rangle - \langle(\phi(\gamma_2) - \eta_2) \cup \delta_2\rangle$.

(2.10) Let $f_0, f_1: X^4E \to E$ be combinatorial A-functions. Suppose $x_1, x_2 \in \Lambda(A)$. Suppose that for (γ_1, γ_2) in a dense subset D of $X^2\mathscr{P}(E)$, $f_{0\Lambda}(x_1, x_2, \langle\gamma_1\rangle, \langle\gamma_2\rangle) \neq f_{1\Lambda}(x_1, x_2, \langle\gamma_1\rangle, \langle\gamma_2\rangle)$. Then the set of all $(\beta_1, \beta_2) \in X^2\mathscr{P}(E)$ such that $f_0(x_1, x_2, \langle\beta_1\rangle, \langle\beta_2\rangle) = f_1(x_1, x_2, \langle\beta_1\rangle, \langle\beta_2\rangle)$ is of first category.

PROOF. Let ϕ_0, ϕ_1 be combinatorial A-operators inducing f_0, f_1 respectively. Let $x_1 = \langle\alpha_1\rangle$, $x_2 = \langle\alpha_2\rangle$. Let p be an A-equivalence. It is sufficient by (2.5) to show that the following set N is nowhere dense: all $(\beta_1, \beta_2) \in X^2\mathscr{P}(E)$ such that $\phi_0(\alpha_1, \alpha_2, \beta_1, \beta_2)$ is a subset of the domain of p and

$$p(\phi_0(\alpha_1, \alpha_2, \beta_1, \beta_2)) = \phi_1(\alpha_1, \alpha_2, \beta_1, \beta_2) .$$

Let $p_0 = p$, $p_1 = p^{-1}$. Let $U = U^{(\delta_1, \delta_2)(\delta_1', \delta_2')}$ be a non-empty neighborhood in $X^2\mathscr{P}(E)$. Due to the density of D, there is a (γ_1, γ_2) in $D \cap U$. By assumption $f_{0\Lambda}(\langle\alpha_1\rangle, \langle\alpha_2\rangle, \langle\gamma_1\rangle, \langle\gamma_2\rangle) \neq f_{1\Lambda}(\langle\alpha_1\rangle, \langle\alpha_2\rangle, \langle\gamma_1\rangle, \langle\gamma_2\rangle)$. Thus in particular (γ_1, γ_2) is not in N. This means that one of the following four cases holds.

Case i ($i = 0$, or $i = 1$). There is an $n \in \phi_i(\alpha_1, \alpha_2, \gamma_1, \gamma_2)$ such that either n is not in the domain of p_i; or n is in the domain of p_i, but $(p_in)_{\phi_{1-i}}$ does not exist; or n is in the domain of p_i, $(p_in)_{\phi_{1-i}}$ does exist, but either $((p_in)_{\phi_{1-i}})_1$ is not a subset of α_1 or $((p_in)_{\phi_{1-i}})_2$ is not a subset of α_2. Then define $\bar\delta_1 = \delta_1 \cup (n_{\phi_i})_3$, $\bar\delta_2 = \delta_2 \cup (n_{\phi_i})_4$ so that $(\gamma_1, \gamma_2) \in V = U^{(\bar\delta_1, \bar\delta_2)(\delta_1', \delta_2')}$. If $(\beta_1, \beta_2) \in V$, then $(n_{\phi_i})_3 \subseteq \bar\delta_1 \subseteq \beta_1$, $(n_{\phi_i})_4 \subseteq \bar\delta_2 \subseteq \beta_2$; and $n \in \phi_i(\alpha_1, \alpha_2, \gamma_1, \gamma_2)$ implies $(n_{\phi_i})_1 \subseteq \alpha_1$, $(n_{\phi_i})_2 \subseteq \alpha_2$. Thus $n_{\phi_i} \leqq (\alpha_1, \alpha_2, \beta_1, \beta_2)$, or $n \in \phi_i(\alpha_1, \alpha_2, \beta_1, \beta_2)$. But the assumption on n means either p_in is undefined or $p_in \notin \phi_{1-i}(\alpha_1, \alpha_2, \beta_1, \beta_2)$.

Thus V is disjoint from N and is a nonempty subneighborhood of U.

Case $2 + i$ ($i = 0$ or $i = 1$). Assume neither Case 0 nor Case 1 holds. Then $(\gamma_1, \gamma_2) \notin N$ implies that there exists an $n \in \phi_i(\alpha_1, \alpha_2, \gamma_1, \gamma_2)$ such that $(p_i n)_{\phi_{1-i}}$ exists, $((p_i n)_{\phi_{1-i}})_1 \subseteq \alpha_1, ((p_i n)_{\phi_{1-i}})_2 \subseteq \alpha_2$, but either $((p_i n)_{\phi_{1-i}})_3 \not\subseteq \gamma_1$ or $((p_i n)_{\phi_{1-i}})_4 \not\subseteq \gamma_2$. As before, define $\bar{\delta}_1 = \delta_1 \cup (n_{\phi_i})_3$, $\bar{\delta}_2 = \delta_2 \cup (n_{\phi_i})_4$. If there is an $a \in ((p_i n)_{\phi_{1-i}})_3 - \gamma_1$, let $\bar{\delta}_1' = \delta_1' \cup \{a\}$ and $\bar{\delta}_2' = \delta_2'$; otherwise there is an $a \in ((p_i n)_{\phi_{1-i}})_4 - \gamma_2$ and we may let $\bar{\delta}_1' = \delta_1', \bar{\delta}_2' = \delta_2' \cup \{a\}$. Then $(\gamma_1, \gamma_2) \in V = U^{(\bar{\delta}_1, \bar{\delta}_2), (\bar{\delta}_1', \bar{\delta}_2')}$. Suppose $(\beta_1, \beta_2) \in V$. Then as before $n \in \phi_i(\alpha_1, \alpha_2, \beta_1, \beta_2)$; and in addition the choice of $\bar{\delta}_1', \bar{\delta}_2'$ assures that $((p_i n)_{\phi_{1-i}})_3 \not\subseteq \beta_1$ or $((p_i n)_{\phi_{1-i}})_4 \not\subseteq \beta_2$, so $(p_i n)_{\phi_{1-i}} \not\leq (\alpha_1, \alpha_2, \beta_1, \beta_2)$ or $p_i n \notin \phi_{1-i}(\alpha_1, \alpha_2, \beta_1, \beta_2)$. Thus $(\beta_1, \beta_2) \notin N$ and V is a nonempty subneighborhood of U disjoint from N.

(2.11) Let $h_i, k_i: X^2E^* \to E^*$ be A-functions, $i = 1, 2$. Suppose that $x, y_1, y_2 \in \Lambda^*(A)$ and for $i = 1, 2$, $h_{i_{\Lambda^*}}(x, y_i) \neq k_{i_{\Lambda^*}}(x, y_i)$. Then there is a $y \in \Lambda^*(A)$ such that $h_{i_{\Lambda^*}}(x, y) \neq k_{i_{\Lambda^*}}(x, y)$, $i = 1, 2$.

Proof. By Baire's category theorem it suffices to show that the following three subsets of $X^2\mathscr{P}(E)$ are of first category: for $i = 1, 2$ the set of all (β_1, β_2) such that β_1, β_2 are A-isolated and $h_{i_{\Lambda^*}}(x, \langle\beta_1\rangle - \langle\beta_2\rangle) = k_{i_{\Lambda^*}}(x, \langle\beta_1\rangle - \langle\beta_2\rangle)$; and the set of all (β_1, β_2) such that either β_1 or β_2 is not A-isolated. The first two are of first category by (2.9), (2.10) above applied to $f_{i_0} = h_i^+ + k_i^-$ and $f_{i_1} = k_i^+ + h_i^-$. As for the third, by the countability of A-enumerable sets it suffices to show that for each pair $(\bar{\beta}_1, \bar{\beta}_2)$ of infinite A-enumerable sets, the set N of all (β_1, β_2) such that $\beta_1 \supseteq \bar{\beta}_1$ or $\beta_2 \supseteq \bar{\beta}_2$ is nowhere dense. If $U = U^{(\delta_1, \delta_2)(\delta_1', \delta_2')}$ is a neighborhood and $x \in \bar{\beta}_1 - (\delta_1 \cup \delta_1')$, $y \in \bar{\beta}_2 - (\delta_2 \cup \delta_2')$ and $\bar{\delta}_1' = \delta_1' \cup \{x\}$, $\bar{\delta}_2' = \delta_2' \cup \{y\}$, then $V = U^{(\bar{\delta}_1, \bar{\delta}_2)(\bar{\delta}_1' \bar{\delta}_2')}$ is a nonempty subneighborhood of U disjoint from N.

3. Elementary theory of $\Lambda^*(A)$. Throughout the remainder of this paper A is arithmetically closed. Let L_μ be a first order functional calculus with identity symbol and function symbols, and with $v_0, v_1, \cdots$ as individual variables. For this calculus ($\exists$) (there exists) and ($|$) (neither-nor) will be taken as primitive, other logical symbols being used as abbreviations; i.e., ($\vee$) (or), ($\wedge$) (and), ($\sim$) (not). Suppose with E^* as domain of individuals an interpretation is specified in which each function symbol denotes an A-function. By the corresponding interpretation with $\Lambda^*(A)$ as domain of individuals is meant the one in which function symbols denote corresponding extensions to $\Lambda^*(A)$. Let θ be a formula of L_μ all of whose free variables are among $v_0, \cdots, v_n$. Then θ defines an A-relation $R(n) \subseteq X^{n+1}E^*$ consisting of all $(x_0, \cdots, x_n) \in X^{n+1}E^*$ such that θ is satisfied in E^* when $v_0, \cdots, v_n$ are assigned respective values $x_0, \cdots, x_n$. Let $\chi^{R(n)}: X^{n+1}E^* \to E^*$ be the characteristic function of $R(n)$, 1 on $R(n)$ and 0 on $(X^{n+1}E^*) - R(n)$. Then $\chi^{R(n)}$ is an A-function. Further, for $n \leq n'$ and all $x_0, \cdots, x_{n'}$ in E^*, $\chi^{R(n)}(x_0, \cdots, x_n) = \chi^{R(n')}(x_0, \cdots, x_{n'})$. Consequently by Corollary 2.3 the corresponding identity holds in $\Lambda^*(A)$ between extensions to $\Lambda^*(A)$. It is thus legitimate to define a function $K_\theta: \Lambda^*(A)^\omega \to \Lambda^*(A)$ by $K_\theta(x) = \chi^{R(n)}_{\Lambda^*}(x_0, \cdots, x_n)$ for $x = (x_0, x_1, x_2, \cdots) \in \Lambda^*(A)^\omega$.

(3.1) Suppose that θ_1, θ_2 are formulas of L_μ and $x \in \Lambda^*(A)^\omega$. Then $((K_{\theta_1}(x))^2 =$

$K_{\theta_1}(x)$, $K_{\theta_1|\theta_2}(x) = (1 - K_{\theta_1}(x)) \cdot (1 - K_{\theta_2}(x))$. Hence if $\wedge, \vee, ', |$ are defined by $a \vee b = a + b - ab$, $a \wedge b = a \cdot b$, $a' = 1 - a$, $a | b = (1 - a) \cdot (1 - b)$, one has

$$K_{\theta_1 \vee \theta_2}(x) = K_{\theta_1}(x) \vee K_{\theta_2}(x),\ K_{\theta_1 \wedge \theta_2}(x) = K_{\theta_1}(x) \wedge K_{\theta_2}(x)\,,$$
$$K_{\sim\theta_1}(x) = (K_{\theta_1}(x))',\ K_{\theta_1|\theta_2}(x) = K_{\theta_1}(x) \,|\, K_{\theta_2}(x)\,.$$

PROOF. Let $v_0, \cdots, v_n$ contain all variables occurring free in either θ_1 or θ_2. Let $R, S \subseteq X^{n+1}E^*$ be the relations defined by θ_1, θ_2 respectively. Then for all $x_0, \cdots, x_n$ in E^*,

$$(\chi^R(x_0, \cdots, x_n))^2 = \chi^R(x_0, \cdots, x_n),\ \chi^{(X^nE^*-R)\cap(X^nE^*-S)}(x_0, \cdots, x_n)$$
$$= (1 - \chi^R(x_0, \cdots, x_n)) \cdot (1 - \chi^S(x_0, \cdots, x_n))\,.$$

Thus by Corollary 2.3 the corresponding identities are true in $\Lambda^*(A)$ and yield the desired conclusion.

(3.2) Suppose that $\theta_0, \cdots, \theta_m$ are formulas of L_μ. Suppose that for $x \in E^{*\omega}$, x satisfies $\theta_0 \vee \cdots \vee \theta_m$; and also satisfies $\sim(\theta_i \wedge \theta_j)$ for $0 \leqq i < j \leqq m$. Suppose that $y_0, \cdots, y_m \in \Lambda^*(A)$, that $x \in \Lambda^*(A)^\omega$, and that $k \in \omega = \{0, 1, 2, \cdots\}$. Then there exists a $z \in \Lambda^*(A)^\omega$ such that $z_i = x_i$ for $i \neq k$ and $K_{\theta_i}(z) = y_i$ for $i = 0, \cdots, m$ if and only if $y_0 \vee \cdots \vee y_m = 1$, $y_i \wedge y_j = 0$ for $0 \leqq i < j \leqq m$, $y_i^2 = y_i$ and $y_i \cdot K_{\exists v_k \theta_i}(x) = y_i$ for $i = 0, \cdots, m$.

PROOF. In one direction, suppose $z \in \Lambda^*(A)^\omega$ is such that $z_i = x_i$ for $i \neq k$ and $K_{\theta_i}(z) = y_i$ for $i = 0, \cdots, m$. Combine the assumption that $\theta_0 \vee \cdots \vee \theta_m$ and $\sim(\theta_i \wedge \theta_j)$ for $0 \leqq i < j \leqq m$ are satisfied by all $x \in E^{*\omega}$ with the fact that the extension of a constant function to $\Lambda^*(A)$ is a constant function to obtain from (3.1) that

$$K_{\theta_0}(z) \vee \cdots \vee K_{\theta_m}(z) = K_{\theta_0 \vee \cdots \vee \theta_m}(z) = 1\,,$$
$$K_{\theta_i}(z) \wedge K_{\theta_j}(z) = K_{\theta_i \wedge \theta_j}(z) = 1 - K_{\sim(\theta_i \wedge \theta_j)}(z) = 1 - 1 = 0\,.$$

Thus from these assumptions and (3.1), $y_i^2 = y_i$ for $i = 0, \cdots, m$, $y_0 \vee \cdots \vee y_m = 1$, $y_i \cdot y_j = 0$ for $0 \leqq i < j \leqq m$. To obtain the final conclusion, let $v_0, \cdots, v_n$ contain all variables occurring free in any of $\theta_0, \cdots, \theta_m$. Let R^i, S^i be the relations defined by $\theta_i, \exists v_k \theta_i$ respectively. Due to Corollary 2.3 applied to a corresponding Horn sentence true in E^*, $x_j = z_j$ for $j \neq k$ implies that for $i = 0, \cdots, m$, $\chi_{\Lambda^*}^{R_i}(z_0, \cdots, z_n) \cdot \chi_{\Lambda^*}^{S_i}(x_0, \cdots, x_n) = \chi_{\Lambda^*}^{R_i}(z_0, \cdots, z_n)$. This proves that $y_i \cdot K_{\exists v_k \theta_i}(x) = y_i$ for $i = 0, \cdots, m$.

For the converse, if $0 \leqq i \leqq m$ define $f^i \colon X^{n+1}E^* \to E^*$ as follows: $f^i(a_0, \cdots, a_n) = 0$ if $(a_0, \cdots, a_n) \notin S^i$; $f^i(a_0, \cdots, a_n) = a$ if a is the least non-negative integer such that when $b_i = a_i$ for $i \neq k$ and $b_k = a$, we have $(b_0, \cdots, b_n) \in R^i$; and in the remaining case $f^i(a_0, \cdots, a_n)$ is the largest negative integer with the same property. Suppose $0 \leqq i_1 \leqq m$. It suffices to prove that from the suppositions $y_0 \vee \cdots \vee y_m = 1$, $y_i \cdot \chi_{\Lambda^*}^{S_i}(x_0, \cdots, x_n) = y_i$, $y_i^2 = y_i$ for $i = 0, \cdots, m$, $y_i \cdot y_j = 0$ for $0 \leqq i < j \leqq m$, $z_i = x_i$ for $i \neq k$, $z_k = f_{\Lambda^*}^0(x_0, \cdots, x_n) \cdot y_0 \vee \cdots \vee f_{\Lambda^*}^m(x_0, \cdots, x_n) \cdot y_m$; we may conclude $y_{i_1} = \chi_{\Lambda^*}^{R_{i_1}}(z_0, \cdots, z_n)$. This whole assertion is a Horn sentence. Hence by Corollary 2.3 it will be true in $\Lambda^*(A)$ if the corresponding assertion is true in E^*. The assertion in E^* to be proved is then the following. Suppose $a_0, \cdots, a_m, b_0, \cdots, b_n, c_0, \cdots, c_n$ are in E^*.

Suppose $a_0 \vee \cdots \vee a_m = 1, a_i^2 = a_i$ and $a_i \cdot \chi^{S_i}(b_0, \cdots, b_n) = a_i$ for $i = 0, \cdots, m$, $a_i \cdot a_j = 0$ for $0 \leqq i < j \leqq m, c_i = b_i$ for $i \neq k$; $c_k = f^0(b_0, \cdots, b_n) \cdot a_0 \vee \cdots \vee f^m(b_0, \cdots, b_n) \cdot a_m$. Then $a_{i_1} = \chi^{R_{i_1}}(c_0, \cdots, c_n)$. This is true since $1 = a_0 \vee \cdots \vee a_m$ and $a_0^2 = a_0, \cdots, a_m^2 = a_m$ imply that exactly one $a_{i_2} = 1$, all the other $a_i = 0$. Thus $c_k = f^{i_2}(b_0, \cdots, b_n)$ and $\chi^{S_{i_2}}(b_0, \cdots, b_n) = 1$. By the definition of f^{i_2} it then follows that $\chi^{R_{i_2}}(c_0, \cdots, c_n) = 1$. Since the R^{i}'s form a partition of $X^{n+1}E^*$, $\chi^{R_i}(c_0, \cdots, c_n) = 0$ for $i \neq i_2$. Thus if $i_1 = i_2$, $a_{i_1} = 1 = \chi^{R_{i_1}}(c_0, \cdots, c_n)$; and if $i_1 \neq i_2$, $a_{i_1} = 0 = \chi^{R_{i_1}}(c_0, \cdots, c_n)$.

Let L_σ be a first order functional calculus with identity symbol, function symbols $\vee, \wedge, ', |$, constant 0, and individual variables $X_0, X_1, \cdots$. The interpretation intended for L_σ has the Boolean algebra $B(A)$ of idempotents of $\Lambda^*(A)$ as domain of individuals with $\wedge, \vee, ', |, 0$ denoting respectively infimum, supremum, complement, intersection of complements, and the zero element. For each finite set S of natural numbers let a term $\bigvee_{k\in S} X_k$ be defined by the following inductive requirement: if S is empty, $\bigvee_{k\in S} X_k$ is 0; if S has a largest number n, $\bigvee_{k\in S} X_k$ is $(\bigvee_{k\in S-\{n\}} X_k) \vee X_n$. A sequence $(\phi, \theta_0, \cdots, \theta_m)$ consisting of a formula ϕ of L_σ and formulas $\theta_0, \cdots, \theta_m$ of L_μ is called *acceptable* if the free variables of ϕ are among $X_0, \cdots, X_m$; and is called a *partitioning sequence* if in addition $\theta_0 \vee \cdots \vee \theta_m$ and $\sim(\theta_j \wedge \theta_{j'})$ for $0 \leqq j < j' \leqq m$ are truth table tautologies.

(3.3) There is a uniform effective procedure correlating with each acceptable sequence $(\phi', \theta'_0, \cdots, \theta'_{m'})$ a partitioning sequence $(\phi, \theta_0, \cdots, \theta_m)$ such that for $x \in \Lambda^*(A)^\omega$, $K_{\theta'_0}(x), \cdots, K_{\theta'_{m'}}(x)$ satisfy ϕ' when assigned as values to $X_0, \cdots, X_{m'}$ if and only if $K_{\theta_0}(x), \cdots, K_{\theta_m}(x)$ satisfy ϕ when assigned as values to $X_0, \cdots, X_m$.

Proof (Adaptation of Feferman-Vaught [2]). Let $m = 2^{m'+1}$, let $r_0, \cdots, r_m$ be a list of all subsets of $\{0, \cdots, m'\}$. For $0 \leqq k \leqq m$ define θ_k as $\bigwedge_{j\in r_k} \theta'_j \wedge \bigwedge_{j\in\{0,\cdots,m'\}-r_k} \sim\theta'_j$. Let S_l consist of all k such that $0 \leqq k \leqq m$ and $l \in r_k$. Let ϕ be the result of substituting $\bigvee_{k\in S_0} X_k$ for $X_0, \cdots, \bigvee_{k\in S_{m'}} X_k$ for $X_{m'}$ throughout ϕ'. Since $x \in E^{*\omega}$ satisfies θ'_l if and only if x satisfies $\bigvee_{k\in S_l} \theta_k$, (3.1) implies that for $x \in \Lambda^*(A)^\omega$, $K_{\theta_l'}(x) = \bigvee_{k\in S_l} K_{\theta_k}(x)$. This yields the conclusion.

Theorem 3.1. *Let A be arithmetically closed. Then there is a uniform effective procedure assigning to each formula Γ of L_μ a partitioning sequence $(\phi, \theta_0, \cdots, \theta_m)$ such that $\bigvee_{0\leqq j\leqq m} \theta_j$ has its free variables among those of Γ and an $x \in \Lambda^*(A)^\omega$ satisfies Γ if and only if $K_{\theta_0}(x), \cdots, K_{\theta_m}(x)$ satisfy ϕ when assigned as values to $X_0, \cdots, X_m$.*

Proof (Adaptation of Feferman-Vaught [2]). If Γ is atomic, $(X_0 = 0', \Gamma, \sim\Gamma)$ will do. If Γ is $\Gamma' \mid \Gamma''$ and $(\phi', \theta'_0, \cdots, \theta'_{m'})$, $(\phi'', \theta''_0, \cdots, \theta''_{m''})$ are partitioning sequences correlated with Γ', Γ'', then the partitioning sequence obtained from $(\phi' \mid \phi''(X_{m'+1}, \cdots, X_{m'+m''+1}), \theta'_0, \cdots, \theta'_{m'}, \theta''_0, \cdots, \theta''_{m''})$ by (3.3) will do.

Finally (and crucially) if Γ is of the form $\exists v_k \Gamma'$ and $(\phi', \theta'_0, \cdots, \theta'_{m'})$ is the partitioning sequence associated with Γ', let ϕ be $(\exists Y_0) \cdots (\exists Y_{m'})(\bigvee_{j\leqq m'}(Y_j = 1) \wedge \bigwedge_{i<j\leqq m'}(Y_i \wedge Y_j = 0) \wedge \bigwedge_{j\leqq m'}(Y_j \wedge X_j = Y_j) \wedge \phi'(Y_0, \cdots, Y_{m'}))$. Then by (3.2) the partitioning sequence obtained using (3.3) from $(\phi, \exists v_k \theta'_0, \cdots, \exists v_k \theta'_{m'})$ will

do.

If Γ is a statement with associated partitioning sequence $(\phi, \theta_0, \cdots, \theta_m)$, then $\theta_0, \cdots, \theta_m$ are statements; and θ_i is true or false in E^* according as $K_{\theta_i}(x)$ has the constant value 1 or 0. Thus

COROLLARY 3.2. *Let A be arithmetically closed. Let $B(A)$ be the Boolean algebra of idempotents of $\Lambda^*(A)$. Then the decision problem for the truth of statements in L_μ about $\Lambda^*(A)$ reduces (by unbounded truth tables) to the decision problem for the truth of statements in L_μ about E^* and the decision problem for the truth of statements in L_σ about $B(A)$.*

Thus for A arithmetically closed, the elementary theory of $\Lambda^*(A)$ depends only on that of $B(A)$; while if A is countable, Theorem 2.4 asserts that $B(A)$ is atomless. It is, however, an ancient theorem that all atomless Boolean algebras have the same elementary theory, which in fact has an elimination of quantifiers procedure and is hence decidable.

Thus if A is the set of all arithmetic or all hyperarithmetic or all analytic functions, the elementary theory of $\Lambda^*(A)$ is the same and is reducible by truth tables to that of E^*.

It is perhaps worthwhile to mention two of the (many) proofs of the required fact about atomless Boolean algebras. First, any two countably infinite atomless Boolean algebras are isomorphic, hence by Vaught's test (an immediate consequence of the completeness theorem) the theory of atomless Boolean algebras is complete: being axiomatizable, it is hence decidable. Second, the straightforward (Skolem) elimination of quantifiers procedure shows that with each formula ϕ of L_σ one can effectively associate an equivalent (quantifier-free) formula of the form

$$\bigvee_j \bigwedge_{r_0 \subseteq \{0, \cdots, m\}} \phi^{r_0, j}\Bigl(\bigwedge_{k \in r_0} X_k \wedge \bigwedge_{k \in \{0, \cdots, m\} - r_0} X_k'\Bigr),$$

where $\phi^{r_0, j}(X_0)$ is either $X_0 = 0$ or $\sim(X_0 = 0)$.

Finally, with notation as in § 1, observe that the same reduction procedure works for Q and for Q', in each case an atomless Boolean algebra of idempotents being obtained. Consequently these two have the same elementary theory as $\Lambda^*(A)$ when A is countable and arithmetically closed.

COROLLARY 3.3. *Suppose A is arithmetically closed and countable. Then the decision problem for the truth of statements in L_μ about $\Lambda^*(A)$ reduces to the decision problem for the truth of statements in L_μ about E^*. Further $\Lambda^*(A)$ has the same elementary theory as Q or Q'.*

4. $\Lambda^*(A)/P$ as nonstandard models.

THEOREM 4.1. *Let A be arithmetically closed. Let P be a minimal prime ideal in $\Lambda^*(A)$. Then all A-functions are well-defined* mod P. *Moreover, define an interpretation of L_μ with $\Lambda^*(A)/P$ as domain of individuals in the natural way; namely each function symbol denotes f_{Λ^*}* mod P *if it denoted f with E^* as domain of individuals. Then the set of true statements of L_μ with E^* as domain of individuals is the same as the set of true statements of L_μ with $\Lambda^*(A)/P$ as* *domain of individuals.*

PROOF. By Corollary 2.3 and the truth of a corresponding Horn sentence in E^*, $x^n = 0$ implies $x = 0$ for $x \in \Lambda^*(A)$. Thus there are no nilpotents in $\Lambda^*(A)$. Then it follows for trivial algebraic reasons (holding in any commutative ring with unit) that the intersection of all minimal prime ideals in $\Lambda^*(A)$ is $\{0\}$. In particular, we will use the fact that $\Lambda^*(A)$ itself can thus not be a minimal prime ideal of $\Lambda^*(A)$.

That A-functions are well-defined mod P is (4.2) below. The remaining assertions follow from (4.3) below applied to statements θ.

If $e(x) = 1$ for $x \neq 0$, $e(0) = 0$, let $x^0 = e_{\Lambda^*}(x)$.

(4.1) For $x \in \Lambda^*(A)$, $x \in P$ if and only if $x^0 \in P$.

PROOF. Let $P' = [x \in \Lambda^*(A) \mid x^0 \in P]$. It suffices to show that P' is a subset of P which is a prime ideal. To see that $P' \subseteq P$, suppose $x \in P'$. Then $x^0 \in P$ implies $1 - x^0 \notin P$; otherwise $1 = (1 - x_0) + x_0 \in P$ or $P = \Lambda^*(A)$, which is impossible. Due to Corollary 2.3 and a corresponding identity in E^*, $x(1 - x^0) = 0$. Thus $(1 - x^0) \notin P$ and $x(1 - x^0) \in P$ implies (due to the primeness of P) that $x \in P$.

To see that P' is an ideal, suppose $x \in P'$ and $y \in \Lambda^*(A)$. Due to a corresponding identity in E^*, $(xy)^0 = x^0 y^0$. Since P is an ideal and $x^0 \in P$, $(xy)^0 \in P$. Thus $xy \in P'$. Next, suppose $x, y \in P'$. Then $x^0, y^0 \in P$ and $x^0 \vee y^0 = x^0 + y^0 - x^0 y^0 \in P$. Due to a corresponding identity in E^*, $(x - y)^0 = (x - y)^0(x^0 \vee y^0)$. Since P is an ideal, $(x - y)^0 \in P$ or $x - y \in P'$. Finally, to see that P' is prime, suppose $xy \in P'$. Then $x^0 y^0 = (xy)^0 \in P$. Since P is prime, $x^0 \in P$ or $y^0 \in P$. Thus $x \in P'$ or $y \in P'$.

(4.2) If $f: X^k E^* \to E^*$ is an A-function, then f_{Λ^*} is well-defined mod P. Hence identities holding in E^* yield identities holding in $\Lambda^*(A)/P$.

PROOF. Suppose $x_1 \equiv y_1 \bmod P, \cdots, x_k \equiv y_k \bmod P$. By (4.1), $(x_1 - y_1)^0 \in P, \cdots, (x_k - y_k)^0 \in P$. If $x \vee y = x + y - xy$, the fact that P is an ideal implies P is closed under $\vee$. Thus $(x_1 - y_1)^0 \vee \cdots \vee (x_k - y_k)^0 \in P$. From a corresponding identity in E^*,

$$(f_{\Lambda^*}(x_1, \cdots, x_k) - f_{\Lambda^*}(y_1, \cdots, y_k))^0((x_1 - y_1)^0 \vee \cdots \vee (x_k - y_k)^0) = (f_{\Lambda^*}(x_1, \cdots, x_k) - f_{\Lambda^*}(y_1, \cdots, y_k))^0 .$$

Thus since P is an ideal, $(f_{\Lambda^*}(x_1, \cdots, x_k) - f_{\Lambda^*}(y_1, \cdots, y_k))^0 \in P$. Hence by (4.1), $f_{\Lambda^*}(x_1, \cdots, x_k) - f_{\Lambda^*}(y_1, \cdots, y_k) \in P$. Finally,

$$f_{\Lambda^*}(x_1, \cdots, x_k) \equiv f_{\Lambda^*}(y_1, \cdots, y_k) \bmod P .$$

If $x \in \Lambda^*(A)^\omega$, let $x \bmod P$ be that element of $(\Lambda^*(A)/P)^\omega$ such that $(x \bmod P)_i = x_i + P$ for all $i \in \omega$.

(4.3) If θ is a formula of L_μ and $x \in \Lambda^*(A)^\omega$, then $x \bmod P$ satisfies θ with $\Lambda^*(A)/P$ as domain of individuals if and only if $K_\theta(x) \equiv 1 \bmod P$.

PROOF. The proof will proceed by induction on the length of formulas. If θ is an atomic formula, using (4.2) the assertion about θ reduces to the following. Suppose $f, g: X^k E^* \to E^*$ are A-functions and R is the set of all $(x_1, \cdots, x_k)$

$\in X^k E^*$ such that $f(x_1, \cdots, x_k) = g(x_1, \cdots, x_k)$. Then for $x_1, \cdots, x_k \in \Lambda^*(A)$, $f_{\Lambda^*}(x_1, \cdots, x_k) \equiv g_{\Lambda^*}(x_1, \cdots, x_k) \bmod P$ if and only if $\chi^R_{\Lambda^*}(x_1, \cdots, x_k) \equiv 1 \bmod P$. Equivalently, it must be shown that $1 - \chi^R_{\Lambda^*}(x_1, \cdots, x_k) \in P$ if and only if $f_{\Lambda^*}(x_1, \cdots, x_k) - g_{\Lambda^*}(x_1, \cdots, x_k) \in P$. But due to a corresponding identity in E^*, $1 - \chi^R_{\Lambda^*}(x_1, \cdots, x_k) = (f_{\Lambda^*}(x_1, \cdots, x_k) - g_{\Lambda^*}(x_1, \cdots, x_k))^0$. Hence the conclusion follows from (4.1).

Now suppose that the conclusion is known for θ_1 and for θ_2. Due to the fact that P is prime and values of K_θ are idempotents ((3.1)), $K_\theta \bmod P$ takes on only values 0 and 1. By (3.1), for $x \in \Lambda^*(A)^\omega$, $K_{\theta_1 | \theta_2}(x) \equiv 1 \bmod P$ if and only if $K_{\theta_1}(x) \mid K_{\theta_2}(x) \equiv 1 \bmod P$; or if and only if $K_{\theta_1}(x) \equiv 0 \bmod P$ and $K_{\theta_2}(x) \equiv 0 \bmod P$; or by inductive hypothesis if and only if $x \bmod P$ satisfies neither θ_1 nor θ_2; or if and only if $x \bmod P$ satisfies $\theta_1 | \theta_2$. This proves the desired conclusion for $\theta_1 | \theta_2$.

Finally, suppose the conclusion is known for θ and consider $\exists v_k \theta$. In one direction, suppose $x \bmod P$ satisfies $\exists v_k \theta$. Let $y \in \Lambda^*(A)^\omega$ be such that $y_i = x_i$ for $i \neq k$ and $y \bmod P$ satisfies θ. By inductive assumption $K_\theta(y) \equiv 1 \bmod P$. Using (3.2), $K_\theta(y) \cdot K_{\exists v_k \theta}(x) = K_\theta(y)$. Reducing mod P, $K_{\exists v_k \theta}(x) \equiv 1 \bmod P$. For the other direction, suppose $K_{\exists v_k \theta}(x) \equiv 1 \bmod P$. By (3.2) there is a y with $K_{\exists v_k \theta}(x) = K_\theta(y)$ and $y_i = x_i$ for $i \neq k$. Thus $K_\theta(y) \equiv 1 \bmod P$. By inductive hypothesis $y \bmod P$ satisfies θ. Since $(x \bmod P)_i = (y \bmod P)_i$ for $i \neq k$, $x \bmod P$ satisfies $\exists v_k \theta$.

Finally, we remark that if A is countable, by an argument of [**6**], $\Lambda^*(A)/P$ is actually uncountable. Of course, if A is the set of all functions, $\Lambda^*(A)$ is E^*, P is necessarily $\{0\}$, and $\Lambda^*(A)/P$ is E^* itself.

In conclusion, the fact that the intersection of all minimal prime ideals is $\{0\}$ corresponds by universal algebra to $\Lambda^*(A)$ being a subdirect sum of non-standard models of E^*; so $\Lambda^*(A)$ is an associate ring in the sense of Sussman [**7**].

References

1. J. C. E. Dekker and J. Myhill, *Recursive equivalence types*, Univ. California Publ. Math. (N.S.) vol. 3 (1960) pp. 67–214.

2. S. Feferman and R. L. Vaught, *The first order properties of products of algebraic systems*, Fund. Math. vol. 47 (1959) pp. 57–103.

3. J. Myhill, *Elementary properties of the group of isolic integers*, Math. Z., to appear.

4. ———, *Recursive equivalence types and combinatorial functions*, Bull. Amer. Math. Soc. vol. 64 (1958) pp. 373–376.

5. A. Nerode, *Extensions to isols*, Ann. of Math. vol. 73 (1961) pp. 362–403.

6. ———, *Extensions to isolic integers*, to appear in Ann. of Math.

7. Irving Sussman, *A generalization of Boolean rings*, Math. Ann. vol. 136 (1958) pp. 326–338.

Cornell University,
Ithaca, New York

ALGEBRAS OF SETS BINUMERABLE IN COMPLETE EXTENSIONS OF ARITHMETIC

BY

DANA SCOTT

A set of integers is usually called *arithmetically definable* if there is a formula $B(x)$ with one free variable (in the notation of elementary first-order arithmetic) such that the set in question consists of just those integers n for which the corresponding instance $B(\Delta_n)$ of the formula is true in the domain of integers. Here the notation Δ_n denotes the nth formal digit in the formalized language of arithmetic. The class of all arithmetically definable sets of integers is of course a denumerable Boolean algebra (or field) of sets. Now aside from the set of all true sentences of arithmetic, there are many other complete and consistent extensions of the axiomatic theory of first-order arithmetic in view of the well-known incompleteness of these axioms. Clearly each such complete extension leads to a denumerable Boolean algebra of sets obtained by using the instances of formulas valid in the complete extension rather than using the true instances. The purpose of this paper is to answer the question: *Which denumerable Boolean algebras of sets are obtainable from complete extensions of arithmetic?* A mathematical characterization of these algebras will be given using some simple notions from recursive function theory.

All formal theories used here are formalized in an applied first-order logic with identity with the individual constants 0 and 1 and the binary operation symbols $+$ and $\cdot$ as the only nonlogical constants. The logical symbols are $\wedge, \vee, \neg, \rightarrow, \longleftrightarrow, \forall, \exists, =$ with $x, y, z, w, x', y', \cdots$ etc. as individual variables. the formal digits Δ_n are defined by recursion so that Δ_0 is 0 and Δ_{n+1} is $(\Delta_n + 1)$. Capital Roman letters will generally denote formulas. Free variables and the substitution of digits for variables will be indicated by an informal parenthesis notation as in $B(x)$ and $B(\Delta_n)$. The standard axiomatic theory of first-order arithmetic is denoted by $\mathbf{P}$ (as in [6, p. 52]). A theory in general is simply a set $\mathbf{T}$ of sentences (without free variables) closed under the rules of deduction of first-order logic. We shall be concerned with those theories $\mathbf{T}$ such that $\mathbf{P} \subseteq \mathbf{T}$ and for every sentence A either $A \in \mathbf{T}$ or $\neg A \in \mathbf{T}$ but not both; these are the complete (and consistent) extensions of arithmetic. A set S of integers is *numerable* in such a theory $\mathbf{T}$ if there is a formula $B(x)$ such that $n \in S$ if and only if $B(\Delta_n) \in \mathbf{T}$. Note that the formula $\neg B(x)$ can be used to show that the complement of S is also numerable in $\mathbf{T}$, and hence S is *binumerable* in $\mathbf{T}$ (cf. [1, p. 51]). The Boolean algebra of all sets of integers *binumerable* in $\mathbf{T}$ will be denoted by $\mathfrak{B}[\mathbf{T}]$.

Even though our main interest here is in the algebras $\mathfrak{B}[\mathbf{T}]$, it seems to be

Received by the editor May 12, 1961.

a little more convenient to replace sets of integers by their characteristic functions. The class of characteristic functions of the sets in $\mathfrak{B}[\mathbf{T}]$ will be denoted by $\mathfrak{F}[\mathbf{T}]$. To be more precise, let ω denote the set of (non-negative) integers and let 2^ω denote the class of all $\{0, 1\}$-valued functions defined on ω. If A is a formula of arithmetic, let A^1 be A itself, but let A^0 be $\neg A$. If $\mathbf{T}$ is a complete extension of arithmetic, then a function $f \in \mathfrak{F}[\mathbf{T}]$ if and only if there is a formula $B(x)$ such that $B(\Delta_n)^{f(n)} \in \mathbf{T}$, for all $n \in \omega$. The theorem that will be given below characterizes those subsets of 2^ω that are of the form $\mathfrak{F}[\mathbf{T}]$ for some complete extension $\mathbf{T}$. Obviously we could transform the result to give a characterization of the algebras $\mathfrak{B}[\mathbf{T}]$, but this transformation will be left to the reader.

Let 2^n denote the class of all $\{0, 1\}$-valued sequences (functions) with domain $\{0, \cdots, n-1\}$. If $s \in 2^m$ and $n \leqq m$, then $s \restriction n$ denotes the restriction of s to $\{0, \cdots, n-1\}$. If $f \in 2^\omega$, then $f \restriction n$ is the restriction of f to $\{0, \cdots, n-1\}$. To each $s \in 2^n$ we make correspond a unique integer

$$||s|| = p_0^{s(0)} \cdot \cdots \cdot p_{n-1}^{s(n-1)} \cdot p_n ,$$

where p_k is the kth prime number starting with $p_0 = 2$. By a *tree* we shall understand a set $\mathscr{T}$ of finite sequences (with arbitrary domains) closed under the formation of restrictions; that is, if $s \in \mathscr{T}$ and $s \in 2^m$, then $s \restriction n \in \mathscr{T}$ for all $n \leqq m$. A tree $\mathscr{T}$ is called *recursive within* a class $\mathscr{F} \subseteq 2^\omega$, if the set of integers of the form $||s||$ for $s \in \mathscr{T}$ is recursive in some finite number of functions $f_0, \cdots, f_{k-1}$ in $\mathscr{F}$ (in the sense of [**2**, p. 275]). A *path* of a tree $\mathscr{T}$ is a function $f \in 2^\omega$ such that $f \restriction n \in \mathscr{T}$ for all $n \in \omega$. As is well-known, every infinite tree has at least one path. It is now possible to state the main theorem.

THEOREM. *For there to exist a complete extension* $\mathbf{T}$ *of arithmetic such that* $\mathscr{F} = \mathfrak{F}[\mathbf{T}]$ *it is necessary and sufficient that:*
(i) $\mathscr{F}$ *is a denumerable subclass of* 2^ω;
(ii) *every infinite tree recursive within* $\mathscr{F}$ *has a path belonging to* $\mathscr{F}$.

The proof of the necessity of conditions (i) and (ii) will not be given here, but it will be included in [**5**] along with the proofs of the results mentioned in [**4**]. For the proof of sufficiency, we shall employ a lemma which seems to be of some independent interest. This lemma slightly generalizes a result of A.Mostowski given in [**3**]. For the statement of the lemma certain additional terminology is convenient. Let us say that a formula $B(x)$ is *independent* of a formula $A(x)$ if whenever $f \in 2^\omega$ and the set of all sentences consisting of those in $\mathbf{P}$ together with sentences $A(\Delta_n)^{f(n)}$, for $n \in \omega$, is consistent, then the set remains consistent upon the adjunction of sentences $B(\Delta_n)^{g(n)}$, for $n \in \omega$, no matter which $g \in 2^\omega$ is chosen. In other words, on the basis of $\mathbf{P}$ it is impossible to deduce any fact about the relations among the instances of $B(x)$ from any consistent assumption about the instances of $A(x)$. We can define in a similar way the notion of $B(x)$ being independent of a finite number of formulas $A_0(x), \cdots, A_{k-1}(x)$.

LEMMA. *Given a finite number of formulas* $A_0(x), \cdots, A_{k-1}(x)$ *one can effectively find a formula* $B(x)$ *which is independent of them.*

The proof of the lemma will be outlined at the end of the paper. Let us now see how the lemma can be used for the proof of sufficiency. Suppose that $\mathscr{F}$ satisfies conditions (i) and (ii) of the theorem. Enumerate the elements of $\mathscr{F}$ in a sequence $g_0, \cdots, g_k, \cdots$. Let $A_0(x), \cdots, A_k(x), \cdots$ be a (recursive) sequence containing *all* the formulas of arithmetic with one free variable x. Evoke the lemma to introduce (by a recursion) a sequence $B_0(x), \cdots, B_k(x), \cdots$ of formulas such that for each $k \in \omega$ the formula $B_k(x)$ is independent of the finite number of formulas $A_0(x), \cdots, A_k(x), B_0(x), \cdots, B_{k-1}(x)$. By construction it is clear that the set of sentences of the form $B_k(\Delta_n)^{g_k(n)}$, for $k, n \in \omega$, is consistent with $\mathbf{P}$. Hence, there is at least one complete extension $\mathbf{T}$ of $\mathbf{P}$ which contains all these sentences. Obviously for such a $\mathbf{T}$ we have $\mathscr{F} \subseteq \mathfrak{F}[\mathbf{T}]$. What we need to show is that there is at least one $\mathbf{T}$ for which the inclusion is an equality.

To obtain the desired complete extension a new sequence $f_0, \cdots, f_k, \cdots$ of functions in $\mathscr{F}$ will be introduced so that the following set of sentences is consistent:

$$\mathbf{P} \cup \{A_k(\Delta_n)^{f_k(n)} : k, n \in \omega\} \cup \{B_k(\Delta_n)^{g_k(n)} : k, n \in \omega\} .$$

Suppose that $f_0, \cdots, f_{k-1} \in \mathscr{F}$ have already been obtained so that the set

$$\mathbf{U}_m = \mathbf{P} \cup \{A_k(\Delta_n)^{f_k(n)} : k < m; n \in \omega\} \cup \{B_k(\Delta_n)^{g_k(n)} : k < m; n \in \omega\}$$

is consistent. Let $\mathscr{T}_m$ be the set of all functions s such that $s \in 2^r$ for some $r \in \omega$ and such that among the first r proofs (in some standard enumeration of all the proofs of first-order logic) there is no proof establishing the inconsistency of the set of sentences:

$$\mathbf{U}_m \cup \{A_m(\Delta_n)^{s(n)} : n < r\} .$$

It is obvious that $\mathscr{T}_m$ is a tree. Since $\mathbf{U}_m$ is consistent, it is easy to prove that $\mathscr{T}_m$ is an infinite tree. Notice that the predicate of Gödel numbers of proofs which tells which proofs establish the inconsistency of a set of sentences is recursive in the set of Gödel numbers of the sentences. Hence, we see that $\mathscr{T}_m$ is recursive in $f_0, \cdots, f_{m-1}, g_0, \cdots, g_{m-1}$, all of which are in $\mathscr{F}$. Thus from condition (ii) there must be a path of $\mathscr{T}_m$ which is in $\mathscr{F}$; let this path be f_m. By the construction of $B_m(x)$ and the choice of f_m, it follows at once that the set $\mathbf{U}_{m+1}$ is consistent; therefore we can continue obtaining functions. Let $\mathbf{T}$ be the deductive closure of the set $\bigcup_{m \in \omega} \mathbf{U}_m$ of sentences. $\mathbf{T}$ is consistent because each $\mathbf{U}_m$ is consistent and $\mathbf{U}_m \subseteq \mathbf{U}_{m+1}$ for $m \in \omega$. On the other hand $\mathbf{T}$ is complete. For if A is any sentence, then there is a $k \in \omega$ such that $A_k(x)$ is the formula $[A \wedge x = x]$. But $A_k(\Delta_0)^{f_k(0)}$ is equivalent to $A^{f_k(0)}$; hence, this sentence must be in $\mathbf{T}$. Thus $\mathbf{T}$ is a complete extension for which $\mathscr{F} = \mathfrak{F}[\mathbf{T}]$ in view of the construction of the functions f_k.

Finally we must return to the proof of the lemma. It will be enough to show how to obtain a formula $B(x)$ which is independent of a given formula $A(x)$. First of all, let $C_0, \cdots, C_k, \cdots$ be the usual list of all sentences of arithmetic; that is, the sentence with Gödel number k is C_k. The Gödel numbering should be chosen in one of the standard ways so that we can introduce by definition functions $\dot{\to}$ and $\dot{\neg}$ into the formal theory of arithmetic where $\Delta_k = (\Delta_m \dot{\to} \Delta_n)$ is provable in **P** if and only if C_k is the formula $[C_m \to C_n]$, and where $\Delta_k = \dot{\neg}\Delta_m$ is provable in **P** if and only if C_k is $\neg C_m$. Further, we need to introduce by definition a formula $Pf(x, y)$ with the following meaning: *x is the Gödel number of a proof from the axioms of* **P** *of a sentence of the form*

$$[[A(\Delta_0)^{s(0)} \wedge \cdots \wedge A(\Delta_{z-1})^{s(z-1)}] \to C_y] ,$$

where s is a $\{0, 1\}$*-valued sequence of length z such that whenever* $w < z$, *then* $s(w) = 1$ *if and only if* $A(w)$ *holds.* The predicate $Pf(x, y)$ is of course not a recursive predicate since it involves $A(x)$; however, it can easily be obtained as a modification of the usual recursive proof predicate for the theory **P**. The particular predicate used here will have the property that if $f \in 2^\omega$; if $\mathbf{P} \cup \{A(\Delta_n)^{f(n)} : n \in \omega\}$ is consistent; and if $\mathbf{T}'$ is the deductive closure of this set of sentences, then $Pf(\Delta_p, \Delta_q) \in \mathbf{T}'$ implies $C_q \in \mathbf{T}'$, and $C_q \in \mathbf{T}'$ implies that $Pf(\Delta_p, \Delta_q) \in \mathbf{T}'$ for some $p \in \omega$. Using this proof predicate we may construct a recursive function d such that for $n \in \omega$ the sentence

$$C_{d(n)} \longleftrightarrow \forall x[Pf(x, \Delta_n \dot{\to} \Delta_{d(n)}) \to \exists y \exists z[x = y + (z + 1) \wedge Pf(y, \Delta_n \dot{\to} \dot{\neg}\Delta_{d(n)}]]$$

is provable in theory **P** (using, e.g., the method of [1, p. 65]). This function d will have the additional property that if $\mathbf{T}'$ is a theory of the type mentioned above, and if either $[C_n \to C_{d(n)}]$ or $[C_n \to \neg C_{d(n)}]$ is in $\mathbf{T}'$, then $\neg C_n$ is in $\mathbf{T}'$.

We can now define by recursion a sequence $B_0, \cdots, B_k, \cdots$ of sentences. Suppose $B_0, \cdots, B_{k-1}$ have already been introduced. Corresponding to each $s \in 2^k$ let $e(s)$ be an integer such that $C_{e(s)}$ is the formula

$$[B_0^{s(0)} \wedge \cdots \wedge B_{k-1}^{s(k-1)}].$$

We let B_k be the conjunction of all sentences $[C_{e(s)} \to C_{d(e(s))}]$ where $s \in 2^k$. Notice that $[C_{e(s)} \to [B_k \longleftrightarrow C_{d(e(s))}]]$ is a tautology. It will now be very easy to establish by induction on k that if $\mathbf{T}'$ is a theory as mentioned above, then $\mathbf{T}' \cup \{C_{e(s)}\}$ is consistent for all $s \in 2^k$. In other words, if whenever $f \in 2^\omega$ and $\mathbf{U} = \mathbf{P} \cup \{A(\Delta_n)^{f(n)} : n \in \omega\}$ is consistent, then $\mathbf{U} \cup \{B_n^{g(n)} : n \in \omega\}$ will be consistent for any choice of $g \in 2^\omega$. If you examine the choice of sentences B_n, you will see that each of them is a Boolean combination of sentences of the form $C_{d(m)}$ using many different m's. But the sentences $C_{d(m)}$ are essentially substitution instances of the same formula; or at least on the basis of **P** we can find a formula $D(x)$ such that $[C_{d(m)} \longleftrightarrow D(\Delta_m)]$ is in **P** for all $m \in \omega$, making use of the fact that d is a recursive function. Now the fact that B_n is obtained in a recursive way from the sentences $C_{d(m)}$ by Boolean combinations can be used to construct a formula $B(x)$ such that $[B_n \longleftrightarrow B(\Delta_n)]$ is in **P** for all $n \in \omega$. The formula $B(x)$ is the desired formula which is independent of $A(x)$.

BIBLIOGRAPHY

1. S. Feferman, *Arithmetization of metamathematics in a general setting*, Fund. Math. vol. 49 (1960) pp. 36–92.

2. S. Kleene, *Introduction to metamathematics*, New York, D. Van Nostrand, 1952.

3. A. Mostowski, *A generalization of the incompleteness theorem*, Fund. Math. vol. 49 (1961) pp. 205–232.

4. D. Scott and S. Tennenbaum, *On the degrees of complete extensions of arithmetic*, Abstract 568-3, Notices Amer. Math. Soc. vol. 7 (1960) pp. 242–243.

5. ———, *On the degrees of complete extensions of arithmetic*, in preparation.

6. A. Tarski, A. Mostowski and R. M. Robinson, *Undecidable theories*, Amsterdam, North-Holland Publishing Company, 1953.

UNIVERSITY OF CALIFORNIA,
BERKELEY, CALIFORNIA

SOME PROBLEMS IN HIERARCHY THEORY

BY

J. W. ADDISON

Last August at the Stanford Congress I proposed [**2**] a general *theory of hierarchies*, to include as special cases the hierarchy theories of (1) analysis, (2) recursive function theory, and (3) the theory of models. At that time general definitions were given, and a number of theorems were cited that were general enough to apply to hierarchies arising in each of the three areas; however, for the more interesting theorems proofs could be given only general enough to cover recursive function theory and analysis. Indeed, the consolidated proofs for recursive function theory and analysis tended to be mockingly different from the proofs of corresponding theorems in the theory of models.

Today I want to present a preliminary report on a development that shows promise of leading to some consolidation of proofs for all three areas. To keep this guiding purpose in the forefront and to avoid technicalities not central to our investigation, we shall discuss directly proofs in analysis (which can be viewed—cf. [**1**]—as fully relativized recursive function theory) and the theory of models, without dealing openly with the case of (absolute) recursive function theory. However, both because in the general theory of hierarchies recursive function theory falls between the theory of models and analysis (cf. [**2**]), and because the link between analysis and recursive function theory has been explored (cf. [**1**]), the discussion today has direct implications for (absolute) recursive function theory, the broad outlines of which will I hope be reasonably clear.

1. Preliminaries. The theory of hierarchies deals with the properties of formulas and their models having to do with the kinds of prefixes the formulas have when written in full prenex normal form. For a general discussion of the concepts and notation the reader is referred to [**2**]. We recall here that if $\mathfrak{P}$ is an applied predicate language[1] with equality of order ω then we denote the class of formulas of $\mathfrak{P}$ in full prenex normal form with leading quantifier existential (universal) and of type t and with k blocks of homogeneous quantifiers of type t by "$\bigvee_k^t(\mathfrak{P})$" ("$\bigwedge_k^t(\mathfrak{P})$"). With each formula of $\mathfrak{P}$ we associate the set of its models relative to the interpretation of the constants given in $\mathfrak{P}$. Then we also use "$\bigvee_k^t(\mathfrak{P})$" and "$\bigwedge_k^t(\mathfrak{P})$" to denote the corresponding classes of sets of models.

In the present discussion we shall be interested first in the case where $\mathfrak{P}$ is an applied language for number theory with constants for all number-theoretic predicates (analysis), and second in the case where $\mathfrak{P}$ is the pure language with no constants (the theory of models). Following past practice

Received by the editor, May 16, 1961.

[1] By a *language* we mean a grammar together with an interpretation for it.

we shall abbreviate "$\bigvee_k^t(\mathfrak{P})$", "$\bigwedge_k^t(\mathfrak{P})$" in these two cases, respectively, to "$\boldsymbol{\Sigma}_k^t$", "$\boldsymbol{\Pi}_k^t$" and to "$\bigvee_k^t$", "$\bigwedge_k^t$".

Moreover, throughout the present discussion the free variables in the formulas will be *restricted to types* 0 *and* 1. And in our discussion of analysis, to prevent degeneracy, we will assume that there is at least one free variable of type 1 in each formula considered. Therefore, in our discussion of analysis we shall be working (in effect) in a space $\mathscr{N}$ which is a Cartesian product of finitely many copies of the set N of natural numbers $0, 1, \cdots$ under the discrete topology and finitely many (and more than 0) copies of N^N under the induced product topology. And in our discussion of the theory of models we shall be working in a system space $\mathscr{S}$, each point of which is a relational system of some arbitrary but fixed similarity type. That is, in either theory given sets will be assumed to be in the fixed space $\mathscr{N}$ or $\mathscr{S}$, although during a discussion it may be useful, for example, to move to higher dimensions.

If $\mathscr{C}$ is a class of sets, then, as usual, $\mathscr{C}_\sigma$ ($\mathscr{C}_\delta$) is the class of countable unions (intersections) of sets in $\mathscr{C}$. We abbreviate "$\bigcup_k \bigvee_k^t$" to "$\bigvee^t$" and "$\bigcup_k \bigwedge_k^t$" to "$\bigwedge^t$". Further it is convenient here to let "$\bigwedge_0^0$" denote the class of sets of models of atomic formulas, and to let "$\bigvee_0^0$" denote the class of sets of models of denials of atomic formulas.

2. The separation and construction principles. One of the propositions which I suggested at Stanford might be true on the basis of a study of the analogies was what might be called a "strong separation principle" for $\bigwedge_k^0$ for $k \geqq 2$, viz.

Strong separation principle for $\bigwedge_k^0$. Every two disjoint $\bigwedge_k^0$ sets can be separated by a set in the Boolean algebra over $\bigwedge_{k-1}^0$.

This principle splits, in a manner characteristic of hierarchy theory (cf. [2]), into the following two corollaries, the conjunction of which has as a corollary the principle itself.

First separation principle for $\bigwedge_k^0$. Every two disjoint $\bigwedge_k^0$ sets can be separated by an $\bigvee_k^0 \cap \bigwedge_k^0$ set.

Construction principle for $\bigvee_k^0 \cap \bigwedge_k^0$. $\bigvee_k^0 \cap \bigwedge_k^0 =$ the Boolean algebra over $\bigwedge_{k-1}^0$.

These three principles are analogs of the following three principles of descriptive set theory (i.e. of the hierarchy theory of analysis).

Strong separation principle for $\boldsymbol{\Pi}_k^0$. Every two disjoint $\boldsymbol{\Pi}_k^0$ sets can be separated by a set in the class of sums of countable decreasing alternating series of $\boldsymbol{\Pi}_{k-1}^0$ sets.

First separation principle for $\boldsymbol{\Pi}_k^0$. Every two disjoint $\boldsymbol{\Pi}_k^0$ sets can be separated by a $\boldsymbol{\Sigma}_k^0 \cap \boldsymbol{\Pi}_k^0$ set.

Construction principle for $\boldsymbol{\Sigma}_k^0 \cap \boldsymbol{\Pi}_k^0$. $\boldsymbol{\Sigma}_k^0 \cap \boldsymbol{\Pi}_k^0 =$ the class of sums of countable decreasing alternating series of $\boldsymbol{\Pi}_{k-1}^0$ sets.

Although not originally formulated that way the uniformity of statement of the three principles can be improved by replacing, in the formulation of the strong separation principle for $\bigwedge_k^0$ and the construction principle for $\bigvee_k^0 \cap \bigwedge_k^0$, "the Boolean algebra over $\bigwedge_{k-1}^0$" by "the class of sums of finite decreasing alternating series of $\bigwedge^0_{k-1}$ sets". These two formulations are of course easily shown to be equivalent.

The three principles from analysis have long been known to hold for all $k \geqq 2$. The strong separation principle was known to fail for $\bigwedge^0_1$, but to hold for $\bigvee^0_1$—cf. the discussion of this in [2]—and the construction principle was known to hold for $\bigvee^0_1 \cap \bigwedge^0_1$. Shortly after the Stanford Congress Shoenfield verified (using ultraproducts) that the strong separation principle holds for $\bigwedge^0_k$ for $k \geqq 2$, and since then several proofs have been found, including ones by H. J. Keisler and Melven R. Krom.[2]

The point of departure for the present discussion is a proof of this theorem that was suggested to me by Krom's proof. The present proof has the following features: (i) although not suggested by a proof in descriptive set theory, the proof it yields for the construction principle for $\bigvee^0_2 \cap \bigwedge^0_2$ presents a striking analogy with the proof by the method of residues of the analogous principle in descriptive set theory; (ii) it provides a sharp yet simple model-theoretic criterion for the degree of complexity of $\bigvee^0_k \cap \bigwedge^0_k$ sets and of how easily separable disjoint $\bigwedge^0_k$ sets are; (iii) it suggests a proof of the Craig separation theorem that is a natural limit of the proofs of the strong separation principles for $\bigwedge^0_k$ as $k \to \omega$; and (iv) it suggests a relativization of the notion of residue that yields a generalization of the proof by the method of residues of the construction principle for $\boldsymbol{\Sigma}^0_2 \cap \boldsymbol{\Pi}^0_2$ to a direct proof of the strong separation principle for $\boldsymbol{\Pi}^0_2$.

3. First-order analysis. We shall first recall the classical proof of the construction principle for $\boldsymbol{\Sigma}^0_2 \cap \boldsymbol{\Pi}^0_2$ (cf., e.g., [3], for details). Let $A \in \boldsymbol{\Sigma}^0_2 \cap \boldsymbol{\Pi}^0_2$, and let $B = -A$. We consider the set B_1 of points *adjoining* A, i.e. of points in the closure $\bar{A}$ of A but not in A. We call this set the *adjoin* of A. The adjoin A_1 of B_1, i.e. the adjoin of the adjoin of A, is the *residue* of A. If B_2 is the adjoin of A_1, then the adjoin A_2 of B_2 is the residue of the residue of A, or the *second residue* of A. Continuing in this fashion, and taking intersections at limit ordinals, one develops the sequence of adjoins and residues of A. It is easily seen that the sequence $\lambda\nu A_\nu$ of even adjoins and the sequence $\lambda\nu B_\nu$ of odd adjoins are monotonic nonincreasing. Purely on grounds of cardinality it follows that these sequences must eventually become stationary, and because the space $\mathscr{N}$ is separable it follows that this must happen before Ω. Since A and B are both $\boldsymbol{\Sigma}^0_2$ sets it follows from the Baire category theorem that if $A_\nu = A_{\nu+1}$, $B_\nu = B_{\nu+1}$ then $A_\nu = B_\nu = \varnothing$. In other words the "last residue" of a $\boldsymbol{\Sigma}^0_2 \cap \boldsymbol{\Pi}^0_2$ set is always empty. Since each of the adjoins is the intersection of a $\boldsymbol{\Pi}^0_1$ set with A or B, it is now just a straightforward computation in terms of these $\boldsymbol{\Pi}^0_1$ sets to show that A is the sum of a countable decreasing alternating series of $\boldsymbol{\Pi}^0_1$ sets.

To generalize this argument to yield a proof of the strong separation principle for $\boldsymbol{\Pi}^0_2$ we relativize the notions of adjoin and residue. By the *adjoin* of

[2] Krom's proof is only for the case $k = 2$, but may be capable of generalization to higher k. Rather syntactic in spirit, his proof categorizes the atomic subformulas in a way motivated by Gödel's decision procedure for $\vee\vee\wedge$ formulas.

a set C *relative to* a set D we mean $(\bar{C} - C) \cap D$, i.e. the intersection with D of the adjoin of C. By the *residue* of C *relative to* D we mean the adjoin relative to D of the adjoin of C relative to D. The adjoin and residue of C are thus simply the adjoin and residue, respectively, of C relative to the space.

Now if A and B are two disjoint $\boldsymbol{\Pi}^0_2$ sets we form the sequence of adjoins of A relative to $A \cup B$, i.e. we form the adjoin B_1 of A relative to $A \cup B$, the adjoin A_1 of B_1 relative to $A \cup B$, etc. The argument then proceeds essentially as it did in the absolute case. At the end, upon tallying up the $\boldsymbol{\Pi}^0_1$ sets whose intersections with A and B are the successive relative adjoins, we find a set separating A and B which is the sum of a countable decreasing alternating series of $\boldsymbol{\Pi}^0_1$ sets.

4. The theory of models. The proof of the strong separation principle for $\boldsymbol{\Pi}^0_2$ by the method of (relative) adjoins carries over in general outline to a proof of the strong separation principle for $\bigwedge^0_2$ if topological closure is replaced by closure under subsystems. And the proof can then by extended to $\bigwedge^0_k$ for $k \geqq 3$ by the use of closure under what we will call $\bigwedge^0_{k-2}$-subsystems, and finally to $\bigvee^1_1$ by the use of closure under elementary subsystems (or of closure under $\bigwedge^0_k$-subsystems for increasing k).

First let us give some necessary definitions and conventions. We will use capital German letters for (relational) systems and the corresponding Roman capitals for their domains. We use the words "subsystem", "extension", "reduct", and "expansion" in the way indicated by the following: $\mathfrak{S}$ is a *subsystem* of $\mathfrak{T}$ if and only if $\mathfrak{T}$ is an *extension* of $\mathfrak{S}$; $\mathfrak{S}$ is a *reduct* of $\mathfrak{T}$ if and only if $\mathfrak{T}$ is an *expansion* of $\mathfrak{S}$. If $\mathfrak{S}$ is a system, then $\tilde{\mathfrak{S}}$ is "the" minimum expansion of $\mathfrak{S}$ in which every element of S is a coordinate.

DEFINITION. $\mathfrak{S}$ is a $\mathscr{C}$-subsystem of $\mathfrak{T} \longleftrightarrow [S \subseteq T$ & for any $\bar{s}$ in S^m and any formula Γ in $\mathscr{C}$ with m free individual variables [$\bar{s}$ satisfies Γ over $\mathfrak{S} \longleftrightarrow \bar{s}$ satisfies Γ over $\mathfrak{T}$]].

DEFINITION. $\mathfrak{S}$ is a $\mathscr{C}$-extension of $\mathfrak{T} \longleftrightarrow \mathfrak{T}$ is a $\mathscr{C}$-subsystem of $\mathfrak{S}$.

DEFINITION. The $\mathscr{C}$-description of $\mathfrak{S}$ = the set of sentences in $\mathscr{C}$ and denials of sentences in $\mathscr{C}$ in a grammar for $\tilde{\mathfrak{S}}$ that are true over $\mathfrak{S}$.

It follows easily that the isomorphs of $\mathscr{C}$-extensions of $\mathfrak{S}$ are just the systems having an expansion which is a model of the $\mathscr{C}$-description of $\mathfrak{S}$. (The understanding is, of course, that $\mathscr{C}$-subsystems and $\mathscr{C}$-extensions of $\mathfrak{S}$ are in the same similarity type as $\mathfrak{S}$.) Since the classes of models we will deal with will all be closed under isomorphism we can use $\mathscr{C}$-descriptions as an effective tool in working with $\mathscr{C}$-extensions.

Important special cases of these concepts arise when $\mathscr{C} = \bigwedge^0_0$ and when $\mathscr{C} = \bigwedge^0$. In the first case we have the familiar notions of *subsystem*, *extension*, and *diagram*, and in the latter those of *elementary subsystem*, *elementary extension*, and *description*.

DEFINITION. $\mathbf{S}^0_k(K) = \{\mathfrak{S} \mid \exists\mathfrak{T}[\mathfrak{T} \in K$ & $\mathfrak{S}$ is an $\bigwedge^0_k$-subsystem of $\mathfrak{T}]\}$.
$\mathbf{S}^0(K) = \{\mathfrak{S} \mid \exists\mathfrak{T}[\mathfrak{T} \in K$ & $\mathfrak{S}$ is an $\bigwedge^0$-subsystem of $\mathfrak{T}]\}$.

DEFINITION. The *νth $\mathbf{S}^0_k$-adjoin of A relative to $A \cup B$*, which we denote by

"$\alpha^0_{k,\nu}(A, B)$", is defined by transfinite induction as follows:[3]

$$\alpha^0_{k,\nu}(A, B) = C_\nu \cap \mathbf{S}^0_k\Big(\bigcap_{\mu<\nu;\ \mu+\nu \text{ odd}} \alpha^0_{k,\mu}(A, B)\Big),$$

where $C_\nu = A$ if ν is even and $C_\nu = B$ if ν is odd.

α^0_ν is defined similarly, using $\mathbf{S}^0$ in place of $\mathbf{S}^0_k$.

$\alpha^0_{*,\nu}$ is defined for finite ν as follows:

$$\alpha^0_{*,\nu}(A, B) = C_\nu \cap \mathbf{S}^0_{\nu-1}\Big(\bigcap_{\mu<\nu;\ \mu+\nu \text{ odd}} \alpha^0_{*,\mu}(A, B)\Big),$$

where $\mathbf{S}^0_{-1}(\mathscr{S}) = \mathscr{S}$. Next we proceed to the statement and proof of several lemmas, some of which may have some independent interest. Several are generalizations of known results, of course, and are close at least in spirit to work of H. J. Keisler.

LEMMA 1. *For any system $\mathfrak{S}$, any set K in $\bigvee^1_{1\delta}$, and any k ($k \geqq 0$): $\mathfrak{S} \notin \mathbf{S}^0_k(K) \longleftrightarrow$ $\mathfrak{S}$ can be separated from K by a $\bigvee^0_{k+1}$ set.*

PROOF. $\rightarrow$: $\mathfrak{S} \notin \mathbf{S}^0_k(K)$ implies that the $\bigwedge^0_k$-description of $\mathfrak{S}$ is inconsistent with the set $\mathscr{K}$ of formulas defining K. Hence, applying the compactness theorem, forming a conjunction, putting the conjunction into prenex normal form, and existentially quantifying the "constants" not in the grammar for $\mathfrak{S}$, we obtain an $\bigvee^0_{k+1}$ formula satisfied by $\mathfrak{S}$ but inconsistent with $\mathscr{K}$.

$\leftarrow$: This follows from the fact that $\bigvee^0_{k+1}$ sets are closed under $\bigwedge^0_k$-extensions, which in turn follows easily from the definition of $\bigwedge^0_k$-extension.

LEMMA 2. *For any sets A, B in $\bigvee^1_{1\delta}$ and any k, n ($k, n \geqq 0$): $\alpha^0_{k,2n}(A, B)$ is the intersection of A with an $\bigwedge^0_{(k+1)\delta}$ set and $\alpha^0_{k,2n+1}(A, B)$ is the intersection of B with an $\bigwedge^0_{(k+1)\delta}$ set.*

PROOF. This is a corollary of Lemma 1.

LEMMA 3. *For any k ($k \geqq 0$): if for all n $\mathfrak{S}_n \in \mathbf{S}^0_k(\mathfrak{S}_{n+1})$, then for all n $\mathfrak{S}_n \in \mathbf{S}^0_k(\bigcup \mathfrak{S}_n)$.*

PROOF. This is easily proved by induction on k.

LEMMA 4. *For any k ($k \geqq 0$) and any $\bigwedge^0_{k+2}$ set K: if for all n $\mathfrak{S}_n \in K$ and $\mathfrak{S}_n \in \mathbf{S}^0_k(\mathfrak{S}_{n+1})$, then $\bigcup \mathfrak{S}_n \in K$.*

PROOF. This follows easily from Lemma 3 by a well-known idea.

Although we do not need it in our discussion, it is perhaps worth noting that this property of being closed under unions of what might be called $\mathbf{S}^0_k$-chains characterizes the $\bigwedge^0_{k+2}$ sets among the $\bigwedge^0$ sets.[4]

[3] If $\mathbf{S}^0_k$ were replaced by $\overline{}$ this would give a different but equivalent formulation of the definition of relative adjoin given in § 2, with the following exception: the present definition does not include at limit ordinals the intersections themselves as adjoins. The present definition seems more natural to us than the analog of the relativization of the refinement of the classical definition of residue.

[4] This has been verified by James W. Thatcher.

Now we consider some applications of these lemmas.

Theorem 1. *For any two $\bigvee_1^1$ sets A, B and any k, n $(k, n \geqq 0)$: $\alpha^0_{k,n}(A, B) = \emptyset$ $\longleftrightarrow$ A can be separated from B by the sum of a decreasing alternating series of length n of $\bigwedge^0_{k+1}$ sets.*

Proof. $\rightarrow$: A straightforward computation based on n applications of both Lemma 1 (used in general many times at each application) and the compactness theorem shows that A is separable from B, or B is separable from A (according as n is even or odd), by a set of the form $E_1 - (E_2 - (\cdots - E_n)\cdots)$ for $E_i \in \bigvee^0_{k+1}$. Lemma 2 is needed to assure the applicability of Lemma 1. A straightforward Boolean computation then completes the proof.

$\leftarrow$: This is easily proved by induction on n, using the fact that $\bigwedge^0_{k+1}$ sets are closed under $\mathbf{S}^0_k$.

Corollary 1. *For any $\bigvee_1^1 \cap \bigwedge_1^1$ set A and any k, n $(k, n \geqq 0)$: the nth $\mathbf{S}^0_k$-adjoin of A is empty (i.e. $\alpha^0_{k,n}(A, -A) = \emptyset$) $\longleftrightarrow$ A is the sum of a decreasing alternating series of length n of $\bigwedge^0_{k+1}$ sets.*

The case $k = 0$, $n = 1$ of this corollary is a well-known theorem of Tarski.

Theorem 2. *For any two $\bigvee^1_{1\delta}$ sets A, B, any k $(k \geqq 0)$, and any infinite even ordinal ν:*

$$\bigcap_n \alpha^0_{k,2n}(A, B) = \mathbf{S}^0_k\Big(\bigcap_n \alpha^0_{k,2n+1}(A, B)\Big) = \alpha^0_{k,\nu}(A, B)$$

and

$$\bigcap_n \alpha^0_{k,2n+1}(A, B) = \mathbf{S}^0_k\Big(\bigcap_n \alpha^0_{k,2n}(A, B)\Big) = \alpha^0_{k,\nu+1}(A, B)\,.$$

Proof. If $\mathfrak{S} \in \bigcap_n \alpha^0_{k,2n}(A, B)$, then $\mathfrak{S}$ has an $\bigwedge^0_k$-extension in $\alpha^0_{k,2n+1}(A, B)$ for every n. Hence its $\bigwedge^0_k$-description has a model in $\alpha^0_{k,2n+1}(A, B)$ for every n, so using Lemma 2 we can apply the compactness theorem to conclude $\mathfrak{S}$ has an $\bigwedge^0_k$-extension in $\bigcap_n \alpha^0_{k,2n+1}(A, B)$. Hence $\bigcap_n \alpha^0_{k,2n}(A, B) \subseteqq \mathbf{S}^0_k(\bigcap_n \alpha^0_{k,2n+1}(A, B))$. Arguing similarly, $\bigcap_n \alpha^0_{k,2n+1}(A, B) \subseteqq \mathbf{S}^0_k(\bigcap_n \alpha^0_{k,2n}(A, B))$. The desired equations follow easily from these inclusions and the definitions.

Theorem 3. *For any k $(k \geqq 0)$: for any disjoint $\bigwedge^0_{k+2}$ sets A, B there exists an n such that $\alpha^0_{k,n}(A, B) = \emptyset$.*

Proof. If there were no such n, $\bigcap_n \alpha^0_{k,2n}(A, B)$ would be, by the compactness theorem (which is applicable by Lemma 2), nonempty; so picking an element $\mathfrak{S}_1$ we could find by Theorem 2 an $\bigwedge^0_k$-extension $\mathfrak{T}_1$ of $\mathfrak{S}_1$ in B, an $\bigwedge^0_k$-extension $\mathfrak{S}_2$ of $\mathfrak{T}_1$ in A, etc. Then $\bigcup \mathfrak{S}_i = \bigcup \mathfrak{T}_i \in A \cap B$ by Lemma 4, contradicting the disjointness of A, B.

From Theorems 1 and 3 we now have

Theorem 4. *For any k $(k \geqq 0)$ and any two disjoint $\bigwedge^0_{k+2}$ sets A, B: the least n such that $\alpha^0_{k,n}(A, B) = \emptyset$ and the least n such that A is separable from B by the sum of a decreasing alternating series of length n of $\bigwedge^0_{k+1}$ sets both exist and are equal.*

We call this least n for a given pair of disjoint $\bigwedge_k^0$ sets A, B the *index of separability of A from B* (as $\bigwedge_k^0$ sets).

Corollary 2. *For any k ($k \geqq 0$) and any $\bigvee_{k+2}^0 \cap \bigwedge_{k+2}^0$ set A: the least n such that the nth $\mathbf{S}_k^0$-adjoin of A is empty and the least n such that A is the sum of a decreasing alternating series of length n of $\bigwedge_{k+1}^0$ sets both exist and are equal.*

Corollary 3 (Shoenfield). *For any k ($k \geqq 2$): the strong separation principle and the first separation principle for $\bigwedge_k^0$ hold; the construction principle for $\bigvee_k^0 \cap \bigwedge_k^0$ holds.*

It is perhaps worth noting that the separation theorems, like the Craig separation theorem, can be reformulated as interpolation theorems. Thus if we have $\models \Gamma_1 \supset \Gamma_2$, where $\Gamma_1 \in \bigwedge_k^0$, $\Gamma_2 \in \bigvee_k^0$ ($k \geqq 2$) and Γ_1, Γ_2 contain the same constants, we know we can interpolate between them a formula Δ with the same constants that is a propositional combination of $\bigwedge_{k-1}^0$ formulas, i.e. we can find such a Δ such that $\models \Gamma_1 \supset \Delta$ and $\models \Delta \supset \Gamma_2$.

5. The Craig separation theorem. We now present a treatment of Craig's theorem that follows very closely the pattern developed in the preceding section.

Theorem 1′. *For any two $\bigvee_1^1$ sets A, B and any n ($n \geqq 0$): $\alpha_{*,n}^0(A, B) = \varnothing$ $\longleftrightarrow$ [n is odd and A can be separated from B by an $\bigwedge_n^0$ set] or [n is even and A can be separated from B by an $\bigvee_n^0$ set].*

Proof. Similar to that of Theorem 1. An appropriate variant of Lemma 2, which follows from Lemma 1 (using all k), is used in place of Lemma 2. Because $\bigwedge_{k+1}^0 \supseteqq \bigvee_k^0 \cup \bigwedge_k^0$ the final form of the separating set simplifies in this case.

Corollary 1′. *For any $\bigvee_1^1 \cap \bigwedge_1^1$ set A and any n ($n \geqq 0$): $\alpha_{*,n}^0(A, -A) = \varnothing$ $\longleftrightarrow$ [n is odd and $A \in \bigwedge_n^0$] or [n is even and $A \in \bigvee_n^0$].*

Theorem 2′. *For any two $\bigvee_{1S}^1$ sets A, B: $\bigcap_n \alpha_{*,2n}^0(A, B) = \mathbf{S}^0(\bigcap_n \alpha_{*,2n+1}^0(A, B))$ and $\bigcap_n \alpha_{*,2n+1}^0(A, B) = \mathbf{S}^0(\bigcap_n \alpha_{*,2n}^0(A, B))$.*

Proof. Similar to that of Theorem 2.

Theorem 3′. *For any two disjoint $\bigvee_1^1$ sets A, B there exists an n such that $\alpha_{*,n}^0(A, B) = \varnothing$.*

Proof. Similar to that of Theorem 3. The only difference is that in picking $\mathfrak{S}_1, \mathfrak{T}_1, \mathfrak{S}_2, \cdots$ some care must be exercised. Let the sets A, B be defined by the formulas $\bigvee G \Gamma_1(F, G)$, $\bigvee H \Gamma_2(F, H)$, respectively, where Γ_1, Γ_2 are elementary. Then, for example, $\mathfrak{T}_2$ must be chosen so that it is an elementary extension of $\mathfrak{S}_2$ and so that it has an expansion $\mathfrak{T}_2^*$ which is both a model of Γ_2 and an elementary extension of the (previously fixed) expansion $\mathfrak{T}_1^*$ of $\mathfrak{T}_1$ satisfying Γ_2. Such a $\mathfrak{T}_2^*$ is obtained as a model of the $\bigwedge^0$-descriptions of $\mathfrak{S}_2$ and $\mathfrak{T}_1^*$. That these $\bigwedge^0$-descriptions must have a model follows from the compactness theorem, using the existence of $\mathfrak{T}_1^*$. The argument is concluded by applying the well-known Lemma 4′ (obtained by dropping the subscripts from "$\bigwedge_{k+2}^0$" and "$\mathbf{S}_k^0$" in Lemma 4) to the sequences of expansions $\mathfrak{S}_1^*, \mathfrak{S}_2^*, \cdots$

and $\mathfrak{T}_1^*, \mathfrak{T}_2^*, \cdots$ to obtain a contradiction as before.

The reader will recognize here, of course, the basic idea of the proof of the Robinson consistency theorem (cf. [4]).

From Theorems 1′ and 3′ we now have

THEOREM 4′. *For any two disjoint* $\bigvee_1^1$ *sets* A, B: *the least* n *such that* $\alpha^0_{*,n}(A, B) = \varnothing$ *and the least* n *such that* A, B *are separable by an* $\bigwedge^0_n$ *set* (A *from* B *if* n *is odd,* B *from* A *if* n *is even*) *both exist and are equal.*

We call this least n for a given pair of disjoint $\bigvee_1^1$ sets A, B their *index of separability* (as $\bigvee_1^1$ sets).

COROLLARY 3′ (CRAIG).

(1) *Strong separation theorem for* $\bigvee_1^1$. *Every two disjoint* $\bigvee_1^1$ *sets can be separated by a set in* (*the Boolean algebra over*) $\bigwedge^0$.

(2) *First separation theorem for* $\bigvee_1^1$. *Every two disjoint* $\bigvee_1^1$ *sets can be separated by an* $\bigvee_1^1 \cap \bigwedge_1^1$ *set.*

(3) *Construction theorem for* $\bigvee_1^1 \cap \bigwedge_1^1$. $\bigvee_1^1 \cap \bigwedge_1^1 =$ (*the Boolean algebra over*) $\bigwedge^0$.

In closing we note that the proof of this corollary could have been carried out using α^0 instead of α^0_*. But in this case the derivation of the index of separability is lost.

BIBLIOGRAPHY

1. J. W. Addison, *Separation principles in the hierarchies of classical and effective descriptive set theory*, Fund. Math. vol. 46 (1959) pp. 123-135.

2. ———, *The theory of hierarchies*, Proceedings of the International Congress for Logic, Methodology and Philosophy of Science, Palo Alto, Stanford University Press, forthcoming.

3. F. Hausdorff, *Set theory*, New York, Chelsea, 1957, 352 pp.

4. A. Robinson, *A result on consistency and its application to the theory of definition*, Indag. Math. vol. 18 (1956) pp. 47-58.

THE UNIVERSITY OF MICHIGAN,
ANN ARBOR, MICHIGAN

THE FORM OF THE NEGATION OF A PREDICATE

BY

J. R. SHOENFIELD

The Kleene hierarchies classify certain predicates[1] arising in recursive function theory. Roughly speaking, these are the predicates which may be written in prenex form with a recursive matrix. The forms in the hierarchies then correspond to the forms which the prefix may assume.

The conjunction or disjunction of two predicates in a given form in one of the hierarchies is again in that form; this follows from the rules for bringing an expression to prenex form, together with certain simple principles on contraction of quantifiers. For a negation, the rules for prenex form lead to quite a different result; the negation of a predicate in a given form is in the dual form, where the dual form is obtained by changing universal quantifiers to existential quantifiers and vice versa.

The question then arises: given a form, what predicates in that form also have their negations in that form? Certainly not all predicates in that form; a counterexample is provided by the diagonal argument, using the appropriate enumeration theorem for the given form. The study of the above question has led to some of the most interesting results in hierarchy theory.

The simplest result is the well-known theorem: a recursively enumerable predicate has a recursively enumerable negation if and only if it is recursive. Somewhat deeper is Post's theorem, which answers the question for the entire arithmetical hierarchy:

A predicate in a given $(n+1)$-*quantifier form in the arithmetical hierarchy has its negation in that form if and only if it is recursive in predicates in* n-*quantifier forms.*

In the analytical hierarchy, the problem is much more difficult. The answer at the lowest level is given by the following theorem of Kleene [**3**]:

A predicate in a 1-*quantifier form*[2] *in the analytical hierarchy has its negation in that form if and only if it is hyperarithmetical.*

This suggests a plausible conjecture for the remaining forms in the analytical hierarchy. We obtain it from Post's theorem by replacing "arithmetical" by "analytical" and "recursive in" by "hyperarithmetical in". The if part of this conjecture is verified without difficulty. However, Addison and

Received by the editor April 7, 1961.

[1] By predicate, we shall mean predicate of natural numbers unless otherwise indicated. Many (but not all) of the results discussed have extensions to predicates of function variables or higher type variables.

[2] In counting quantifiers in a form in the analytical hierarchy, we count only the quantifiers on function variables.

Kleene [2] have shown that the only if part fails for $n = 1$. In fact, they prove the following result:

Any predicate which is in a 1-quantifier form relative to prèdicates in both 2-quantifier forms[3] *is itself in both 2-quantifier forms.*

Now the diagonal argument produces a predicate which is in one of the 1-quantifier forms relative to predicates in 1-quantifier form but is not in the other 1-quantifier form relative to predicates in 1-quantifier form. By the above result, this predicate is in both 2-quantifier forms; but it is not hyperarithmetical in predicates in 1-quantifier form.

The Kleene-Addison result can be extended as follows:

Any predicate which is in a n-quantifier form relative to predicates in both $(n+1)$*-quantifier forms is itself in both* $(n+1)$*-quantifier forms.*

For example, let $n = 2$. Suppose that $P(x) \equiv (\alpha)(E\beta)(y)V(\alpha, \beta, y, x)$, where V is recursive in Q and Q is in both 3-quantifier forms. Letting τ be the representing function of Q, we then have $P(x) \equiv (\alpha)(E\beta)(y)R(\tau, \alpha, \beta, y, x)$ with R recursive. From this we obtain

$$P(x) \equiv (\gamma)(\gamma = \tau \supset (\alpha)(E\beta)(y)R(\gamma, \alpha, \beta, y, x)) , \tag{1}$$

$$P(x) \equiv (E\gamma)(\gamma = \tau \ \& \ (\alpha)(E\beta)(y)R(\gamma, \alpha, \beta, y, x)) . \tag{2}$$

Now

$$\gamma = \tau \equiv (x)(\gamma(x) \leqq 1 \ \& \ (\gamma(x) = 0 \equiv Q(x))) .$$

Hence $\gamma = \tau$ is arithmetical in Q, and therefore can be written in the 3-quantifier form with existence first [4, Theorem 5]. Writing it in this form in (1) and (2) and then bringing quantifiers to the front and contracting in the usual manner, we obtain expressions for P in both 3-quantifier forms.

This extension is relevant to conjectures such as the following: for $n \geqq 1$, a predicate in a given $(n+1)$-quantifier form has its negation in that form if and only if it is in both 2-quantifier forms relative to predicates in n-quantifier forms. The extension verifies the if part, and, as above, furnishes a counter-example to the only if part for $n = 2$.

A new possibility for solving our problem in the analytical hierarchy is given by Kleene's theory of recursive functions of higher types [5]. Kleene defines a type 2 object $\mathbf{E}$ by

$$\begin{aligned} \mathbf{E}(\alpha) &= 0 \text{ if } (Ex)(\alpha(x) = 0) , \\ &= 1 \text{ otherwise} . \end{aligned}$$

He then shows that a predicate is hyperarithmetical (or, equivalently, in both 1-quantifier forms) if and only if it is recursive in $\mathbf{E}$.

By analogy, we define a type 2 object $\mathbf{A}$ by

$$\begin{aligned} \mathbf{A}(\alpha) &= 0 \text{ if } (\beta)(Ex)(\alpha(\bar{\beta}(x)) = 0) , \\ &= 1 \text{ otherwise} . \end{aligned}$$

[3] A predicate is in 1-quantifier form relative to P if it has a prenex form with one function quantifier in which the matrix is recursive in P.

Tugué [7] has shown that any predicate recursive in **A** is in both 2-quantifier forms.[4] The converse is a tempting conjecture; but it also proves to be false. To prove this, let $(E\alpha)_{\mathbf{A}}$ represent a quantifier over all functions recursive in **A**. Then we have:

Any predicate of the form $(E\alpha)_{\mathbf{A}}(\beta)(Ey)R(\alpha, \beta, y, x)$ *with R recursive can be written in the form* $(\alpha)(E\beta)(y)S(\alpha, \beta, y, x)$ *with S recursive.*[5]

PROOF. First note that $(E\alpha)_{\mathbf{A}}(\beta)(Ey)R(\alpha, \beta, y, x)$ can be written

(3) $(Ee)((z)(\{e\}(\mathbf{A}, z) \text{ is defined}) \;\&\; (\alpha)[(z)(\{e\}(\mathbf{A}, z) = \alpha(z)) \supset (\beta)(Ey)R(\alpha, \beta, y, x)]$.

Now by [5, XXVII],

$$\{e\}(\mathbf{A}, z) \text{ is defined } \equiv (\beta)(Ey)K(e, z, \mathbf{A}, \beta, y)$$

with K recursive. By Tugué's result, $K(e, z, \mathbf{A}, \beta, y)$, as a predicate of e, z, β, y, can be written in both 2-quantifier forms. Hence "$\{e\}(\mathbf{A}, z)$ is defined" can be written in the 2-quantifier form beginning with a universal quantifier. Now assume $\{e\}(\mathbf{A}, z)$ is defined for all z. By [5, XXVI],

$$\{e\}(\mathbf{A}, z) = \alpha(z) \equiv (E\beta)(y)J(e, z, \mathbf{A}, \alpha(z), \beta, y) ,$$

with J recursive. The same argument as above then shows that $\{e\}(\mathbf{A}, z) = \alpha(z)$ can be written in the 2-quantifier form beginning with an existential quantifier. If we make these two substitutions in (3) and bring quantifiers to the front and contract, we obtain the 2-quantifier form beginning with a universal quantifier.

To get our counterexample, we now need the following result of Addison [1]: the functions recursive in predicates in both 2-quantifier forms are a basis for predicates of the form $(\beta)(Ex)R(\alpha, \beta, x)$ with R recursive. If the conjecture were correct, the functions recursive in **A** would also form such a basis, and hence the subscript A in $(E\alpha)_{\mathbf{A}}(\beta)(Ey)R(\alpha, \beta, y, x)$ could be omitted. The above result would tell us that any predicate in one of the 2-quantifier forms is also in the other 2-quantifier form; and this would contradict the hierarchy theorem.

The problem of characterizing those predicates such that both they and their negations are in a given 2-quantifier form in the analytical hierarchy thus remains open; and we are left without any very plausible conjecture on what such a characterization might be.

We close with a few remarks suggested by the above.

(1) It is now a familiar fact that the analytical hierarchy is an analogue of the projective hierarchy of point set topology. Thus Kleene's theorem quoted above is an analogue of Souslin's theorem stating that an analytic set has an analytic complement if and only if it is a Borel set. The problem considered above is analogous to the following problem: characterize those sets which are both *PCA* sets and *CPCA* sets. This problem has resisted the efforts of many topologists.

(2) The predicates recursive in **A** form a large subclass of the predicates

[4] This result holds also for predicates of function variables.

[5] This result is an analogue of Lemma 1 of [6].

in both 2-quantifier forms. Thus Tugué [7] shows that they include the hierarchy of sets considered by Addison and Kleene [2]. It would be of considerable interest to extend this Addison-Kleene hierarchy to include all the predicates recursive in **A**. In particular, it seems likely that this would lead to an interesting class of ordinals.

(3) The problem discussed here can, of course, be formulated in the Kleene hierarchies based on quantification of higher type variables, and there are again some plausible conjectures. We discuss the hierarchy based on quantification of type 2 variables briefly.

Define a type 3 object $\mathbf{E}^3$ by setting $\mathbf{E}^3(\alpha^2) = 0$ if $(E\beta^1)(\alpha^2(\beta^1) = 0)$ and $\mathbf{E}^3(\alpha^2) = 1$ otherwise. Following a suggestion of Kleene, we call predicates recursive in $\mathbf{E}^3$ *hyperanalytical*. It is not difficult to show that analytical predicates are hyperanalytical, and that hyperanalytical predicates can be written in both 1-quantifier forms (that is, one quantifier over type 2 variables). The converse of the last statement seems a plausible conjecture.

For predicates of number variables, the above methods yield no information about the conjecture. However, the conjecture can equally well be formulated for predicates of function variables (i.e., type 1 variables). Here the same methods as above lead to a counterexample, except that the necessary basis result (also due to Addison) requires the axiom of constructibility. It seems likely that further clarification of the problem must await more information on Kleene's recursive functionals of higher type variables.

REFERENCES

1. J. W. Addison, *Hierarchies and the axiom of constructibility*, Summaries of talks presented at the Summer Institute of Symbolic Logic in 1957 at Cornell University, vol. 3, pp. 355-362.

2. J. W. Addison and S. C. Kleene, *A note on function quantification*, Proc. Amer. Math. Soc. vol. 8 (1957) pp. 1002-1006.

3. S. C. Kleene, *Hierarchies of number-theoretic predicates*, Bull. Amer. Math. Soc. vol. 61 (1955) pp. 193-213.

4. ———, *Arithmetical predicates and function quantifiers*, Trans. Amer. Math. Soc. vol. 79 (1955) pp. 312-340.

5. ———, *Recursive functionals and quantifiers of finite types*. I, Trans. Amer. Math. Soc. vol. 91 (1959) pp. 1-52.

6. ———, *Quantification of number-theoretic functions*, Compositio Math. vol. 14 (1959) pp. 23-40.

7. Tosiyuki Tugué, *Predicates recursive in a type-2 object and Kleene hierarchies*, Comment. Math. Univ. St. Paul vol. 8 (1959) pp. 97-117.

DUKE UNIVERSITY,
DURHAM, NORTH CAROLINA

APPLICATIONS OF RECURSIVE FUNCTION THEORY TO NUMBER THEORY

BY

MARTIN DAVIS

As a sample application of recursive function theory to the theory of numbers, let me state the following theorem:

THEOREM. *There is a function* $P(x_1, \cdots, x_n, t, 2^t)$ *where* P *is a polynomial with integer coefficients whose range* (*that is the set of all values assumed by it as* $x_1, \cdots, x_n, t$ *take on all positive integral values*), *consists precisely of* 0, *the negative integers, and the positive primes.*

Unfortunately it is not to be expected that any deep properties of the prime numbers will follow from this "prime-representing function." In fact, the result has very little to do with the prime numbers and holds as we shall shortly see for every set of positive integers which has the property which in recursive function theory is called: recursive enumerability. A set S of positive integers is called *recursively enumerable* if there is an algorithm or computing procedure which applied to a given positive integer n will eventually terminate if and only if that integer n is a member of the set S. The vagueness resulting from the use here of such apparently ill-defined terms as "algorithm" or "computing procedure" is eliminated in the formal development of the theory of recursive functions where these notions receive precise elucidation.[1] However, for the purposes of this talk, the intuitive notion will do perfectly well. The class of all recursively enumerable sets will be denoted by $\mathscr{R}$. A set S such that $S \in \mathscr{R}$ and $\bar{S} \in \mathscr{R}$, where $\bar{S}$ here and below stands for the complement of S with respect to the set of all positive integers, is called *recursive*. For a recursive set, by making use of the two algorithms for S and $\bar{S}$, once can determine for any given integer n, in a finite number of steps, whether or not n belongs to the set S. Conversely, as is immediately seen, any set S for which this last is possible is recursively enumerable as is its complement. The main result from recursive function theory which will be employed here is the theorem:[2]

THEOREM. *There exists a recursively enumerable set* K *which is not recursive, i.e.* $\bar{K}$ *is not recursively enumerable.*

This last theorem provides the existence of perfectly definite problems which cannot be solved by means of any algorithm, i.e. which are recursively unsolvable. Such results as the unsolvability of the decision problem for quantification theory or the unsolvability of the word problem for groups were obtained, essentially, by making use of just this result. A problem which has attracted considerable interest in this connection is the 10th on

Received by the editor April 7, 1961.

[1] Cf. [**3**],

[2] Cf., e.g. [**3**, p. 75].

Hilbert's famous list, which may be stated as follows:

Give an algorithm for determining whether or not a given polynomial Diophantine equation has a solution.

Efforts to prove this problem recursively unsolvable have tended to reverse the usual way of looking at a Diophantine equation. Ordinarily one thinks of a Diophantine equation as given, and seeks information about its set of solutions. Here we reverse this procedure and think of the solution set, or rather one of its projections, as being given and seek to determine a corresponding equation. Thus, a set S is called *Diophantine* if there is a polynomial $P(n, x_1, \cdots, x_m)$ with integer coefficients such that

$$S = \{n \mid (En, x_1, \cdots, x_m)[P(x_1, \cdots, x_m) = 0]\} \tag{i}$$

that is, S is the projection of the set of positive lattice points on the algebraic manifold determined by $P = 0$, on an axis. If one permits variable exponents in the Diophantine equation $P = 0$, one calls it an exponential Diophantine equation, and then, the corresponding set S will be called *exponential-Diophantine*. The class of all Diophantine sets is denoted by $\mathscr{D}$, and the class of all exponential-Diophantine sets, by $\mathscr{E}$. It is then clear at once that:

$$\mathscr{D} \subset \mathscr{E} \subset \mathscr{R} .$$

The main interest here is in the converse inclusions. In particular, if it could be shown that $\mathscr{D} = \mathscr{R}$, then it would follow that the nonrecursive set K mentioned above, belongs to $\mathscr{D}$ and hence that Hilbert's 10th problem is recursively unsolvable.[3] Unfortunately, this last conjecture remains an open question. However, it is now known that:[4]

THEOREM: $\mathscr{E} = \mathscr{R}$.

This result was first obtained by Hilary Putnam and myself making use of the unproved hypothesis that there are arbitrarily long arithmetic progressions of primes. Later Julia Robinson showed that this hypothesis could be dispensed with. From this last theorem it follows at once that the set K is exponential-Diophantine and that therefore there can be no algorithm for testing exponential-Diophantine equations for solvability. The result mentioned at the beginning of this talk follows from the fact that every exponential-Diophantine set can be represented in the particular form there employed.

The question of the equality of $\mathscr{D}$ and $\mathscr{R}$ thus reduces to that of the equality between $\mathscr{D}$ and $\mathscr{E}$. This question however, was discussed about a decade ago by Julia Robinson.[5] She defined $x * n$ by the equations:

$$x * 1 = x , \quad x * (n + 1) = x^{(x * n)} .$$

She then considered the following hypothesis which we shall call J. R.:

There is a Diophantine set S of ordered pairs such that:

[3] Cf. [**2**], or [**3** Chapter 7].

[4] Cf. [**5**].

[5] Cf. [**9**].

(1) For some n, $(u, v) \in S$ implies $v \leqq (u*n)$

(2) For each k, there are u, v, such that $(u, v) \in S$ and $v > u^k$.

Julia Robinson proved that J. R. implies $\mathscr{D} = \mathscr{E}$. Hence by the principal result stated above, J. R. implies $\mathscr{D} = \mathscr{R}$ and therefore that Hilbert's 10th problem is unsolvable. Moreover it is very easy to see that $\mathscr{D} = \mathscr{R}$ in turn implies J. R. so that the problem of proving $\mathscr{D} = \mathscr{R}$ is seen to be entirely equivalent to that of proving J.R. On the other hand, if one could disprove $\mathscr{D} = \mathscr{R}$ for example by constructing a particular set of integers which was recursively enumerable but not Diophantine, then one would prove that J.R. is false[6] which would be a rather remarkable general result about bounds on the solutions of Diophantine equations. I should like to state various unproved propositions about particular second and third degree Diophantine equations which can be shown to imply J. R. The fact that these assertions do imply J. R. has been proved by Hilary Putnam and myself. Each of these assertions implies the unsolvability of Hilbert's 10th problem and, from another point of view, to prove the existence of a recursively enumerable set which is not Diophantine is to prove the falsity of each of these propositions about Diophantine equations.

PROPOSITION 1. *For each k, there are x, y, m such that $x^3 - my^3 = 1$, and $x > m^k$.*

That Proposition 1 implies J. R. is easily seen. For, Proposition 1 states precisely that the set of ordered pairs (m, x) satisfying the cubic equation $x^3 - my^3 = 1$ obeys condition (2) of J. R. However, by a result of Harvey Cohn[7] condition (1) of J. R. holds with $n = 3$.

PROPOSITION 2. *Let (a_n, a_n') be the successive solutions of a Pell equation $x^2 - dy^2 = 1$. Then, for at least one value of d, a_n' cannot be written in the form $r^2 + ds^2$ if n is not a power of* 2.

That Proposition 2 implies J. R. is proved by showing that Proposition 2 implies that the set of a_n' for which n is a power of 2 is Diophantine and that this in turn implies J.R.

The next proposition involves the Diophantine equation $Ax^3 + By^3 = C$, where $C = 1$ or $C = 3$. Let ξ be the fundamental unit of the algebraic number domain $R(AB^2)^{1/3}$. Then by a result of T. Nagell[8] it follows that for a solution

$$\frac{1}{C}[xA^{1/3} + yB^{1/3}]^3 = \xi^{2^n} \tag{ii}$$

for some value of n.

PROPOSITION 3. *For arbitrarily large values of n, values of A, B, C ($C = 1$, or $C = 3$), x, y, with $Ax^3 + By^3 = C$ exist satisfying relation* (ii).

[6] To disprove J. R., it would, as is easily seen, also suffice to prove that there are Diophantine sets of arbitrarily large dimension. Here the *dimension* of a Diophantine set S is the least value of m for which (i) can be made to hold.

[7] Cf. [**1**].

[8] Cf. [**7**].

I would like to conclude by mentioning a rather striking theorem about the order of growth of solutions of Diophantine equations which is a corollary of the main result stated above and was first proved by Julia Robinson:[9]

THEOREM. *Consider Diophantine equations* $P(x, y, u_1, \cdots, u_m) = 0$ *such that to every* x *there are only a finite number of values of* y *for which* $P = 0$ *has a solution. If to every such equation* $P = 0$, *there are* k *and* n *so that every solution of* $P = 0$ *satisfies* $y < k(x*n)$, *then to every such equation there are* K' *and* N' *so that each solution satisfies* $y < K'x^{N'}$.

BIBLIOGRAPHY

1. Harvey Cohn, *Some algebraic number theory estimates based on the Dedekind eta-function*, Amer. J. Math. vol. 78(1956) pp. 791-796.

2. Martin Davis, *Arithmetical problems and recursively enumerable predicates*, J. Symb. Logic vol. 18 (1953) pp. 33-41.

3. ———, *Computability and unsolvability*, New York, McGraw-Hill, 1958.

4. Martin Davis and Hilary Putnam, *Reduction of Hilbert's tenth problem*, J. Symb. Logic vol. 23 (1958) pp. 183-187.

5. Martin Davis, Hilary Putnam and Julia Robinson, *The decision problem for exponential Diophantine equations*, Ann. of Math., vol. 74 (1961) pp. 425-436.

6. David Hilbert, *Mathematische probleme. Vortrag, gehalten auf dem internationalen Mathematiker-Kongress zu Paris* 1900, Nachrichten von der K. Gesellschaft der Wissenschaften zu Göttingen, Math. -Phys. Kl. 1900, pp. 253-297. Reprinted, Archiv der Mathematik und Physik, vol. 1 (1901) pp. 44-63, 213-237. English translation, Bull. Amer. Math. Soc. vol. 8 (1901-1902) pp. 437-479.

7. T. Nagell, *Solution complète de quelques équations cubiques à deux indéterminées*. Journal de Mathématique 9e serie, vol. 4 (1925) pp. 209-270.

8. Hilary Putnam, *An unsolvable problem in number theory*, submitted to J. Symb. Logic.

9. Julia Robinson, *Existential definability in arithmetic*, Trans. Amer. Math. Soc. vol. 72 (1952) pp. 437-449.

YESHIVA UNIVERSITY,
NEW YORK, NEW YORK

[9] Cf. [5].

SEQUENCE GENERATORS AND DIGITAL COMPUTERS[1]

BY

A. W. BURKS AND J. B. WRIGHT

1. INTRODUCTION

1.1. **Sequence generators.** The basic concept of this paper, that of sequence generator, is a generalization of the concepts of digital computer, finite automaton, logical net, and other information-processing systems. In this subsection, we will define sequence generator and some related concepts and will illustrate them immediately thereafter.

DEFINITIONS. A *sequence generator* $\Gamma = (S, G, R, P^1, \cdots, P^n)$ consists of a set S (whose elements are called *complete states*), a set G (whose elements are called *generators*), a binary relation R (called the *direct transition relation*), and functions $P^1, \cdots, P^n$ (called *projections*), for some $n = 0, 1, 2, 3, \cdots$, satisfying the conditions: (1) S is finite, (2) G is a subset of S, (3) R is defined on S, and each P^i (for $i = 1, 2, \cdots, n$) is also defined on S. The values of the function P^i, which may be entities of any kind, are called P^i-states.

A sequence generator may be represented by a finite-directed graph whose vertices denote complete states and whose arrows indicate when the direct transition relation holds between two states. In our diagrams, we will use rectangles at those vertices which represent generator states and circles at vertices representing complete states which are not also generators; the names of complete states and of P-states are written in the circles and rectangles (see Figs. 1(b), 2(b), 3, etc.). Though our diagrams are closely related to the usual state diagrams (transition diagrams) employed to represent automata (see, for example, Moore [**22**, p. 134]), there are very significant differences. The vertices (nodes) of our diagrams represent complete states, while in the usual state diagrams, the nodes represent internal states. This difference results from the fact that in sequence generators complete states are basic and input and output states are derived from complete states by means of

---

Received by the editor April 7, 1961.

[1] This research was supported by the U.S. Army Signal Corps through Project MICHIGAN (Contract No. DA-36-039-sc-52654), by the U.S. Army Office of Ordnance Research (Contract No. DA-20-018-ORD-16971), and by the U.S. Navy Office of Naval Research (Contract No. Nonr 1224 (21)).

Some of the material in this paper was presented at the International Conference on Information Processing, UNESCO, Paris, 15–20 June 1959. An abstract and report of the discussion was published in the *Proceedings*, Paris, UNESCO, 1960, p. 425.

This paper grew out of some researches on well-behaved nets (see § 3.1); Hao Wang participated in these early investigations and supplied an essential part of the proof of Lemma 3.3-1 for the case of well-behaved nets.

J. Richard Büchi has made many helpful suggestions during the course of our work.

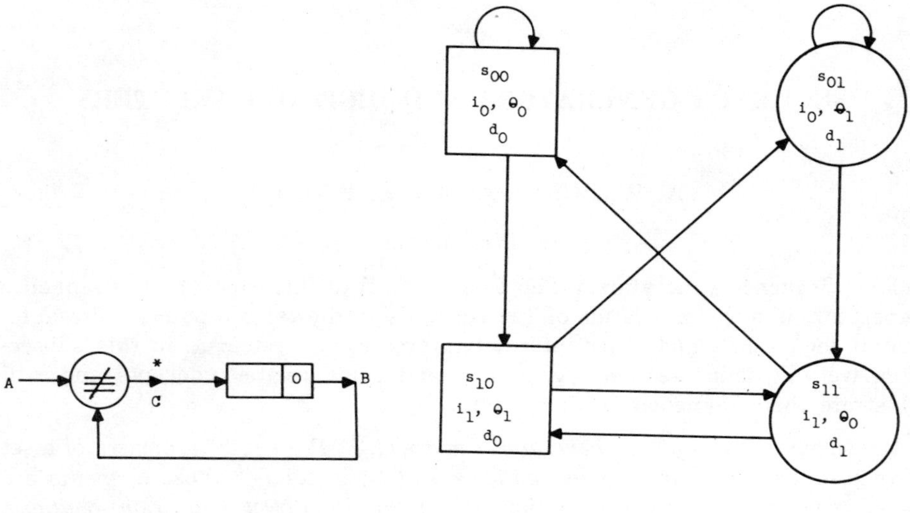

(a) (b)

Figure 1.

(a) Binary counter.

(b) Three-projection sequence generator $\Gamma = (S, G, R, I, \theta, D)$ associated with the binary counter (a).

projections, while in the usual approach complete states are derived by compounding internal states and input states. (The latter process is explained in § 1.2; we will discuss the relation of the two approaches further in § 2.1.)

Some comments and explanations concerning the definition of sequence generator may be helpful. If $n = 0$, then $\Gamma = (S, G, R)$ is a sequence generator with no projections. Though our definition of sequence generator permits any number of projections, in this paper we will be mainly interested in sequence generators with zero, one or two projections. Furthermore, the set of complete states S of a sequence generator may be a null set; in this case the domain of definition of each function P^i will be empty. It is worth noting that essentially (but not quite) the same concept of sequence generator can be obtained without using the set S of complete states in the definition and then defining S to be the union of G and the field of R.

We will use $[\alpha](j, k)$ (where j is a non-negative integer, k is a non-negative integer or $k = \omega$; $j \leqq k$) to denote the sequence $\langle \alpha(j), \alpha(j+1), \cdots, \alpha(k) \rangle$ when k is finite and the sequence $\langle \alpha(j), \alpha(j+1), \alpha(j+2), \cdots \rangle$ when $k = \omega$. If P is a projection $P([\alpha](j, k))$ abbreviates the sequence $\langle P(\alpha(j)), P(\alpha(j+1)), \cdots, P(\alpha(k)) \rangle$ when k is finite and the sequence $\langle P(\alpha(j)), P(\alpha(j+1)), P(\alpha(j+2)), \cdots \rangle$ when $k = \omega$.

DEFINITIONS: Let $\Gamma = (S, G, R, P^1, \cdots, P^n)$ be a sequence generator and let k be a non-negative integer or ω. $[s](0, k)$ is a Γ-sequence if (1) $s(0) \in G$ and (2) for each j, $j < k$, $R(s(j), s(j+1))$. A complete state s is {*Γ-accessible*} [Γ-

(a)

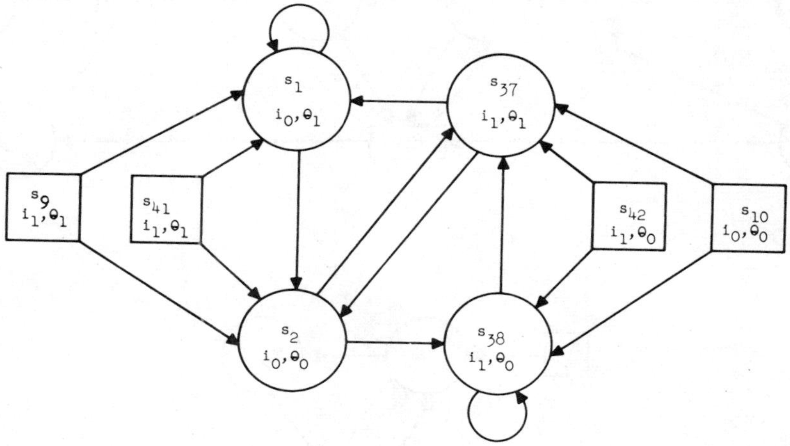

(b)

Figure 2.
(a) Ill-formed net with behavior $F(t) \equiv \sim E(t+1)$.
(b) Sequence generator $\Gamma = (S, G, R, I, \theta)$ associated with net (a).

admissible] if s occurs in some {——} [infinite] Γ-sequence.[2]

These concepts may be illustrated by reference to the direct transition diagram of Fig. 3(a). The sequence $\langle s_7, s_8 \rangle$ is a Γ-sequence, while the sequence $\langle s_3, s_4, s_5, s_6, s_4, s_6, s_4, s_6, s_4, s_6, \cdots \rangle$ is an infinite Γ-sequence. Complete states s_7, s_8, and s_9 are Γ-accessible but not Γ-admissible; complete states s_3, s_4, s_5, and s_6 are Γ-accessible and Γ-admissible, while states s_0, s_1, s_2, and s_{10} are inaccessible (and hence inadmissable).

DEFINITIONS. Let ρ be a binary relation and α a set; we define

$$\rho(\alpha) = \{ y \mid (\exists x)\rho(x, y) \ \& \ x \in \alpha \} .$$

A complete state s of $\Gamma = (S, G, R, P^1, \cdots, P^n)$ is a *terminal state* of Γ if $R(\{s\})$ is null.

A terminal state of Γ is a complete state for which there is no successor by the direct transition relation R. Complete states s_8 and s_{10} are the terminal states of Fig. 3(a). Note that if Γ has no terminal states, every Γ-accessible

[2] In Burks and Wright [**6**, p. 1364], we defined the concept of an admissible state of a net. When a net is converted into a sequence generator (see § 1.2 below), these states will be accessible rather than admissible in the senses of these terms defined above.

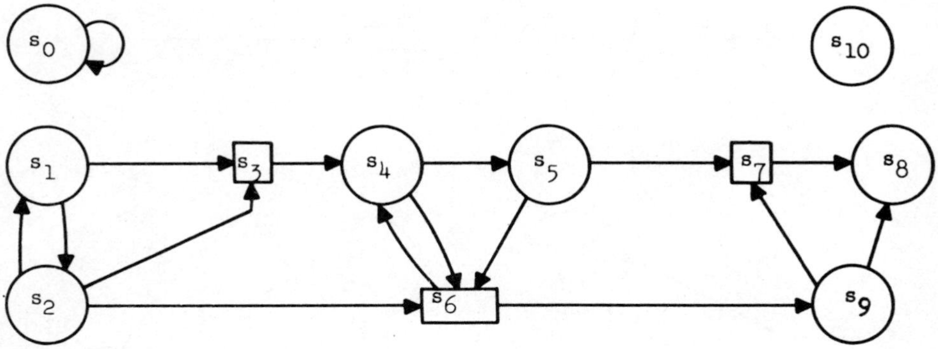

(a)

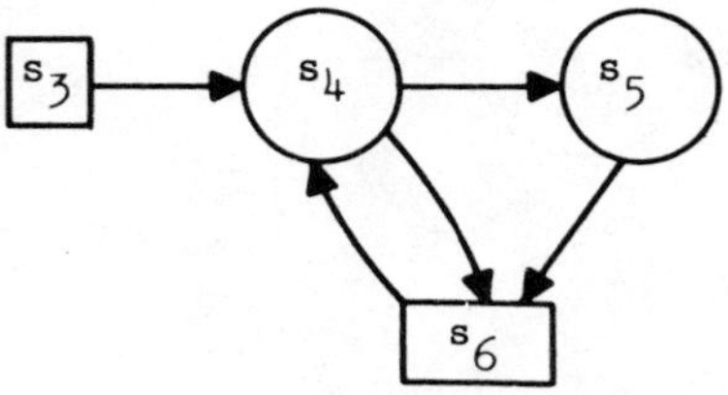

(b)

Figure 3.
(a) Sequence generator $\Gamma = (S, G, R)$.
(b) $\Gamma^{\dagger}$, the reduced form of Γ.

state is Γ-admissible and vice versa.

We will sometimes need to combine several projections to make a composite projection of them. For this we will use the notation

$$P^1 \times P^2 \times \cdots \times P^n$$

which is defined by

$$[P^1 \times P^2 \times \cdots \times P^n](s) = \langle P^1(s), P^2(s), \cdots, P^n(s)\rangle \, .$$

1.2. **Special cases of sequence generators.** Many concepts in the theory of information processing turn out to be special cases of the concept of sequence generator or are closely related to this concept. We will discuss a number of these in the present subsection. Since digital computers (automata) and logical nets are of special interest to us, we will show in detail how the concept of sequence generator applies to them. In later sections we will derive both new and old results about automata and nets from our new theory of sequence generators.

We will begin with well-formed nets, review the method of deriving a finite automaton from a well-formed net, and then show how to derive a sequence generator from a finite automaton. We will use the definition of well-formed nets of Burks and Wright [**6**, p. 1361], modified to allow arbitrary switching elements and delay elements whose initial output states are one, as well as delays whose initial output states are zero.[3] In net diagrams, certain nodes (junctions) are designated as net outputs and are distinguished by stars [see Fig. 1(a)].

A well-formed net (w.f.n.) may be analyzed in terms of its input states, delay output states, and net output states. A digital computer represented by a w.f.n. operates as follows. The "state" of a net at a given time is determined by its input state i and its delay output state d at that time; these pairs $\langle i, d\rangle$ are called the complete states of the net. For each time t ($t = 0, 1, 2, \cdots$) the complete state $\langle i, d\rangle$ determines the net output state θ at the same time (t) in accordance with an output function λ, i.e., $\theta = \lambda(i, d)$. At time 0 the delay output state d_0 is uniquely determined by the initial delay output states of the delay elements. For each time t the complete state $\langle i, d\rangle$ determines the delay output state d_1 at the next moment of time $(t + 1)$ in accordance with a direct transition function τ, i.e., $d_1 = \tau(i, d)$. The net of Fig. 1(a) is a well-formed net which represents a binary counter. A is the input node, the starred node C is its net output node, and B its delay output node (the initial state of B is zero). The state of C at t indicates the binary count, i.e., the number modulo 2 of 1's which have appeared (during the interval of time $0, \cdots, t$) on the input node A. The state analysis is given by the following table, where 0 is the initial delay output state.

$i(t)$	$d(t)$	$d(t+1) = \tau(i, d)$	$\theta(t) = \lambda(i, d)$
A	B	B	C
0	0	0	0
0	1	1	1
1	0	1	1
1	1	0	0

Definition. A *finite automaton* is a sextuple $\langle\{i\}, \{d\}, \{\theta\}, d_0, \tau, \lambda\rangle$ where $\{i\}, \{d\}, \{\theta\}$ are finite nonempty sets (whose elements are called input states, internal states, and output states, respectively), $d_0 \in \{d\}$ (d_0 is called the initial internal state), τ is a function from the Cartesian product $\{i\} \times \{d\}$ into $\{d\}$ (called the direct transition function), and λ is a function from the

[3] Sequence generators may also be derived from automata containing delays whose initial output states are unspecified; these are called "abstract delays" in Burks and Wang [**5**, p. 201] and Burks [**4**, § 3]. But we will not complicate the present discussion by considering automata with such delay elements.

Cartesian product $\{i\} \times \{d\}$ onto $\{\Theta\}$ (called the output function). (This is essentially the definition of Burks and Wang [**5**, p. 203]; see also Moore [**22**, p. 133].) The procedure for analyzing a well-formed net which is described in the preceding paragraph clearly converts a well-formed net into a finite automaton. This procedure is reversible; that is, given a finite automaton, one can construct a well-formed net which realizes it. Thus the concepts of well-formed net and finite automaton are basically equivalent and either can be taken as a formal definition of the concept "finite digital computer." (See Church [**9**], Kleene [**17**, p. 5], and Burks [**4**, § 3] for other definitions of these concepts.)

A three-projection sequence generator $\Gamma = (S, G, R, I, \theta, D)$ may be associated with a finite automaton as follows. The elements of S are the complete states $\langle i, d\rangle$ and the elements of G are the complete states $\langle i, d_0\rangle$. The direct transition relation is defined by

$$R(\langle i_1, d_1\rangle, \langle i_2, d_2\rangle) \equiv [d_2 = \tau(i_1, d_1)]$$

and the input, output, and internal state projections by $I(\langle i, d\rangle) = i$, $\theta(\langle i, d\rangle) = \lambda(i, d)$, and $D(\langle i, d\rangle) = d$, respectively. The sequence generator associated with the binary counter of Fig. 1(a) is represented by Fig. 1(b). As before, the rectangles represent elements of G. The subscripts on the complete states correspond to the nodes of the counter in the order A, B. $\langle s_{00}, s_{10}, s_{11}, s_{00}, s_{10}, s_{01}\rangle$ is an example of a finite Γ-sequence; in it the input sequence $i_0, i_1, i_1, i_0, i_1, i_0$ produces the output sequence $\Theta_0, \Theta_1, \Theta_0, \Theta_0, \Theta_1, \Theta_1$ (and thus three "ones" on the input leave the counter recording "one"). Note that, though the internal states d_0, d_1 are represented in Fig. 1(b), the nodes of the graph correspond to complete states and not to internal states, as is the case with the usual state diagrams used to represent automata.

We have shown how to transform a well-formed net into a finite automaton and vice-versa. We have also shown how to derive a sequence generator from a finite automaton. The latter process is not in general reversible. Only certain sequence generators (those which are deterministic) may be realized by finite automata (see § 2.1).

Our next application of sequence generators is to arbitrary "nets," including nets that are not well-formed. We will use the concept of Burks and Wright [**6**, p. 1353], modified to allow arbitrary switching elements and both kinds of concrete delays. Each switch element translates into a switch equivalence which gives the state of the switch output as a truth function of the switch input, and each delay element translates into two delay equivalences, called the "initial delay equivalence" and the "recursive delay equivalence." The initial delay equivalence gives the initial state of the delay output and the recursive delay equivalence equates the delay output at any time other than 0 to the delay input at the previous time. Hence, each net translates into a conjunction of equivalences. If the net is not well-formed, this conjunction will not directly correspond to (will not give the structure of) a digital computer, but it may specify a computation or behavior condition on a digital computer (see § 4), and on this account is of interest. Fig. 2(a) shows a net with input node E and output node F. The non input switch element driving node A represents the contradictory or "always false" function. The initial state of the delay AB is "true," which for coding reasons we represent by "1"; the initial state of the delay GC is 0. The switch equivalences for this net are $A(t) \equiv 0$, $F(t) \equiv F(t)$, and $C(t) \equiv [E(t) \ \& \ \overline{B(t)}]$. $B(0) \equiv 1$ and $C(0) \equiv 0$ are the initial delay equivalences, while $B(t+1) \equiv A(t)$ and $C(t+1) \equiv G(t)$ are the recursive delay equivalences.

A two-projection sequence generator $\Gamma = (S, G, R, I, \theta)$ may be associated with an arbitrary net in the following way:[4] A complete state s is an assignment of a truth value to each node of the net which makes the switch equivalences of the net true. An element s of S is a generator (element of G) if s assigns to the delay output nodes truth values which make the initial delay equivalences true. $R(s_1, s_2)$, where $s_1, s_2 \in S$, if and only if the truth values which s_1 assigns to the delay input nodes and the truth values which s_2 assigns to the delay output nodes satisfy the recursive delay equivalences. For each complete state s, $\{I(s)\}$ $[\theta(s)]$ is s cut down to the {input} [output] nodes [i.e., $\{I(s)\}$ $[\theta(s)]$ is the net {input} [output] state contained in s]; the input projection will not exist if there are no input nodes.

The sequence generator $\Gamma = (S, G, R, I, \theta)$ associated with Fig. 2(a) is represented by Fig. 2(b). Though there are 6 nodes in the net, there are only 8 complete states. The subscripts on the state symbols s_1, s_{37}, etc., are the decimal codings of the binary representations of the states of the nodes taken in the order E, A, B, C, G, F; e.g., the subscript on s_9 decodes into 001001, showing that in this state nodes B and F are active while the remaining nodes are inactive. The subscript of the input state i is the state of node E and the subscript on the output state θ is the state of node F. $\langle s_{10}, s_{37}, s_2, s_{38}, s_{37}\rangle$ is a Γ-sequence which has a derived input sequence $\langle i_0, i_1, i_0, i_1, i_1\rangle$ and a derived output sequence $\langle \theta_0, \theta_1, \theta_0, \theta_0, \theta_1\rangle$. It can be proved that $F(t) \equiv \sim E(t+1)$; such behavior would not, of course, be possible in a well-formed net.

Our process for associating a sequence generator with an arbitrary net is different from our process for associating a sequence generator with a well-formed net in the following basic respect. In the latter case we first defined input states, delay output states (internal states), and output states for the net, and then compounded complete states from input states and internal states. On the other hand, in associating a sequence generator with an arbitrary net, we first defined states (complete states) over every node, and then derived input and output states by means of projections. It turns out that in general not every assignment of truth values to the input nodes of an arbitrary net is an input state. In fact, we know of no way of defining states for parts of a net (input, internal, and output states) without presupposing states for the whole net (complete states). Indeed, it was our work with arbitrary nets which led us to consider sequence generators (in which complete

[4] If the net is well-formed either the procedure about to be described or the procedure described earlier may be used. The resultant sequence generator will, of course, be different in the two cases.

states are basic; input, internal, and output states derivative).

This completes our discussion of the method of transforming an arbitrary net into a two-projection sequence generator. This process may be reversed; that is, given any sequence generator with two projections, one can find a corresponding net. It is not difficult to construct this procedure (for going from a sequence generator to a net) from the information to be given in § 4.2, so we will not describe it here.

There are other entities besides nets and well-formed nets (digital computers) which are either sequence generators or are closely related to sequence generators. The concept of a nondeterministic automaton of Rabin and Scott [**27**, Definition 9] is quite similar to our concept of a sequence generator. Sequence generators are in a certain sense equivalent to formulas constructed from truth functional connectives, monadic predicates, one individual variable "t" (which ranges over discrete times), the successor function, and zero (see §§ 4.2 and 4.3). The following are special cases of sequence generators: a finite state grammar (Chomsky and Miller [**7**, p. 95]); sequential circuits representable in combinatory logic (Fitch [**13**, p. 263]); incompletely specified automata, i.e., automata in which certain sequences of input states are proscribed (Aufenkamp and Hohn [**1**, § IV]); automata with terminal states [**1**, § VII]; and the flow diagrams used in programming a digital computer. A sequence generator may be used to characterize a class of finite sequences defined by a regular expression (see § 4.4). Finite graphs may be used to analyze certain games (König [**18**] and McKinsey [**20**, Chapter 6]). There is an obvious relation between finite graphs and sequence generators, and hence some problems concerning games may be studied by means of sequence generators; we will give an example in the next subsection. Though he makes no reference to automata theory, Putnam [**26**, pp. 44–49] uses sequence generators to establish some results about satisfiability; he uses the concept of admissibility in connection with "Γ-sequences" which are infinite in both directions. We remark finally that Harary and Paper [**14**], in applying relational logic to linguistics use ideas closely related to the concept of sequence generator.

Though we have noted a number of applications of the concept of sequence generator, we wish to make it clear that we are not attempting in the present paper to solve all the problems that have been considered for these applications. In the next subsection we will establish some results concerning infinite Γ-sequences for sequence generators without projections. In § 2 we will treat some concepts in which a single projection plays an essential role, and in § 3 we will work with concepts in which two projections play a special role. In § 4 we will present some generalizations and further applications of sequence generators.

1.3. **Reduced form algorithm.** Algorithms play a fundamental role in this paper, so we will make a few informal comments about them. An algorithm presupposes a well-defined set of entities, called "the domain of the algorithm." An algorithm is a finite system of rules which may be mechanically applied to any entity of its domain. An algorithm which terminates in a finite number of steps when applied to any entity of its domain is called a "terminating algorithm." The Reduced Form Algorithm to be described soon is a terminating algorithm, since, when it is applied to any sequence generator, it will eventually terminate in a sequence generator. An algorithm with a domain D is called a decision procedure for a class A which is a subset of D, if for every element of D which *belongs* to A, the algorithm terminates in "yes," and for every element of D which *does not belong* to A, the algorithm terminates in "no." The truth table procedure is a decision procedure for the class of tautologies of the propositional calculus.

Before formulating the Reduced Form Algorithm, we will describe informally what it does. Let us call a state s of a sequence generator $\Gamma = (S, G, R, P^1, \cdots, P^n)$ "extendable" if there is an infinite sequence of complete states $\langle s(0), s(1), s(2), \cdots \rangle$ such that $s(0)$ is s and $R[s(i), s(i+1)]$ for $i = 0, 1, 2, \cdots$. (Note that s is not necessarily a generator, and so the infinite sequence of complete states is not necessarily a Γ-sequence.) In Fig. 3 states s_0 and s_1 are extendable, while states s_7 and s_{10} are not. The Reduced Form Algorithm may be applied to any sequence generator Γ. In part (1) of the algorithm the operation of deleting terminal states is iterated until we arrive at a sequence generator $\ddot{\Gamma}$ with no terminal states. Since a sequence generator has nonextendable states if and only if it has terminal states, $\ddot{\Gamma}$ is essentially the result of deleting all nonextendable states from Γ. In part (2) of the algorithm, one begins with the generators of $\ddot{\Gamma}$ (and hence of Γ), and by a succession of steps obtains the accessible states of $\ddot{\Gamma}$. A new sequence generator $\Gamma^\dagger$, called the reduced form of Γ, is defined on the basis of the states so obtained. Since a state is admissible if and only if it is both extendable and accessible, $\Gamma^\dagger$ is just Γ cut down to its admissible states (see Theorem 1.3–1).

Algorithm (*Reduced Form Algorithm*): Consider any sequence generator $\Gamma = (S, G, R, P^1, \cdots, P^n)$.

(1) Form a new sequence generator by deleting all the terminal states of Γ. Iterate this process until you arrive at a sequence generator with no terminal states. Call this final sequence generator $\ddot{\Gamma} = (\ddot{S}, \ddot{G}, \ddot{R}, \ddot{P}^1, \cdots, \ddot{P}^n)$.

(2) Define A_i inductively by

$$A_0 = \ddot{G} \,,$$
$$A_{i+1} = \ddot{R}(A_i) \,.$$

Form the sequence $A_0, A_1, A_2, \cdots$, stopping when $A_{m+1} \subset \bigcup_{i=0}^{m} A_i$. (Note: "$\alpha \subset \beta$" means that α is either included in β or equals β.) Let $\dot{S} = \bigcup_{i=0}^{m} A_i$, $\dot{G} = \ddot{G}$, and let $\{\dot{R}\}$ $[\dot{P}^i]$ be the $\{$relation $\ddot{R}\}$ [projection $\ddot{P}^i$] cut down to $\dot{S}$. Define $\Gamma^\dagger = \dot{\Gamma} = (\dot{S}, \dot{G}, \dot{R}, \dot{P}^1, \cdots, \dot{P}^n)$.

We will illustrate the Reduced Form Algorithm. Let $\Gamma = (S, G, R)$ be the sequence generator represented by Fig. 3(a), with those complete states which belong to G being designated by rectangles. In part (1) of the algorithm we delete states s_8 and s_{10}, then state s_7, and then state s_9. $\ddot{S}$ consists of the remaining complete states, $\ddot{G}$ contains s_3 and s_6, and $\ddot{R}$ is R cut down to $\ddot{S}$. We have at the end of part (1) the sequence generator $\ddot{\Gamma}$ represented by the result of deleting everything to the right of state s_5 in Fig. 3(a). In part (2) of the algorithm we form the sequence $\{s_3, s_6\}[=A_0]$, $\{s_4\}[=A_1]$, $\{s_5, s_6\}[=A_2]$, $\{s_4, s_6\}[=A_3]$. Simultaneously we form the sequence $\{s_3, s_6\}[=\bigcup_{i=0}^{0} A_i]$, $\{s_3, s_4, s_6\}[=\bigcup_{i=0}^{1} A_i]$, $\{s_3, s_4, s_5, s_6\}[=\bigcup_{i=0}^{2} A_i]$, stopping at this point since $\{s_4, s_6\} \subset \{s_3, s_4, s_5, s_6\}$, i.e., $A_3 \subset \bigcup_{i=0}^{2} A_i$. Hence $S = \{s_3, s_4, s_5, s_6\}$ and $\Gamma^\dagger$, the reduced form of Γ, is represented by Fig. 3(b).

THEOREM 1.3–1. *The Reduced Form Algorithm, when applied to any sequence generator Γ, always terminates in a sequence generator $\Gamma^\dagger$. The set of complete states of $\Gamma^\dagger$ equals the set of Γ-admissible complete states.*

As a step toward proving this theorem, we first establish

LEMMA 1.3–2. *Let ρ be a binary relation and δ_0 a set. Define δ_i for $i = 1, 2 \cdots$ inductively by $\delta_{i+1} = \rho(\delta_i)$ and let $\alpha_l = \bigcup_{i=0}^{l} \delta_i$ for $l = 0, 1, 2, \cdots$. If, for some j, $\alpha_j = \alpha_{j+1}$ then for all l, $\alpha_l \subset \alpha_j$.*

PROOF. We note first that since the operator ρ may be distributed over the union, $\alpha_{l+1} = \alpha_l \cup \rho(\alpha_l)$. We now assume $\alpha_j = \alpha_{j+1}$ and prove that $\alpha_l \subset \alpha_j$, proving first by induction that for all $l \geqq j$, $\alpha_l = \alpha_j$. The initial step is covered by the hypothesis that $\alpha_{j+1} = \alpha_j$ For the general step assume $\alpha_k = \alpha_j$, where $k > j$. By the fact noted above, $\alpha_{k+1} = \alpha_k \cup \rho(\alpha_k)$ and $\alpha_{j+1} = \alpha_j \cup \rho(\alpha_j)$. Combining the four preceding equalities, we get $\alpha_{k+1} = \alpha_j$. To conclude the proof, we note that it follows directly from the definition of α_l that for $l < j$, $\alpha_l \subset \alpha_j$.

We turn now to the proof of Theorem 1.3–1. We will use freely the notation of the algorithm. (I) We prove first that the Reduced Form Algorithm, when applied to any sequence generator Γ, always terminates in a sequence generator $\Gamma^\dagger$. Since S is a finite set, the first part of the algorithm terminates in a sequence generator $\ddot{\Gamma}$. The criterion for stopping in part (2) of the algorithm is based on a monotonically increasing sequence of subsets, of $\ddot{S}$, which is a finite set, so the second part of the algorithm will always terminate. Finally, it is clear that a sequence generator $\Gamma^\dagger$ is defined in part (2) of the algorithm.

(II) We prove next that the set of complete states of $\Gamma^\dagger$ equals the set of Γ-admissible complete states. (IIA) We consider a Γ-admissible complete state s_1 and show that $s_1 \in \dot{S}$. Since s_1 is Γ-admissible, there is an infinite sequence $[s](0, \omega)$ of Γ-admissible states such that for some k, $[s](k) = s_1$. A complete state of Γ is deleted by part (1) of the algorithm only if it cannot belong to an infinite Γ-sequence, and so $[s](0, k)$ is a $\ddot{\Gamma}$-sequence. Hence by the definition of A_k in the algorithm, $s_1 \in A_k$ and $s_1 \in \bigcup_{i=0}^{k} A_i$. We now apply Lemma 1.3–2, letting $\rho = \ddot{R}$ and $\delta_0 = \ddot{G}$. By (I) above, part (2) of the algorithm terminates; in the notation of the algorithm $A_{m+1} \subset \bigcup_{i=0}^{m} A_i$. The result of Lemma 1.3–2, put in this notation, is that for all l, $\bigcup_{i=0}^{l} A_i \subset U_{i=0}^{m} A_i$. We have already shown that $s_1 \in \bigcup_{i=0}^{k} A_i$, and so $s_1 \in \bigcup_{i=0}^{m} A_i$. But in the algorithm $\dot{S}$ is defined to be $\bigcup_{i=0}^{m} A_i$ and so $s_1 \in \dot{S}$.

(IIB) We next consider a complete state $s_1 \in \dot{S}$ and show that s_1 is Γ-admissible. In the notation of the algorithm $\dot{S} = \bigcup_{i=0}^{m} A_i$ and so $s_1 \in \bigcup_{i=0}^{m} A_i$.

Hence s_1 is $\ddot{\Gamma}$-accessible. Part (1) of the algorithm terminates in a sequence generator $\ddot{\Gamma}$ with no terminal states. As remarked in § 1.1, every accessible state of a sequence generator with no terminal states is an admissible state. Consequently, there exists an infinite $\ddot{\Gamma}$-sequence $[s](0, \omega)$ such that for some $k, s_1 = [s](k)$. By the nature of part (1) of the algorithm, $[s](0, \omega)$ is also an infinite Γ-sequence, and so s_1 is Γ-admissible.

COROLLARY 1.3–3. (a) *Every complete state of* $\Gamma^\dagger$ *is* $\Gamma^\dagger$*-admissible.*
(b) *The set of infinite* Γ*-sequences equals the set of infinite* $\Gamma^\dagger$*-sequences.*
(c) *Every finite* $\Gamma^\dagger$*-sequence is an initial segment of an infinite* Γ*-sequence.*

We will next discuss the Reduced Form Algorithm and some alternatives to it. Applied to an arbitrary sequence generator Γ, part (1) of the Reduced Form Algorithm produces the set of extendable states of Γ. Applied to an arbitrary sequence generator Γ, part (2) of the algorithm produces the set of Γ-accessible states. Since a complete state is admissible if and only if it is both extendable and accessible, the two parts of the Reduced Form Algorithm applied to a sequence generator Γ in either order produce the same sequence generator $\Gamma^\dagger$. A sequence generator Γ derived from a well-formed net in the way indicated in § 1.2 has no terminal states; consequently, when part (2) of the Reduced Form Algorithm is applied to Γ, it produces $\Gamma^\dagger$.

There is an alternative procedure for finding the Γ-admissible complete states of a sequence generator. Let x be the number of complete states of Γ. Form all Γ-sequences of length $x + 1$; it can be proved that a state is Γ-accessible if and only if it occurs in one of these sequences. To find the Γ-admissible states, we operate on each sequence as follows: proceeding through the sequence $\langle s(0), s(1), \cdots, s(x)\rangle$ check an occurrence of a state whenever that state has occurred earlier in the same sequence; then delete all states which follow the last checked state. It can be shown that a state is Γ-admissible if and only if it occurs in one of the resultant sequences. This method of finding the Γ-admissible states can be made the essence of an alternative reduced form algorithm which is simpler to formulate and easier to prove adequate than our Reduced Form Algorithm. It is less efficient, however: in the example given earlier, $m = 2$ while $x = 11$. These differences seem to result from the following fundamental difference between these two algorithms. In the Reduced Form Algorithm the length of the computation is not specified in advance; rather, parts (1) and (2) each contain an internal "stop criterion:" one proceeds until he is stopped by these criteria. In contrast, the alternative algorithm first establishes the length of the computation on the basis of a general property of the sequence generator (the number of complete states); since this length is established *a priori*, it is of course determined by the worst case, even though in most cases far fewer steps would have sufficed. This is analogous to the contrast between asynchronous circuits, in which completion of an operation is sensed and the next operation begun immediately, and synchronous circuits, in which the same amount of time is allowed for a given operation in every case, and this is, of course the time required for the worst case (plus a "safety factor!"). We have presented the more efficient of these two algorithms, although it is more difficult to formulate and prove adequate, because finding the reduced form is basic to so many automata algorithms; see, for example, §§ 2.3 and 3.4. But though in many later cases we know of more efficient algorithms (see, for example, the alternative to the h-univalence Decision Procedure in § 3.4), we will not present them because we feel that perspicuity of theory and simplicity of exposition are more important there.

We mentioned in § 1.2 that certain puzzles give rise to sequence generators. The so-called "15 puzzle" is a good example since it may be solved by means of our Reduced Form Algorithm. The puzzle consists of a 4×4 array of 15 movable blocks (numbered 1 through 15) and one empty position. A "move" consists in changing a pattern into any one of the (at most) four patterns obtained by shifting a block into the (neighboring) empty space. The problem is to achieve a stipulated pattern by a succession of moves starting from a given pattern. A sequence generator $\Gamma = (S, G, R, P)$ corresponding to the puzzle may be defined as follows. The 4×4 matrices whose entries are the numbers 0 through 15 are the complete states of Γ; there are 16! of these. The starting pattern is the sole generator of Γ. Two states s_1 and s_2 stand in the direct transition relation R if there is a move taking the pattern corresponding to s_1 into the pattern corresponding to s_2. The projection P has the value 1 on the single pattern stipulated to be the goal and 0 otherwise. The problem is solved by constructing a finite Γ-sequence $\langle s(0), s(1), s(2), \cdots, s(t)\rangle$ such that $P[s(t)] = 1$, if such a sequence exists. Clearly this sequence exists if and only if the complete state with a projection of 1 is Γ-accessible. Whether or not this is the case can be determined by applying part (2) of the Reduced Form Algorithm to Γ: if such a sequence exists, it will be found in the course of carrying out the algorithm.[5] It turns out that exactly half of the complete states of Γ are Γ-accessible and that each of these Γ-accessible states is also Γ-admissible.

2. Sequence generators with one projection

2.1. **Definitions.** In the last sub-section we made no particular use of the projections of a sequence generator. In this section we shall define some concepts which apply primarily to sequence generators with one projection and will prove some theorems about these concepts. In most applications this single projection is an input projection, an output projection, or a combined input-output projection. In the next section we will work with sequence generators having two projections. These two projections will usually be an input and an output projection.

Definition. The *behavior* of $\Gamma = (S, G, R, P^1, P^2, \cdots, P^n)$, where $n > 0$, is the set $P([s](0, k))$, where $P = P^1 \times P^2 \times \cdots \times P^n$ and $[s](0, k)$ is a Γ-sequence (finite or infinite). "$\mathscr{B}(\Gamma)$" denotes the behavior of Γ. The *infinite behavior*

[5] There is a much simpler algorithm for finding the Γ-accessible states of this particular sequence generator. See W. W. R. Ball, *Mathematical recreations and essays,* Macmillan, London 1940, pp. 299–303.

of Γ, denoted by "$\mathscr{B}^{\omega}(\Gamma)$", is the set of infinite sequences in $\mathscr{B}(\Gamma)$. Corollary 1.3–3b clearly yields

COROLLARY 2.1–1. $\mathscr{B}^{\omega}(\Gamma) = \mathscr{B}^{\omega}(\Gamma^{\dagger})$.

It is worth noting that in general it is not true that for a sequence generator $\Gamma = (S, G, R, P)$ there exists a sequence generator $\dot{\Gamma}$ such that the set of $\dot{\Gamma}$-sequences equals the behavior of Γ. This may be shown by a simple example. Let $S = \{s_0, s_1, s_2\}$, $G = \{s_0\}$, $R = \{\langle s_0, s_1\rangle, \langle s_1, s_2\rangle, \langle s_2, s_0\rangle\}$, and $P(s_0) = P(s_1) = p_0$, $P(s_2) = p_1$. There is one infinite Γ-sequence $\langle s_0, s_1, s_2, s_0, s_1, s_2, s_0, s_1, s_2, \cdots\rangle$ and the behavior of Γ consists of the sequence $\langle p_0, p_0, p_1, p_0, p_0, p_1, \cdots\rangle$ and all its initial segments, and does not include the infinite sequence $\langle p_0, p_0, p_0, \cdots\rangle$. Consider a sequence generator $\dot{\Gamma} = (\dot{S}, \dot{G}, \dot{R})$ such that $\{p_0, p_1\} \subset \dot{S}$ and such that $\langle p_0, p_0, p_1, p_0, p_0, p_1, \cdots\rangle$ is an infinite $\dot{\Gamma}$-sequence. It follows from the existence of this sequence that $\dot{R}(p_0, p_0)$ and $p_0 \in \dot{G}$, and hence that $\langle p_0, p_0, p_0, \cdots\rangle$ is an infinite $\dot{\Gamma}$-sequence.

Some remarks about the application of the concept of behavior to nets will be appropriate. By the methods of § 1.2, we can associate with every net (well-formed or not) a sequence generator $\Gamma = (S, G, R, I, \theta)$, where I is the input projection and θ is the output projection. The behavior of a digital computer (w.f.n.) consists of the relationship between its inputs and its outputs, and similarly for an arbitrary net. The behavior of a net may be regarded as the set of sequences (finite and infinite) of pairs $\langle i(0), \mathfrak{o}(0)\rangle$, $\langle i(1), \mathfrak{o}(1)\rangle$, $\langle i(2), \mathfrak{o}(2)\rangle, \cdots$ for which there is a Γ-sequence $[s](0, k)$ such that $i(t) = I\{s(t)\}$ and $\mathfrak{o}(t) = \theta\{s(t)\}$ for every t. This is clearly $\mathscr{B}(\Gamma)$, the behavior of the sequence generator $\Gamma = (S, G, R, I, \theta)$. In § 2.3 we present a Behavior Inclusion Procedure to be applied to a pair $\langle \Gamma, \dot{\Gamma}\rangle$ to decide whether the behavior of Γ is included in the behavior of $\dot{\Gamma}$; when applied to the pair $\langle \dot{\Gamma}, \Gamma\rangle$ as well as to the pair $\langle \Gamma, \dot{\Gamma}\rangle$, this tells us whether the behaviors of Γ and $\dot{\Gamma}$ are equal. Thus, through these considerations, the Behavior Inclusion Procedure can be used to decide whether the behavior of an arbitrary net N is included in or equal to the behavior of a net $\dot{N}$. In the case of well-formed nets, however, a much more efficient algorithm for deciding equality of behaviors is known (Burks and Wang [**5**, § 2.2]); a basic part of this algorithm consists essentially of finding the reduced form (§ 1.3) of a sequence generator associated with the combined nets. Actually this algorithm applies to any deterministic sequence generator (this concept is defined below); moreover, if Γ and $\dot{\Gamma}$ are both deterministic and $\mathscr{B}(\Gamma) \subset \mathscr{B}(\dot{\Gamma})$, then $\mathscr{B}(\dot{\Gamma}) \subset \mathscr{B}(\Gamma)$, so this algorithm also answers the question as to whether $\mathscr{B}(\Gamma) \subset \mathscr{B}(\dot{\Gamma})$ for the case of deterministic sequence generators.

The following lemma will be needed in subsequent proofs. It is a classical interpretation of Brouwer's Fan theorem (Heyting [**15**, pp. 42–43]) and is closely related to König's Infinity Lemma concerning infinite graphs (König [**18**, p. 81]). Our lemma, however, is stronger than König's Infinity Lemma in that it does not require that the α's be pairwise disjoint; because of this difference, we present a proof of it here.

LEMMA 2.1–2. *Let $\langle \alpha_0, \alpha_1, \alpha_2, \cdots\rangle$ be an ω-sequence of finite nonempty sets* *and let* ρ *be a binary relation. If for every* $x \in \alpha_{i+1}$ *there is a* $y \in \alpha_i$ *such that* $\rho(y, x)$, *then there is an infinite sequence* $\langle z_0, z_1, z_2, \cdots \rangle$ *such that for each* i, $z_i \in \alpha_i$ *and* $\rho(z_i, z_{i+1})$.

PROOF. Let β_i consist of all finite sequences $\langle x_i, x_{i+1}, \cdots, x_{i+k} \rangle$ where $k = 0, 1, 2, \cdots, x_j \in \alpha_j$ for $i \leqq j \leqq i + k$, and $\rho(x_j, x_{j+1})$ for $i \leqq j < i + k$. It follows from the requirement on ρ in the hypothesis of the lemma that for each i, k, and element y_{i+k} of α_{i+k} there is an element of β_i $\langle y_i, y_{i+1}, \cdots, y_{i+k} \rangle$. Since this is so for any k, each β_i is infinite. We will now define by induction the desired sequence $\langle z_0, z_1, z_2, \cdots \rangle$.

INITIAL STEP. Since β_0 is infinite while α_0 is finite there will be some element z_0 of α_0 such that an infinite number of elements of β_0 begin with z_0. Let δ_0 be the subset of β_0 all of whose elements begin with z_0.

GENERAL STEP. Assume given a sequence $\langle z_0, z_1, \cdots, z_i \rangle$ (where $i = 0, 1, 2, \cdots$) which belongs to β_0 and satisfies the condition that the set δ_i of elements of β_i which begin with z_i is infinite. The result δ'_{i+1} of deleting the first element of each member of β_i is an infinite subset of β_{i+1}. Since α_{i+1} is finite, there will be some element z_{i+1} of α_{i+1} such that $\rho(z_i, z_{i+1})$ and an infinite number of elements of δ'_{i+1} begin with z_{i+1}. Let δ_{i+1} be the subset of δ'_{i+1}, all of whose elements begin with z_{i+1}; δ'_{i+1} is a subset of β_{i+1} and hence δ_{i+1} is also. Hence $\langle z_0, z_1, \cdots, z_i, z_{i+1} \rangle$ belongs to β_0 and satisfies the condition that the set δ_{i+1} of elements of β_{i+1} which begin with z_{i+1} is infinite. Thus the inductive hypothesis has been established for the sequence $\langle z_0, z_1, \cdots, z_i, z_{i+1} \rangle$. This completes the proof of Lemma 2.1–2.

It may be shown by means of this lemma that a sequence of P-states is an element of the behavior of a sequence generator if and only if every initial segment is.

THEOREM 2.1–3 (INFINITY THEOREM). *Let* $\Gamma = (S, G, R, P)$ *be a sequence generator with behavior* $\mathscr{B}(\Gamma)$ *and let* $[p](0, k)$, $k = 0, 1, 2, \cdots, \omega$, *be a sequence of* P-*states* $[p](0, k) \in \mathscr{B}(\Gamma)$ *if and only if for every finite* $i \leqq k$, $[p](0, i) \in \mathscr{B}(\Gamma)$.

PROOF. The proof of the theorem for finite k is obvious. It is also obvious that $[p](0, \omega) \in \mathscr{B}(\Gamma)$ implies that for every finite i, $[p](0, i) \in \mathscr{B}(\Gamma)$. It remains to be proved that if $[p](0, i) \in \mathscr{B}(\Gamma)$ for every finite i, then $[p](0, \omega) \in \mathscr{B}(\Gamma)$. We define α_i by $s_1 \in \alpha_i$ if there exists a Γ-sequence $[s](0, i)$ such that $[p](0, i) = P\{[s](0, i)\}$ and $s_1 = s(i)$. It is clear that each α_i is finite and nonempty. We let $\rho = R$ and show that the hypothesis of Lemma 2.1–2 is satisfied. Suppose $s_2 \in \alpha_{i+1}$. By definition of α_{i+1} there exists a Γ-sequence $[s_3](0, i + 1)$ such that $[p](0, i + 1) = P\{[s_3](0, i + 1)\}$ and $s_2 = [s_3](i + 1)$. Let $s_4 = s_3(i)$. By the definition of α_i, $s_4 \in \alpha_i$ and by the definition of a Γ-sequence, $R(s_1, s_2)$. By Lemma 2.1–2 there is an infinite Γ-sequence $[s_5](0, \omega)$ and by the definition of α_i we have $P\{[s_5](0, \omega)\} = [p](0, \omega)$. Hence $[p](0, \omega) \in \mathscr{B}(\Gamma)$.

The following is a corollary of the Infinity Theorem. Let $\Gamma = (S, G, R, P)$ and $\dot{\Gamma} = (\dot{S}, \dot{G}, \dot{R}, \dot{P})$ be two sequence generators and suppose that for every finite k, if $[p](0, k) \in \mathscr{B}(\Gamma)$ then $[p](0, k) \in \mathscr{B}(\Gamma)$; then $\mathscr{B}(\Gamma) \subset \mathscr{B}(\dot{\Gamma})$. For consider any infinite sequence $[p](0, \omega) \in \mathscr{B}(\Gamma)$. By the Infinity Theorem, for each k, $[p](0, k) \in \mathscr{B}(\Gamma)$. Then by hypothesis, for each k, $[p](0, k) \in \mathscr{B}(\dot{\Gamma})$. Finally, by the Infinity Theorem $[p](0, \omega) \in \mathscr{B}(\dot{\Gamma})$. This result holds for Γ and $\dot{\Gamma}$ interchanged, of course, so we have: if, for every finite k, $[p](0, k) \in \mathscr{B}(\Gamma) \equiv [p](0, k) \in \mathscr{B}(\dot{\Gamma})$, then $\mathscr{B}(\Gamma) = \mathscr{B}(\dot{\Gamma})$. Thus the Infinity Theorem shows that the "finite" behavior of a sequence generator determines its (complete) behavior.

Definitions. Let $\Gamma = (S, G, R, P)$ be a sequence generator. Γ is *solvable* if every infinite sequence of P-states belongs to its behavior. Γ is {*semi-deterministic*} [*deterministic*] if it satisfies the conditions:

(1) For any P-state p, there is {at most one} [exactly one] complete state s of Γ such that $s \in G$ and $P(s) = p$.

(2) For any complete state s_1 and any P-state p of Γ, there is {at most one} [exactly one] complete state s_2 such that $R(s_1, s_2)$ and $P(s_2) = p$.

It is obvious from the definition of {semi-determinism} [determinism] that there is a decision procedure for the class of {semi-deterministic} [deterministic] sequence generators. The problem of solvability is not so simple, but we will later develop a decision procedure for solvability (see Theorem 2.3–2).

Let us illustrate these concepts. The sequence generator of Fig. 1(b) less its last two projections, is clearly deterministic. The sequence generator of Fig. 4(a) is semi-deterministic but not solvable, while the sequence generator of Fig. 4(b) is neither semi-deterministic nor solvable.

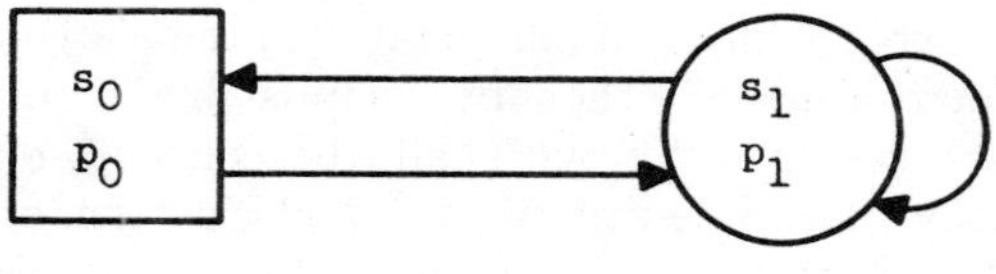

(a)

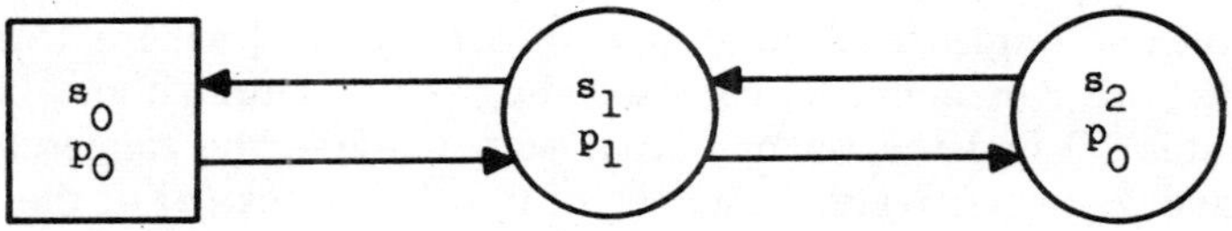

(b)

Figure 4.
(a) Semi-deterministic nonsolvable sequence generator.
(b) Sequence generator neither solvable nor semi-deterministic.

By simple inspection it can be ascertained that the sequence generator (S, G, R, P) of Fig. 10(a) is neither semi-deterministic nor deterministic. It is, however, solvable, as the following considerations show. Given any sequence of P-states, divide it into a sequence (finite or infinite) of subsequences (finite or infinite), where each subsequence is either an iteration of p_0 or an iteration of p_1 and the two types of subsequences alternate. Now a Γ-sequence $s_0, s_0, \cdots, s_0, s_1$ produces a P-state sequence $p_0, p_0, \cdots, p_0, p_0$ followed by at least one occurrence of p_1, while a Γ-sequence $s_3, s_3, \cdots, s_3, s_2$ produces a P-state sequence $p_1, p_1, \cdots, p_1, p_1$ followed by at least one occurrence of p_0. Hence for any sequence of P-states $[p](0, k)$ one can construct a Γ-sequence $[s](0, k)$ such that $[p](0, k) = P([s](0, k))$, and so (S, G, R, P) is solvable. Consider next (S, G, R, I) of Fig. 2(b). (S, G, R, I) is not semi-deterministic, since the input sequence $\langle i_0, i_0, i_0 \rangle$ is the projection of both $\langle s_9, s_1, s_1 \rangle$ and $\langle s_9, s_1, s_2 \rangle$. But (S, G, R, I) is solvable, as may be shown by an analysis like that just given for Fig. 10(a); indeed, except for labeling, the behavior of Fig. 2(b) is the same as the behavior of Fig. 10(a).

The following lemma may be established by simple mathematical inductions with reference to the appropriate definitions.

LEMMA 2.1–4. *Let* $\Gamma = (S, G, R, P)$.

(a) *If* Γ *is {semi-deterministic} [deterministic] (solvable), then* $\Gamma^\dagger$ *is {semi-deterministic} [deterministic] (solvable).*

(b) *If every complete state of* Γ *is* Γ*-accessible, then* Γ *is {semi-deterministic} [deterministic] if and only if for every finite sequence of* P*-states* $[p](0, t)$ *there exists {at most one} [exactly one]* Γ*-sequence* $[s](0, t)$ *such that* $P\{[s](0, t)\} = [p](0, t)$.

(c) *If* Γ *is deterministic, then* Γ *is solvable.*

Other senses of semi-determinism and of determinism may be obtained by replacing every occurrence of "complete state" in the above definition of semi-determinism and determinism either by "admissible complete state" or by "accessible complete state." We will call the concepts obtained by making the latter substitution "semi-determinism$_1$" and "determinism$_1$." It may be shown that these two concepts are equivalent to the conditions stated in the consequent of part (b) of Lemma 2.1–4. In the case of arbitrary nets determinism$_1$ becomes the determinism of Burks and Wright [**6**, p. 1359].

The process described in § 1.2 associates with a finite automaton a sequence generator $\Gamma = (S, G, R, I, \theta, D)$ such that (S, G, R, I) is deterministic. Conversely, given a sequence generator $\Gamma = (S, G, R, I, \theta)$, where (S, G, R, I) is deterministic, we can define a corresponding finite automaton. Let the set of input states $\{i\}$ and the set of output states $\{\text{o}\}$ be the ranges of the projections I and θ, respectively. The set of internal states $\{d\}$ of the automaton is a set of sets of complete states of Γ defined as follows:

$$\alpha \in \{d\} \,.\!\equiv\!.\, \alpha = G \,.\vee.\, (\exists s)(\alpha = R(s)) \,,$$

where α ranges over non-null subsets of S. Let the initial internal state $d_0 = G$. The direct transition function τ is given by

$$\tau(i, d) = R(\iota s \mid s \in d \;\&\; I(s) = i) \,,$$

where "$\iota s \mid \cdots$" means "the complete state s satisfying the condition "$\cdots$". Finally, the output function λ of the net is defined by

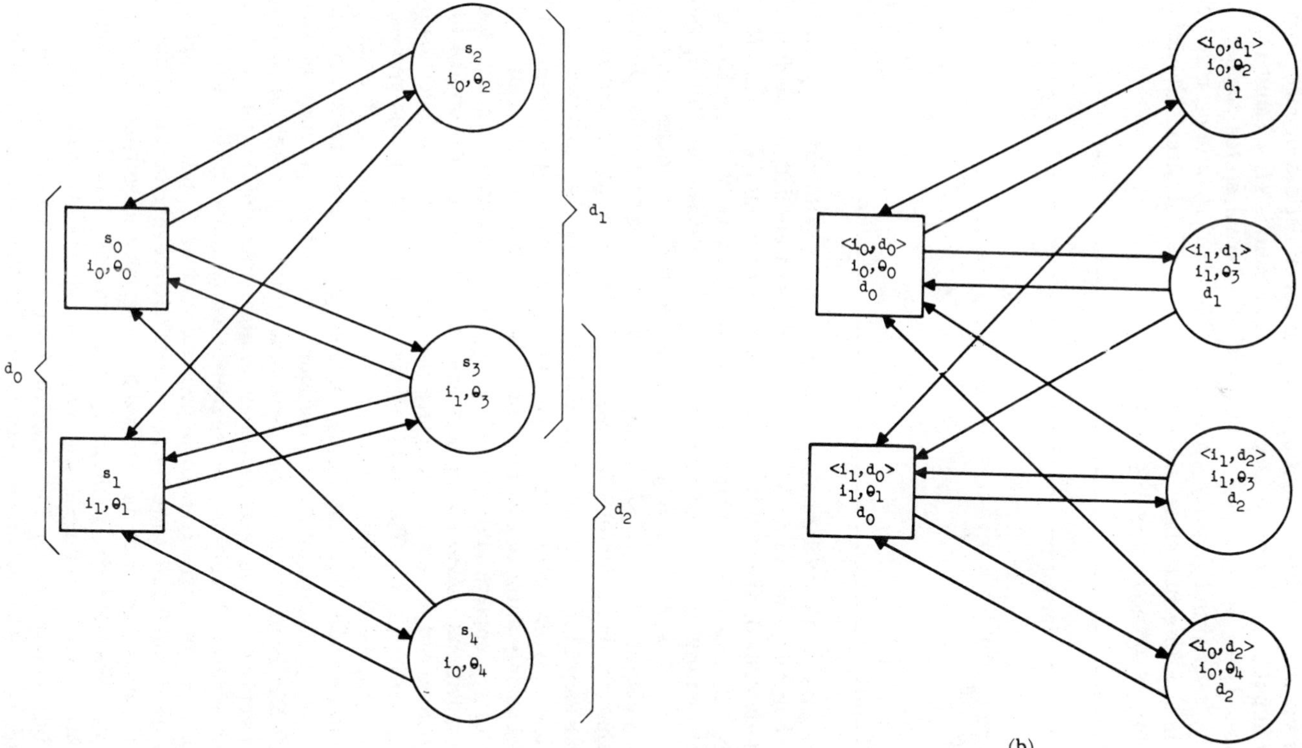

(b)

Figure 5.

(a) Deterministic sequence generator $\Gamma = (S, G, R, I, \theta)$.
(b) Internal state sequence generator $\dot{\Gamma} = (\dot{S}, \dot{G}, \dot{R}, \dot{I}, \dot{\theta}, \dot{D})$ corresponding to (a).

$$\lambda(i, d) = \theta(\iota s \mid s \in d \;\&\; I(s) = i) \,.$$

We will give an example. Fig. 5(a) is a deterministic sequence generator. The set of input states for the associated automaton is $\{i_0, i_1\}$ and the set of output states is $\{\theta_0, \theta_1, \theta_2, \theta_3, \theta_4\}$. The set of internal states consists of the sets $\{s_0, s_1\}$, $\{s_2, s_3\}$, and $\{s_3, s_4\}$, which we will call d_0, d_1, and d_2, respectively. d_0 is the initial internal state since $\{s_0, s_1\} = G$. The direct transition and output functions are given by the table below.

$i(t)$	$d(t)$	$d(t+1) = \tau(i, d)$	$\theta(t) = \lambda(i, d)$
i_0	d_0	d_1	θ_0
i_1	d_0	d_2	θ_1
i_0	d_1	d_0	θ_2
i_1	d_1	d_0	θ_3
i_0	d_2	d_0	θ_4
i_1	d_2	d_0	θ_3

In § 1.2 we gave a process for converting a finite automaton into a three-projection sequence generator. When this process is applied to the finite automaton just described, the result is the sequence generator $\dot{\Gamma} = (\dot{S}, \dot{G}, \dot{R}, \dot{I}, \dot{\theta}, \dot{D})$ of Fig. 5(b). It should be noted that $\mathscr{B}(\Gamma) = \mathscr{B}(\dot{S}, \dot{G}, \dot{R}, \dot{I}, \dot{\theta})$. Hence when the two procedures just described are applied successively to a sequence generator $\Gamma = (S, G, R, I, \theta)$, where (S, G, R, I) is deterministic, the result is a sequence generator $\dot{\Gamma} = (\dot{S}, \dot{G}, \dot{R}, \dot{I}, \dot{\theta}, \dot{D})$ with an internal state projection $\dot{D}$ and such that the behavior of $(\dot{S}, \dot{G}, \dot{R}, \dot{I}, \dot{\theta})$ is the same as the behavior of Γ.

This is an opportune time to compare the state diagrams that we have been using, which may be called "complete state graphs," with those ordinarily used in discussing automata, which may be called "internal state graphs." The nodes of a complete state graph represent complete states and the lines represent transitions between complete states; these lines are unlabeled since the input states, output states, etc., are derived from the complete states by means of the projections. The nodes of an internal state graph represent internal states; the labeled lines represent transitions between internal states, with the labels indicating the inputs that cause the transitions and the outputs which are produced by the transitions. Since the definition of sequence generator (§ 1.1) is in terms of complete states, complete state graphs give a more direct representation of sequence generators than do internal state graphs. We have considered definitions of sequence generators in terms of internal states, but none of these is both as general and as simple to formulate and work with as the definition we have given. However, certain kinds of sequence generators can best be analyzed in terms of internal states and are more simply represented by internal state graphs than by complete state graphs. For example, deterministic sequence generators can be analyzed in terms of internal states in the way we have just shown; moreover, the resulting internal state diagram is always simpler than the corresponding complete state diagram. Whenever the practicality of a technique of analysis is of interest and the internal state diagram is simpler than the complete state diagram, the former should, of course, be used.

Any property of a one-projection sequence generator and any operation applicable to a one-projection sequence generator can be extended to a sequence generator $\Gamma = (S, G, R)$ without projections by adjoining to it a constant projection P (i.e., a projection with only one P-state); Γ is solvable, deterministic, etc., if (S, G, R, P) is solvable, deterministic, etc. For example, a well-formed net without input nodes has associated with it (by either of the techniques of § 1.2) a sequence generator (S, G, R) with one infinite (periodic) Γ-sequence. For constant P, (S, G, R, P) is solvable and semi-deterministic, and hence deterministic, and so is (S, G, R); see, for example, Fig. 6(a).

2.2. **Subset sequence generator operation.** We will next define an operation, denoted by "*", called "the subset sequence generator operation." This operation may be applied to any sequence generator Γ to obtain its subset sequence generator Γ^*. The complete states of Γ^* are sets of complete states

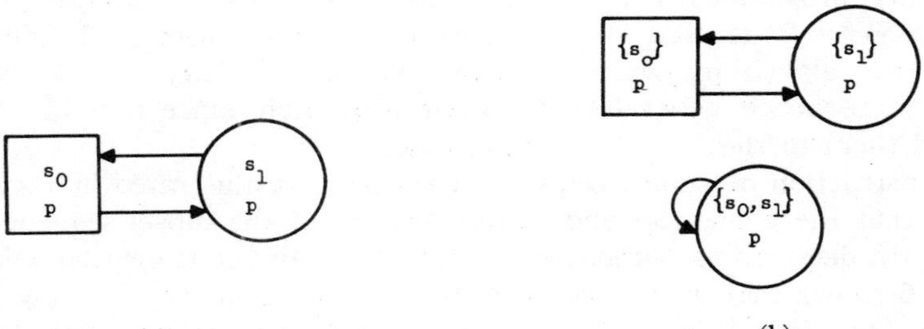

(a) (b)

Figure 6. The construction of a subset sequence generator.
(a) $\Gamma = (S, G, R, P)$.
(b) Γ^*, the subset sequence generator of Γ.

of Γ. The generators, the direct transition relation, and the projections of Γ^* are defined in terms of Γ in such a way that Γ^* has the same behavior as Γ (Theorem 2.2–3 below) and Γ^* is always semi-deterministic, even though Γ may not be (Lemma 2.2–1 below).

Definition. The *subset sequence generator operation*, denoted by "*", applies to any sequence generator $\Gamma = (S, G, R, P^1, P^2, \cdots, P^n)$, where $n > 0$, and produces a sequence generator $\Gamma^* = \dot{\Gamma} = (\dot{S}, \dot{G}, \dot{R}, \dot{P}^1, \dot{P}^2, \cdots, \dot{P}^n)$. Let $P = P^1 \times P^2 \times \cdots \times P^n$.

(1) A subset x of S is an element of $\dot{S}$ if and only if x is non-null and P has the same value for all elements of x. This definition can be expressed symbolically as follows, where Λ is the null set, and the variable x ranges over subsets of S:

$$x \in \dot{S} :\equiv: x \neq \Lambda \;\&\; (s_1, s_2)\{[(s_1 \in S) \;\&\; (s_2 \in S) \;\&\; (s_1 \in x) \;\&\; (s_2 \in x) \supset [P(s_1) = P(s_2)]\} .$$

(2) The elements of $\dot{G}$ are maximal subsets of G which are elements of $\dot{S}$.

Formally,

$$\dot{s} \in \dot{G} :\equiv: \dot{s} \in \dot{S} \ \& \ \dot{s} \subset G \ \& \ (\dot{s}_1)\{[\dot{s}_1 \in \dot{S} \ \& \ (\dot{s} \subset \dot{s}_1 \subset G)] \supset (\dot{s} = \dot{s}_1)\} .$$

(3) Two complete states $\dot{s}_1$ and $\dot{s}_2$ of $\dot{S}$ stand in the direct transition relation $\dot{R}$ if and only if $\dot{s}_2$ is a maximal set of direct successors (by R) of elements of $\dot{s}_1$. Formally,

$$\dot{R}(\dot{s}_1, \dot{s}_2) :\equiv: \dot{s}_1 \in \dot{S} \ \& \ s_2 \subset R(\dot{s}_1) \ \& \ (\dot{s}_3)\{[\dot{s}_3 \in \dot{S} \ \& \ (\dot{s}_2 \subset \dot{s}_3 \subset R(\dot{s}_1))] \supset (\dot{s}_2 = \dot{s}_3)\} .$$

(4) All the elements of a state $\dot{s}$ of $\dot{S}$ have the same P^i-state (for $i = 1, 2, \cdots, n$), and we take this common-value to be the $\dot{P}^i$-state of $\dot{s}$. Formally,

$$\dot{P}^i(\dot{s}) = P^i(s) \qquad \text{where } s \in \dot{s}, \ \dot{s} \in \dot{S} .$$

(Γ^* is called "the subset sequence generator" of Γ. Our concept of a subset sequence generator is similar to concepts used by Myhill [**24**, p. 122], Medvedev [**21**, p. 13] and Rabin and Scott [**27**, Definition 11].)

It should be noted that the concepts of behavior (Section 2.1) and subset sequence generator are essentially one-projection concepts in the sense that when many projections $P^1, P^2, \cdots, P^n$ are given, the composite projection $P^1 \times P^2 \times \cdots \times P^n$ is used in the definitions of the concepts. In subsequent theorems and algorithms we will, for the sake of simplicity, usually state our results for sequence generators with one projection, since it is obvious how to extend them to the many-projection case.

The construction of subset sequence generators is illustrated in Figs. 6 and 7. Note that the generators and complete states of the subset sequence generator Γ^* are determined without reference to the direct transition relation of Γ. Fig. 6 shows that even though Γ is in reduced form, Γ^* may not be, though in fact, if Γ is in reduced form, then Γ^* has no terminal states and so $\mathscr{B}(\Gamma^*) = \mathscr{B}(\Gamma^{*\dagger})$. In Fig. 7 we begin with a semi-deterministic sequence generator, add to it in various ways to obtain three sequence generators, $\dot{\Gamma}, \ddot{\Gamma}, \dddot{\Gamma}$ which are not semi-deterministic, and then derive the subset sequence generator of each of these. All the subset sequence generators $\Gamma^*, \dot{\Gamma}^*, \ddot{\Gamma}^*, \dddot{\Gamma}^*$ are semi-deterministic, as they must be by the next lemma. None of the sequence generators $\Gamma, \dot{\Gamma}, \ddot{\Gamma}, \dddot{\Gamma}$ is solvable; $\Gamma^*, \dot{\Gamma}^*, \ddot{\Gamma}^*$, and $\dddot{\Gamma}^*$ are not solvable either (cf. Corollary 2.2–4).

LEMMA 2.2–1. *For any sequence generator* $\Gamma = (S, G, R, P)$, Γ^* *is semi-deterministic.*

PROOF. Let $\Gamma^* = \dot{\Gamma} = (\dot{S}, \dot{G}, \dot{R}, \dot{P})$. It follows from the construction of $\dot{G}$ that for any $\dot{s}_1, \dot{s}_2$, if $\dot{s}_1 \in \dot{G}$, $\dot{s}_2 \in \dot{G}$, and $\dot{P}(\dot{s}_1) = \dot{P}(\dot{s}_2)$, then $\dot{s}_1 = \dot{s}_2$, and it follows from the definition of $\dot{R}$ that for any $\dot{s}, \dot{s}_1, \dot{s}_2$, if $\dot{R}(\dot{s}, \dot{s}_1)$, $\dot{R}(\dot{s}, \dot{s}_2)$, and

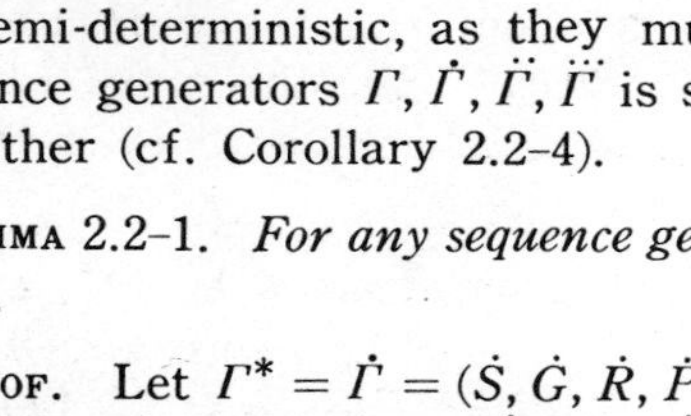

(a)

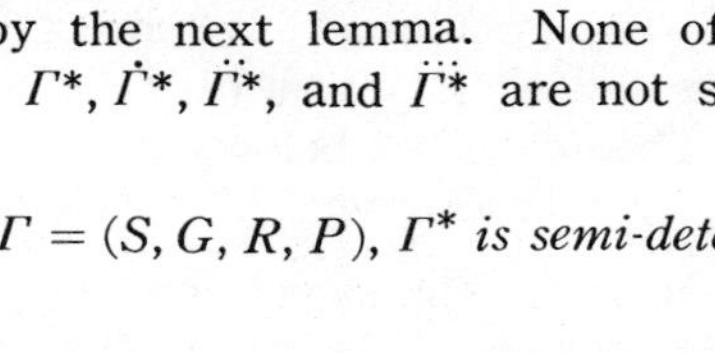

(a′)

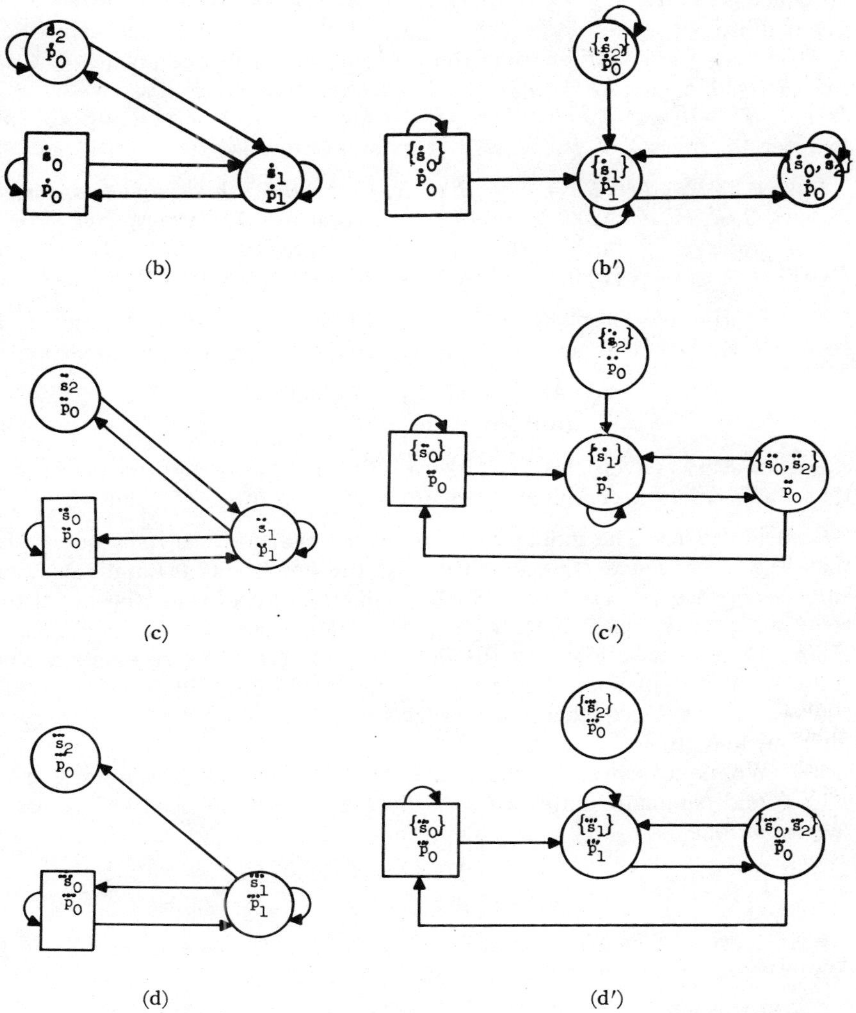

Figure 7. Examples which illustrate Lemma 2.2-1. For any sequence generator $\Gamma = (S, G, R, P)$, Γ^* is semi-deterministic.

(a) Sequence generator $\Gamma = (S, G, R, P)$. Γ is semi-deterministic.
(a′) Γ^*, the subset sequence generator of Γ. Γ^* is semi-deterministic.
(b) Sequence generator $\dot{\Gamma} = (\dot{S}, \dot{G}, \dot{R}, \dot{P})$. $\dot{\Gamma}$ is not semi-deterministic.
(b′) $\dot{\Gamma}^*$, the subset sequence generator of $\dot{\Gamma}$. $\dot{\Gamma}^*$ is semi-deterministic.
(c) $\ddot{\Gamma} = (\ddot{S}, \ddot{G}, \ddot{R}, \ddot{P})$. $\ddot{\Gamma}$ is not semi-deterministic.
(c′) $\ddot{\Gamma}^*$, the subset sequence generator of $\ddot{\Gamma}$. $\ddot{\Gamma}^*$ is semi-deterministic.
(d) $\dddot{\Gamma} = (\dddot{S}, \dddot{G}, \dddot{R}, \dddot{P})$. $\dddot{\Gamma}$ is not semi-deterministic.
(d′) $\dddot{\Gamma}^*$, the subset sequence generator of $\dddot{\Gamma}$. $\dddot{\Gamma}^*$ is semi-deterministic.

$\dot{P}(\dot{s}_1) = \dot{P}(\dot{s}_2)$, then $\dot{s}_1 = \dot{s}_2$.

Given a sequence generator $\Gamma = (S, G, R, P)$, by the above lemma its subset sequence generator $\Gamma^* = \dot{\Gamma} = (\dot{S}, \dot{G}, \dot{R}, \dot{P})$ is semi-deterministic. Hence for any given finite sequence of P-states $[p](0, t)$ there is at most one complete state $\dot{s}_1$ satisfying the condition that there exists a $\dot{\Gamma}$-sequence $[\dot{s}](0, t-1)$, $\dot{s}_1$ such that $P\{[\dot{s}](0, t-1), \dot{s}_1\} = [p](0, t)$. Moreover, this state $\dot{s}_1$ is a set of states of Γ. We will use the locution "the state $\dot{s}_1 \in \dot{S}$ corresponding to $[p](0, t)$" to refer to this set of states $\dot{s}_1$ if it exists, otherwise to the null set.

LEMMA 2.2-2. *Let $\Gamma^* = \dot{\Gamma} = (\dot{S}, \dot{G}, \dot{R}, \dot{P})$ be the subset sequence generator of $\Gamma = (S, G, R, P)$, let $[p](0, t)$ be any finite sequence of P-states, and let α be the set of states s_1 for which there exists a Γ-sequence $[s](0, t-1)$, s_1 such that $P\{[s](0, t-1), s_1\} = [p](0, t)$. Then α is the state $\dot{s}_1 \in \dot{S}$ corresponding to $[p](0, t)$.*

PROOF. (I) We first prove by an induction on t that for any finite Γ-sequence $[s](0, t)$ there is a $\dot{\Gamma}$-sequence $[\dot{s}](0, t)$ satisfying the conditions

$$\text{(a)} \quad P\{[s](0, t)\} = \dot{P}\{[\dot{s}](0, t)\},$$
$$\text{(b)} \quad s(t) \in \dot{s}(t).$$

INITIAL STEP. Given the Γ-sequence $s(0)$, it follows by the definition of $\dot{G}$ that there exists a complete state $\dot{s}(0)$ such that $s(0) \in \dot{s}(0)$ and $\dot{s}(0) \in \dot{G}$.

GENERAL STEP. The inductive hypothesis is that for every Γ-sequence $[s](0, k)$ that is a $\dot{\Gamma}$-sequence $[\dot{s}](0, k)$ satisfying the conditions (a) and (b). Consider any $\dot{\Gamma}$-sequence $[\dot{s}](0, k+1)$. By the inductive hypothesis there exists a $\dot{\Gamma}$-sequence $[\dot{s}](0, k)$ such that $[s](0, k)$ and $[\dot{s}](0, k)$ satisfy (a) and (b). Since $R(s(k), s(k+1))$ it follows by the definition of $\dot{R}$ that there exists a complete state $\dot{s}_1$ such that $s(k+1) \in \dot{s}_1$ and $\dot{R}(\dot{s}(k), \dot{s}_1)$. Hence $P\{s(k+1)\} = \dot{P}(\dot{s}_1)$ and $[\dot{s}](0, k)$, $\dot{s}_1$ is a Γ-sequence, and so $[s](0, k+1)$ and $[\dot{s}](0, k)$, $\dot{s}_1$ satisfy conditions (a) and (b).

(II) We next prove by an induction on t that for any finite $\dot{\Gamma}$-sequence $[\dot{s}](0, t)$ and complete state $s_1 \in \dot{s}(t)$ there is a Γ-sequence $[s](0, t)$ satisfying the conditions

$$\text{(c)} \quad P\{[s](0, t)\} = \dot{P}\{[\dot{s}](0, t)\},$$
$$\text{(d)} \quad s(t) = s_1.$$

INITIAL STEP. Given a $\dot{\Gamma}$-sequence $\dot{s}(0)$ and a state $s_1 \in \dot{s}(0)$, it follows by the definition of $\dot{G}$ that s_1 is the desired Γ-sequence.

GENERAL STEP. The inductive hypothesis is that for every $\dot{\Gamma}$-sequence $[\dot{s}](0, k)$ and state $s_1 \in \dot{s}(k)$ there is a Γ-sequence $[s](0, k)$ satisfying conditions (c) and (d). Consider any $\dot{\Gamma}$-sequence $[\dot{s}](0, k+1)$ and a state $s_2 \in \dot{s}(k+1)$. By the definition of $\dot{R}$ there exists a state s_1 satisfying the conditions $s_1 \in \dot{s}(k)$ and $R(s_1, s_2)$. By the inductive hypothesis there is a Γ-sequence $[s](0, k)$ such that $P\{[s](0, k)\} = \dot{P}\{[\dot{s}](0, k)\}$ and $s(k) = s_1$. $[s](0, k)$, s_2 is the desired Γ-sequence. This completes the proof of Lemma 2.2-2.

THEOREM 2.2-3. *For any sequence generator $\Gamma = (S, G, R, P)$ with behavior $\mathscr{B}(\Gamma)$, $\mathscr{B}(\Gamma) = \mathscr{B}(\Gamma^*)$.*

PROOF. By the preceding Lemma 2.2–2, for every finite t, $[p](0, t) \in \mathscr{B}(\Gamma)$ if and only if $[p](0, t) \in \mathscr{B}(\Gamma^*)$. The theorem to be proved now follows by the Infinity Theorem (2.1–3). It should be noted in this connection that the proof of the Infinity Theorem makes implicit use of Γ^*, the subset sequence generator of Γ. In fact, the sequence $\langle \alpha_0, \alpha_1, \alpha_2, \cdots \rangle$ employed in the proof is an infinite Γ^*-sequence.

COROLLARY 2.2–4. *For any sequence generator $\Gamma = (S, G, R, P)$, Γ is solvable if and only if Γ^* solvable.*

2.3. **Decision procedures.** *Behavior Inclusion Procedure*: Consider two sequence generators $\Gamma = (S, G, R, P)$ and $\dot{\Gamma} = (\dot{S}, \dot{G}, \dot{R}, \dot{P})$, and let $\{a\}[\dot{a}]$ be the number of states in $\{S\}[\dot{S}]$. Form all $\{\Gamma$-sequences$\}[\dot{\Gamma}$-sequences$]$ of length $1 + a2^{\dot{a}}$ or less and form the set $\{\alpha\}[\dot{\alpha}]$ of their $\{P$-projections$\}[\dot{P}$-projections$]$. Write "yes" or "no" as $\alpha \subset \dot{\alpha}$ or not.

THEOREM 2.3–1. *Let A be the class of pairs of one-projection sequence generators $\langle \Gamma, \dot{\Gamma} \rangle$ such that $\mathscr{B}(\Gamma) \subset \mathscr{B}(\dot{\Gamma})$, i.e., such that the behavior of Γ is included in that of $\dot{\Gamma}$. The Behavior Inclusion Procedure is a decision procedure for A.*

PROOF. (I) It is obvious that if $\mathscr{B}(\Gamma) \subset \mathscr{B}(\dot{\Gamma})$ then $\alpha \subset \dot{\alpha}$, i.e., that the algorithm yields "yes."

(II) We assume that $\alpha \subset \dot{\alpha}$, i.e., that the algorithm yields "yes," and prove that $\mathscr{B}(\Gamma) \subset \mathscr{B}(\dot{\Gamma})$. Let $\ddot{\Gamma} = (\ddot{S}, \ddot{G}, \ddot{R}, \ddot{P}) = \dot{\Gamma}^*$ and let $\ddot{\alpha}$ be the set of $\ddot{P}$-projections of $\ddot{\Gamma}$-sequences of length $1 + a2^{\dot{a}}$ or less. By Theorem 2.2–3 $\dot{\alpha} = \ddot{\alpha}$ and $\mathscr{B}(\dot{\Gamma}) = \mathscr{B}(\ddot{\Gamma})$, so we will assume that $\alpha \subset \ddot{\alpha}$ and prove that $\mathscr{B}(\Gamma) \subset \mathscr{B}(\ddot{\Gamma})$.

(IIA) Let $\{\alpha_t\}[\ddot{\alpha}_t]$ be the subset of sequences of $\{\mathscr{B}(\Gamma)\}[\mathscr{B}(\ddot{\Gamma})]$ of length $t + 1$ or less. We will now establish by induction that for every t, $\alpha_t \subset \ddot{\alpha}_t$.

INITIAL STEP. Since by assumption $\alpha \subset \ddot{\alpha}$ and by definition $\alpha = \alpha_{a\cdot 2^{\dot{a}}}$ and $\ddot{\alpha} = \ddot{\alpha}_{a\cdot 2^{\dot{a}}}$, so $\alpha \subset \ddot{\alpha}$ for $k = a\cdot 2^{\dot{a}}$.

GENERAL STEP. We assume that $\alpha_k \subset \ddot{\alpha}_k$ for $k \geqq a\cdot 2^{\dot{a}}$ and prove that $\alpha_{k+1} \subset \ddot{\alpha}_{k+1}$. Consider an arbitrary Γ-sequence $[s](0, k + 1)$ for $k \geqq a\cdot 2^{\dot{a}}$. Let

$$[p](0, k + 1) = P\{[s](0, k + 1)]\} .$$

By the inductive hypothesis there exists a $\ddot{\Gamma}$-sequence $[\ddot{s}](0, k)$ such that

$$\ddot{P}\{[\ddot{s}](0, k)\} = [p](0, k) .$$

We will show that there exists a complete state $\ddot{s}_0$ satisfying the conditions

$$\text{(1)} \qquad \ddot{R}\{\ddot{s}(k), \ddot{s}_0\} ,$$

$$\text{(2)} \qquad \ddot{P}(\ddot{s}_0) = p(k + 1) ,$$

i.e., that there exists a $\ddot{\Gamma}$-sequence $\langle [\ddot{s}](0, k), \ddot{s}_0 \rangle$ such that $\ddot{P}\{[\ddot{s}](0, k), \ddot{s}_0\} = [p](0, k + 1)$. It follows from the nature of the subset sequence generator construction that $\ddot{\Gamma}$ has no more than $2^{\dot{a}}$ complete states, and hence the number of pairs of complete states $\langle s, \ddot{s} \rangle$ is no more than $a\cdot 2^{\dot{a}}$. Since $k \geqq a\cdot 2^{\dot{a}}$ there exist t_1, t_2 such that $0 \leqq t_1 < t_2 \leqq a2^{\dot{a}} \leqq k$, $s(t_1) = s(t_2)$ and $\ddot{s}(t_1) = \ddot{s}(t_2)$.

Let

(3) $$[s_1](0, l+1) = \langle [s](0, t_1), [s](t_2+1, k+1)\rangle ,$$

(4) $$[\ddot{s}_1](0, l) = \langle [\ddot{s}](0, t_1), [\ddot{s}](t_2+1, k)\rangle ,$$

(5) $$[p_1](0, l+1) = \langle [p](0, t_1), [p](t_2+1, k+1)\rangle$$

where $l = k - (t_2 - t_1)$. Note that

(6) $$P\{[s_1](0, l+1)\} = [p_1](0, l+1) ,$$

(7) $$P\{[\ddot{s}_1](0, l)\} = [p_1](0, l) .$$

Because $\{s(t_1) = s(t_2)\}$ $[\ddot{s}(t_1) = \ddot{s}(t_2)]$ we have that $\{[s_1](0, l+1)$ is a Γ-sequence$\}$ $[[\ddot{s}_1](0, l)$ is a $\ddot{\Gamma}$-sequence$]$. And since $l + 1 \leqq k$ there exists by the inductive hypothesis a $\ddot{\Gamma}$-sequence $[\ddot{s}_2](0, l+1)$ such that

(8) $$\ddot{P}\{[\ddot{s}_2](0, l+1)\} = [p_1](0, l+1) .$$

It follows from (7) and (8) that

(9) $$\ddot{P}\{[\ddot{s}_2](0, l)\} = \ddot{P}\{[\ddot{s}_1](0, l)\} = [p_1](0, l)$$

and since $\ddot{\Gamma}$ is semi-deterministic (Lemma 2.2–1) we have that

(10) $$[\ddot{s}_2](0, l) = [\ddot{s}_1](0, l) .$$

It follows from (4) that

(11) $$\ddot{s}_1(l) = \ddot{s}(k)$$

and hence by (10)

(12) $$\ddot{s}_2(l) = \ddot{s}(k) .$$

Since $[\ddot{s}_2](0, l+1)$ is a $\ddot{\Gamma}$-sequence we have that $\ddot{R}\{\ddot{s}_2(l), \ddot{s}_2(l+1)\}$ and hence by (12) that

(13) $$\ddot{R}\{\ddot{s}(k), \ddot{s}_2(l+1)\} .$$

Now by (8) and (5)

(14) $$\ddot{P}\{\ddot{s}_2(l+1)\} = p(k+1) .$$

Conditions (13) and (14) show that $\ddot{s}_2(l+1)$ satisfies conditions (1) and (2) and hence that $\ddot{s}_2(l+1)$ is the desired state $\ddot{s}_0$.

(IIB) We have shown that if $\alpha \subset \ddot{\alpha}$ then for every t, $\alpha_t \subset \ddot{\alpha}_t$. It follows by the Infinity Theorem (Theorem 2.1–3) that if $\alpha \subset \ddot{\alpha}$, then $\mathscr{B}(\Gamma) \subset \mathscr{B}(\ddot{\Gamma})$. As remarked earlier, this is equivalent to: if $\alpha \subset \dot{\alpha}$ then $\mathscr{B}(\Gamma) \subset \mathscr{B}(\dot{\Gamma})$. This completes the proof of Theorem 2.3–1.

In formulating the Behavior Inclusion Procedure we have not attempted to minimize the computation required. Many simplifications will occur to anyone who uses this algorithm. For example, since any two elements of a complete state $\ddot{s} \in \ddot{S}$ must have the same projection, the bound $1 + a \cdot 2^{\dot{a}}$ may be greatly reduced. Note also that if $\dot{\Gamma}$ is already semi-deterministic, it is not necessary to make use of $\dot{\Gamma}^*$ in the proof, and the bound $1 + a \cdot 2^{\dot{a}}$ may be replaced by $1 + a\dot{a}$.

The Behavior Inclusion Procedure may be used as the basis of a decision procedure for solvability. Let $\Gamma = (S, G, R, P)$ be given. By definition Γ is solvable if every infinite sequence of P-states belongs to its behavior (§ 2.1). $\dot{\Gamma} = (S, S, \dot{R}, P)$, where $\dot{R}(s_1, s_2)$ for all $s_1, s_2 \in S$, has as its behavior the set of all sequences of P-states. Hence the behavior of $\dot{\Gamma}$ includes all infinite sequences of P-states, so Γ is solvable if and only if $\mathscr{B}(\dot{\Gamma}) \subseteq \mathscr{B}(\Gamma)$. By Theorem 2.3–1 the Behavior Inclusion Procedure is a decision procedure for behavior inclusion, so we have proved the following theorem.

THEOREM 2.3–2. *Let $\Gamma = (S, G, R, P)$ be a sequence generator and let $\dot{R}(s_1, s_2)$ for all $s_1, s_2 \in S$. The Behavior Inclusion Procedure applied to the pair $\langle (S, S, \dot{R}, P), \Gamma \rangle$ is a decision procedure for the solvability of Γ.*

When the Behavior Inclusion Procedure is applied first to the pair $\langle \Gamma, \dot{\Gamma} \rangle$ and then to the pair $\langle \dot{\Gamma}, \Gamma \rangle$ the result is "yes" in both cases if and only if $\mathscr{B}(\Gamma) = \mathscr{B}(\dot{\Gamma})$. This "behavior equivalence procedure" may be used to reduce the number of complete states of a sequence generator so as to obtain a behaviorally equivalent sequence generator with fewer states. Consider $\Gamma = (S, G, R, P)$. We will say that two complete states s_1 and s_2 are "behaviorally equivalent" if

$$\mathscr{B}[(S, \{s_1\}, R, P)] = \mathscr{B}[(S, \{s_2\}, R, P)] .$$

Let $\dot{\Gamma}$ be the result of identifying all behaviorally equivalent states of Γ. $\dot{\Gamma}$ will in general have fewer states than Γ and yet $\mathscr{B}(\dot{\Gamma}) = \mathscr{B}(\Gamma)$. Moore's concept of two sequential machines being indistinguishable by any experiment is a special case of our concept of behavioral equivalence, and the above process of identifying behaviorally equivalent states is analogous to the "reduction procedure" of Moore [**22**], and Mealy [**19**]. We have examples to show that the procedure we described does not always lead to a behaviorally equivalent sequence generator with a minimal number of complete states and that the procedure of Moore and Mealy does not always lead to a behaviorally equivalent sequential machine with a minimum number of internal states.[6]

3. SEQUENCE GENERATORS WITH TWO PROJECTIONS

3.1. **Definitions.** The results of the last section concern primarily one projection of a sequence generator. In the present section we will work mainly with two-projection sequence generators.

DEFINITION. $\Gamma = (S, G, R, P, Q)$ is *h-univalent* ($h = 0, 1, 2, \cdots; \omega$) if for every two infinite Γ-sequences $[s_1](0, \omega)$, $[s_2](0, \omega)$ and any time t, if $P([s_1](0, t + h)) = P([s_2](0, t + h))$ then $Q(s_1(t)) = Q(s_2(t))$. (By definition, $t + \omega = \omega$.)

Note that h-univalence is essentially a property of a set of infinite sequence

[6] Moore and Mealy showed that identifying behaviorally equivalent states of a sequential machine does lead to a minimum number of internal states if either the sequential machine is "strongly connected" (Moore) or it has exactly one generator (Mealy). The counter-examples mentioned above are not strongly connected and have more than one generator.

of pairs $\langle p, q\rangle$, and hence a property of $\mathscr{B}^{\omega}(\Gamma)$, the infinite behavior of Γ. As a consequence the following lemma holds.

LEMMA 3.1–1. *Let* $\Gamma = (S, G, R, P, Q)$ *and* $\dot{\Gamma} = (\dot{S}, \dot{G}, \dot{R}, \dot{P}, \dot{Q})$. *If* $\mathscr{B}^{\omega}(\Gamma) = \mathscr{B}^{\omega}(\dot{\Gamma})$, *then* Γ *is h-univalent if and only if* $\dot{\Gamma}$ *is h-univalent.*

This lemma, together with Corollary 2.1–1 and Theorem 2.2–3, immediately yields

LEMMA 3.1–2. *Let* $\Gamma = (S, G, R, P, Q)$. *The following three conditions are equivalent*: (1) Γ *is h-univalent,* (2) Γ^* *is h-univalent, and* (3) $\Gamma^{\dagger}$ *is h-univalent.*

There is a close connection between zero-univalence and semi-determinism which is brought out by the following lemma.

LEMMA 3.1–3. (a) *Let* $\Gamma = (S, G, R, P, Q)$ *and* $\Gamma = \Gamma^{\dagger}$. Γ *is* 0*-univalent if and only if for any two finite* Γ*-sequences* $[s_1](0, t)$ *and* $[s_2](0, t)$, *if* $P([s_1](0, t)) = P([s_2](0, t))$ *then* $Q([s_1](0, t)) = Q([s_2](0, t))$.

(b) *Let* $\Gamma = (S, G, R, P)$ *and* $\Gamma = \Gamma^{\dagger}$. Γ *is semi-deterministic if and only if for any two finite* Γ*-sequences,* $[s_1](0, t)$ *and* $[s_2](0, t)$, *if* $P([s_1](0, t)) = P([s_2](0, t))$ *then* $[s_1](0, t) = [s_2](0, t)$.

Note that the sequence generator of part (a) of the lemma has two projections, while that of part (b) has one projection. Part (a) may be established by using Corollary 1.3–3 and the definition of univalence; part (b) follows from Corollary 1.3–3 and Lemma 2.1–4b. It follows from Lemma 3.1–3 that for any projection Q, if (S, G, R, P) is semi-deterministic then (S, G, R, P, Q) is 0-univalent.

The converse is not in general true, but the following lemma asserts a connection between the 0-univalence of a sequence generator and the semi-determinism of a related sequence generator.

LEMMA 3.1–4. *Let* $\Gamma = (S, G, R, P, Q)$ *and let* $\Gamma^{*\dagger} = \dot{\Gamma} = (\dot{S}, \dot{G}, \dot{R}, \dot{P}, \dot{Q})$. Γ *is zero-univalent if and only if* $(\dot{S}, \dot{G}, \dot{R}, \dot{P})$ *is semi-deterministic.*

It might seem that since $\dot{\Gamma}$ is the reduced form of the subset sequence generator of Γ, it would follow immediately by Lemma 2.2–1 that if Γ is 0-univalent then $(\dot{S}, \dot{G}, \dot{R}, \dot{P})$ is semi-deterministic. This is by no means the case. It can be shown from Lemma 2.2–1 by means of the definition of the subset sequence generator operation that $(\dot{S}, \dot{G}, \dot{R}, \dot{P} \times \dot{Q})$ is semi-deterministic, while the conclusion of Lemma 3.1–4 is that $(\dot{S}, \dot{G}, \dot{R}, \dot{P})$, which is a different sequence generator, is semi-deterministic. For any projection $\dot{Q}$, if a sequence generator (S, G, R, P) is semi-deterministic, then $(\dot{S}, \dot{G}, \dot{R}, \dot{P} \times \dot{Q})$ is semi-deterministic, but the converse is not in general true.

PROOF OF LEMMA 3.1–4 (*"Only if"* part). We assume that Γ is 0-univalent and prove that $(\dot{S}, \dot{G}, \dot{R}, \dot{P})$ is semi-deterministic.

(I) We will use three sequence generators in the proof besides Γ. These are $\dot{\Gamma} = (\dot{S}, \dot{G}, \dot{R}, \dot{P}, \dot{Q})$, $\ddot{\Gamma} = (\dot{S}, \dot{G}, \dot{R}, \dot{P} \times \dot{Q})$, and $\dddot{\Gamma} = (\dot{S}, \dot{G}, \dot{R}, \dot{P})$. We will first establish some results that will enable us to use Lemma 3.1–3a on $\dot{\Gamma}$ and Lemma 3.1–3b on $\ddot{\Gamma}$ and $\dddot{\Gamma}$. (A) By construction $\dot{\Gamma} = \dot{\Gamma}^{\dagger}$ and, by Lemma 3.1–2, $\dot{\Gamma}$ is 0-univalent. (B) By construction $\ddot{\Gamma} = \ddot{\Gamma}^{\dagger}$ and by Lemma 2.1–1

$\ddot{\Gamma}$ is semi-deterministic. (C) By construction $\dddot{\Gamma} = \ddot{\Gamma}^{\dagger}$. Our task is to prove that $\dddot{\Gamma}$ is semi-deterministic.

(II) Since $\dot{\Gamma}$, $\ddot{\Gamma}$, and $\dddot{\Gamma}$ have $\dot{S}, \dot{G}, \dot{R}$ in common, the sets of $\dot{\Gamma}$-sequences, $\ddot{\Gamma}$-sequences, and $\dddot{\Gamma}$-sequences are identical with one another. Consider now any two finite $\dot{\Gamma}$-sequences $[\dot{s}_1](0, t)$ and $[\dot{s}_2](0, t)$; these are also arbitrary $\ddot{\Gamma}$-sequences and arbitrary $\dddot{\Gamma}$-sequences. Using (IA) and applying Lemma 3.1–3a to $\dot{\Gamma}$, we obtain

(1) If $\dot{P}([\dot{s}_1](0, t)) = \dot{P}([\dot{s}_2](0, t))$, then $\dot{Q}([\dot{s}_1](0, t)) = \dot{Q}([\dot{s}_2](0, t))$.

Using (IB) and applying Lemma 3.1–3b to $\ddot{\Gamma}$, we obtain

(2) If $\dot{P}([\dot{s}_1](0, t)) = \dot{P}([\dot{s}_2](0, t))$ and $\dot{Q}([\dot{s}_1](0, t)) = \dot{Q}([\dot{s}_2](0, t))$, then $[\dot{s}_1](0, t) = [\dot{s}_2](0, t)$.

Combining (1) and (2) and noting that $[\dot{s}_1](0, t)$ and $[\dot{s}_2](0, t)$ are arbitrary $\dddot{\Gamma}$-sequences, we get

(3) For any two finite $\dddot{\Gamma}$-sequences $[\dot{s}_1](0, t)$ and $[\dot{s}_2](0, t)$, if $\dot{P}([\dot{s}_1](0, t)) = \dot{P}([\dot{s}_2](0, t))$, then $[\dot{s}_1](0, t) = [\dot{s}_2](0, t)$.

Using (3) and (IC) and applying Lemma 3.1–3b to $\dddot{\Gamma}$, we obtain

(4) $\dddot{\Gamma}$ is semi-deterministic,

which completes the proof of the "only if" part of the lemma.

("*If*" part). We assume that $(\dot{S}, \dot{G}, \dot{R}, \dot{P})$ is semi-deterministic and prove that Γ is 0-univalent. Since every complete state of $\dot{\Gamma}$ is $\dot{\Gamma}$-accessible, by Lemma 2.1–4b we have that for every finite sequence of $\dot{P}$-states $[\dot{p}](0, t)$ there exists at most one $\dot{\Gamma}$-sequence $[\dot{s}](0, t)$ such that $\dot{P}\{[\dot{s}](0, t)\} = [\dot{p}](0, t)$. Hence for any two $\dot{\Gamma}$-sequences $[\dot{s}_1](0, \omega)$, $[\dot{s}_2](0, \omega)$ and any time t, if $\dot{P}([\dot{s}_1](0, t)) = \dot{P}([\dot{s}_2](0, t))$, then $\dot{s}_1(t) = \dot{s}_2(t)$. Since a projection is a (single-valued) function, we have that if $\dot{P}([\dot{s}_1](0, t + 0)) = \dot{P}([\dot{s}_2](0, t + 0))$ then $\dot{Q}(\dot{s}_1(t)) = \dot{Q}(\dot{s}_2(t))$, so $\dot{\Gamma}$ is 0-univalent. $\dot{\Gamma} = \Gamma^{*\dagger}$, and by Lemma 3.1–2 Γ is 0-univalent. This completes the proof of Lemma 3.1–4.

We next apply this lemma to an example. Consider $\Gamma = (S, G, R, P, Q)$ of Fig. 8(a). Note that the complete states s_2 and s_4 have the same projections (p_0 and q_2) and stand in the same relation to state s_0. Thus the two Γ-sequences

$$s_0, s_4, s_0$$

$$s_0, s_2, s_0$$

have the same sequence of P-projections

$$p_0, p_0, p_0$$

and hence (S, G, R, P) is not semi-deterministic. These two Γ-sequences do have the same sequence of Q-projections

$$q_0, q_2, q_0$$

and in fact Γ is 0-univalent. By Lemma 3.1–4 $(\dot{S}, \dot{G}, \dot{R}, \dot{P})$ of Fig. 8(b) must be semi-deterministic. An examination of the states $(\dot{S}, \dot{G}, \dot{R}, \dot{P})$ shows that it is deterministic, so *a fortiori* it is semi-deterministic. (The determinism of $(\dot{S}, \dot{G}, \dot{R}, \dot{P})$ will be discussed after Lemma 3.2–3 below.) Note that the main difference between Γ and $\Gamma^{*\dagger}$ in Fig. 8 is that the two states s_2, s_4 of Γ have become a single state $\{s_2, s_4\}$ of $\Gamma^{*\dagger}$.

DEFINITION. $\Gamma = (S, G, R, P, Q)$ is *uniquely solvable* if (1) (S, G, R, P) is solvable and (2) (S, G, R, P, Q) is ω-univalent. We remarked earlier that h-univalence is essentially a property of the infinite behavior of a sequence generator, and this remark applies to unique solvability as well. Thus Γ is uniquely

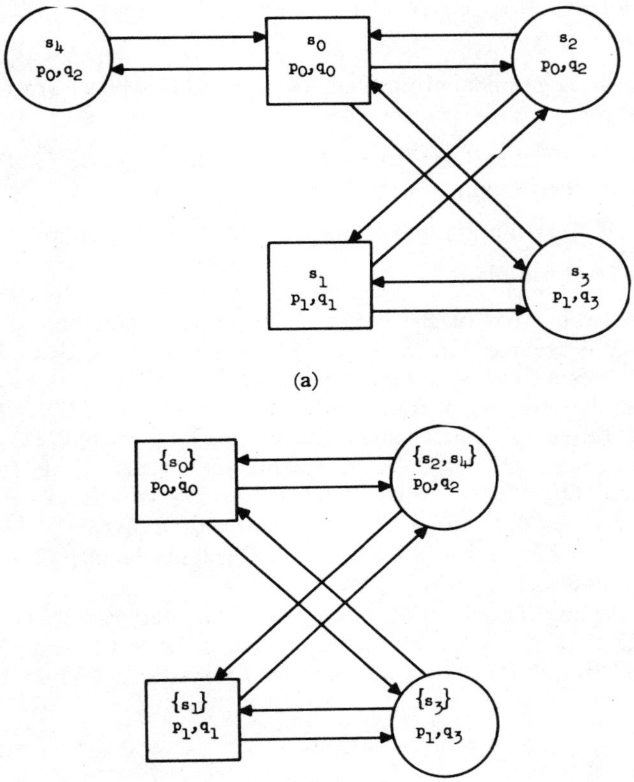

(a)

(b)

Figure 8. Illustration of Lemma 3.1-4: If $\Gamma = (S, G, R, P, Q)$ and $\Gamma^{*\dagger} = \dot{\Gamma} = (\dot{S}, \dot{G}, \dot{R}, \dot{P}, \dot{Q})$, then Γ is zero-univalent if and only if $(\dot{S}, \dot{G}, \dot{R}, \dot{P})$ is semi-deterministic.

(a) $\Gamma = (S, G, R, P, Q)$. Γ is zero-univalent and uniquely solvable, but (S, G, R, P) is not semi-deterministic.

(b) $\Gamma^{*\dagger} = \dot{\Gamma} = (\dot{S}, \dot{G}, \dot{R}, \dot{P}, \dot{Q})$. $(\dot{S}, \dot{G}, \dot{R}, \dot{P})$ is deterministic, and *a fortiori* semi-deterministic.

solvable if and only if for any infinite sequence of P-states $[p](0, \omega)$ there is exactly one sequence of Q-states $[q](0, \omega)$ such that the sequence

$$\langle p(0), q(0)\rangle, \langle p(1), q(1)\rangle, \cdots$$

belongs to $\mathscr{B}(\Gamma)$. To put the point in another way: a sequence generator $\Gamma = (S, G, R, P, Q)$ is uniquely solvable if and only if its behavior defines a single-valued function (transformation) from the set of all infinite sequences of P-states into the set of all infinite sequences of Q-states. Various consequences follow from this fact. The result of replacing "h-univalence" by "uniquely solvable" in Lemma 3.1–1 is also a lemma. A similar remark holds for Lemma 3.1–2 except that $\Gamma^{\dagger}$ may have fewer P-states (values of p) than Γ.

It was shown in § 2.1 that well-formed nets and deterministic sequence generators are equivalent in a certain sense: for every w.f.n. there is a corresponding deterministic sequence generator and vice-versa. The w.f.n. gives the structure of an automaton while the associated deterministic sequence generator gives the corresponding complete state diagram. An analogous relation holds between the well-behaved nets of Burks and Wright [**6**, p. 1358], and uniquely solvable sequence generators. Consider any net and label all its noninput nodes as output nodes. The procedure of § 1.2 will associate with this net a sequence generator which is uniquely solvable if and only if the original net is well-behaved.

3.2. **The displacement operator and the l-shift operation.** We will first define a displacement operator $\mathscr{D}^k$ which applies to sets composed of finite sequences of pairs and/or ω-sequences of pairs. Roughly speaking, $\mathscr{D}^k$ has the effect of leaving the first element of each pair where it is and displacing the second element of each pair k places to the right. Displacing the second element of the first pair k places to the right will leave k gaps, since the first pair is not preceded by any pair. It will be convenient always to fill these gaps with the same element; we will use the null set Λ for this purpose.

DEFINITION. Let the universe of discourse V consist of all finite sequences of pairs and all ω-sequences of pairs and let Λ be the null set. The operator $\mathscr{D}$ (without superscript) is defined to apply to any sequence of V as follows:

$$\begin{aligned}\mathscr{D}(\langle x_0, y_0\rangle, \langle x_1, y_1\rangle, \langle x_2, y_2\rangle, \langle x_3, y_3\rangle, \cdots)\\ = (\langle x_0, \Lambda\rangle, \langle x_1, y_0\rangle, \langle x_2, y_1\rangle, \langle x_3, y_2\rangle, \cdots)\,,\\ \mathscr{D}(\langle x_0, y_0\rangle, \langle x_1, y_1\rangle, \cdots, \langle x_{n-1}, y_{n-1}\rangle, \langle x_n, y_n\rangle)\\ = (\langle x_0, \Lambda\rangle, \langle x_1, y_0\rangle, \cdots, \langle x_{n-1}, y_{n-2}\rangle, \langle x_n, y_{n-1}\rangle)\,.\end{aligned}$$

The operator $\mathscr{D}$ is extended to apply to an arbitrary set α of V by

$$\mathscr{D}(\alpha) = \{v \mid (\exists u)[u \in \alpha \;\&\; v = \mathscr{D}(u)]\}\,,$$

where v and u range over elements of V. Finally, we define $\mathscr{D}^k$, $k = 0, 1, 2, \cdots$, to apply to an arbitrary set α of V by the induction

$$\mathscr{D}^0(\alpha) = \alpha ,$$
$$\mathscr{D}^{i+1}(\alpha) = \mathscr{D}(\mathscr{D}^i(\alpha)) .$$

$\mathscr{D}^l$ is called the *displacement operator.*

We next define a shifting operator which may be applied to an arbitrary sequence generator $\Gamma = (S, G, R, P, Q)$ to produce the l-shifted sequence generator $\Gamma^l = \dot{\Gamma} = (\dot{S}, \dot{G}, \dot{R}, \dot{P}, \dot{Q})$. The effect of this operation is to displace the behavior of Γ, so that the behavior of Γ^l, i.e., $\mathscr{B}(\Gamma^l)$, equals the displaced behavior of Γ, i.e., $\mathscr{D}^l[\mathscr{B}(\Gamma)]$, as is shown in Lemma 3.2–1 below. To help make clear the definition of Γ^l, we will make some remarks about Γ^1. Extend Q to apply to Λ, so that $Q(\Lambda) = \Lambda$. The generators of Γ^1 are the pairs $\langle \Lambda, s\rangle$, where s belongs to G. Suppose

$$s_0, s_1, s_2, s_3, s_4$$

is a Γ-sequence with the resulting behavior element

$$\langle p_0, q_0\rangle, \langle p_1, q_1\rangle, \langle p_2, q_2\rangle, \langle p_3, q_3\rangle, \langle p_4, q_4\rangle .$$

Then

$$\langle \Lambda, s_0\rangle, \langle s_0, s_1\rangle, \langle s_1, s_2\rangle, \langle s_2, s_3\rangle, \langle s_3, s_4\rangle$$

is the corresponding Γ^1-sequence with the resulting behavior element

$$\langle p_0, \Lambda\rangle, \langle p_1, q_0\rangle, \langle p_2, q_1\rangle, \langle p_3, q_3\rangle, \langle p_4, q_3\rangle .$$

Γ^l may be obtained by shifting Γ l times in this way.

DEFINITION. The *unit-shift operation*, denoted by "$\diamond$," applies to any sequence generator $\Gamma = (S, G, R, P, Q)$ and produces a sequence generator $\Gamma^\diamond = \dot{\Gamma} = (\dot{S}, \dot{G}, \dot{R}, \dot{P}, \dot{Q})$ defined as follows. (We wish to assume throughout that Λ is not an element of S; if it is, S should first be redefined so it is not.)

(1) The elements of $\dot{G}$ are all the pairs $\langle \Lambda, s\rangle$ where s belongs to G:

$$\langle \Lambda, s\rangle \in \dot{G} \equiv s \in G .$$

(2) The elements of $\dot{S}$ are all the pairs $\langle s_1, s_2\rangle$ which either belong to $\dot{G}$ or are connected by the direct transition relation R:

$$\dot{S} = \{\langle s_1, s_2\rangle \mid \langle s_1, s_2\rangle \in \dot{G} \vee R(s_1, s_2)\} .$$

(3) Two complete states $\langle s_1, s_2\rangle$ and $\langle s_3, s_4\rangle$ of $\dot{S}$ stand in the direct transition relation $\dot{R}$ if and only if $s_2 = s_3$:

$$\dot{R}(\langle s_1, s_2\rangle, \langle s_3, s_4\rangle) \equiv [\langle s_1, s_2\rangle, \langle s_3, s_4\rangle \in \dot{S} \;\&\; s_2 = s_3] .$$

(4) The $\dot{P}$-projection of a complete state $\langle s_1, s_2\rangle$ of $\dot{S}$ is the P-projection of its *second* element s_2:

$$\dot{P}(\langle s_1, s_2\rangle) = P(s_2), \text{ where } \langle s_1, s_2\rangle \in \dot{S} .$$

(5) Extend Q to apply to Λ, stipulating that $Q(\Lambda) = \Lambda$. The $\dot{Q}$-projection of the complete state $\langle s_1, s_2\rangle$ of $\dot{S}$ is the Q-projection of its *first* element:

$$\dot{Q}(\langle s_1, s_2\rangle) = Q(s_1), \text{ where } \langle s_1, s_2\rangle \in \dot{S} .$$

The *l-shift operation*, denoted by "l," applies to any sequence generator $\Gamma = (G, S, G, R, P, Q)$ and produces a sequence generator Γ^l; it is defined in terms of the unit-shift operation by means of an induction:

$$\Gamma^0 = \Gamma \,,$$
$$\Gamma^{l+1} = (\Gamma^l)^{\diamond} \,.$$

The l-shift construction is illustrated in Fig. 9. Part (a) shows a sequence generator Γ with two projections, while part (b) shows Γ^1, the result of shifting Γ one unit of time. Note that both Γ and Γ^1 are in reduced form; this is a special case of the general fact that if Γ is in reduced form then Γ^h is in reduced form. Note next that the generator $\langle \Lambda, s_0 \rangle$ can only occur as the first state of a Γ^1-sequence. Since Γ^l is defined by induction on $\Gamma^{\diamond}$, it follows by induction that if a Γ^l-admissible complete state occurs in some Γ^l-sequence at a time $\tau < l$ then all occurrences of this state are at time τ. Thus the Γ^l-admissible complete states are partitioned into two sets, those that occur only before time l, and those that occur only at time l or later.

The l-shift operation is of interest because it shifts behavior in the same way the displacement operator $\mathscr{D}^l$ does. Compare the behaviors of Γ and Γ^1. The behavior element [an element of $\mathscr{B}(\Gamma)$]

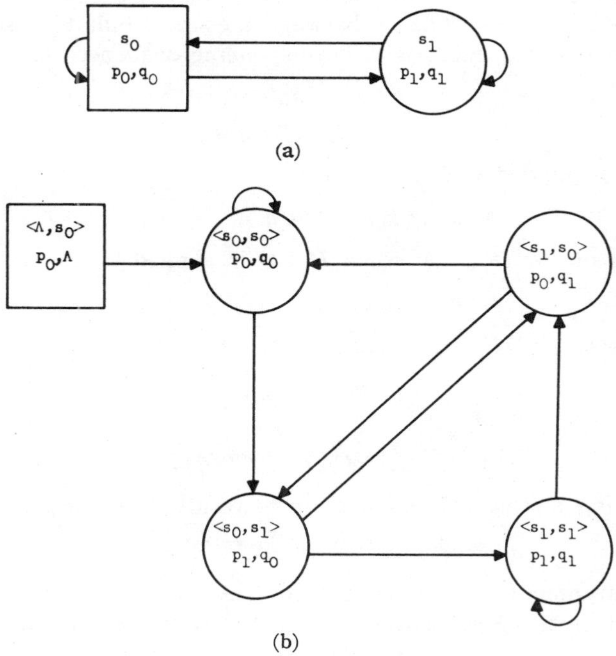

Figure 9. Illustration of the l-shift construction, for $l = 1$. In accordance with Lemma 3.2-1, $\mathscr{D}^1[\mathscr{B}(\Gamma)] = \mathscr{B}(\Gamma^1)$.

(a) $\Gamma = (S, G, R, P, Q)$;

(b) Γ^1 (unit-shifted sequence generator of Γ).

$$(1) \qquad \langle p_0, q_0\rangle, \langle p_0, q_0\rangle, \langle p_1, q_1\rangle, \langle p_0, q_0\rangle$$

is derived from the Γ-sequence

$$s_0, s_0, s_1, s_1, s_0 \ .$$

The corresponding Γ^1-sequence is

$$\langle \Lambda, s_0\rangle, \langle s_0, s_0\rangle, \langle s_0, s_1\rangle, \langle s_1, s_1\rangle, \langle s_1, s_0\rangle \ ,$$

which gives rise to the behavior element [an element of $\mathscr{B}(\Gamma^1)$]

$$(2) \qquad \langle p_0, \Lambda\rangle, \langle p_0, q_0\rangle, \langle p_1, q_0\rangle, \langle p_1, q_1\rangle, \langle p_0, q_1\rangle \ .$$

Note that this last sequence (2) is the result of displacing sequence (1) by one unit. This is an example of the general fact that $\mathscr{B}(\Gamma^1) = \mathscr{D}^1[\mathscr{B}(\Gamma)]$, which is a special case of the following lemma.

LEMMA 3.2-1. *Let* $\Gamma = (S, G, R, P, Q)$. *Then*

$$\text{(a)} \qquad \mathscr{D}^l[\mathscr{B}(\Gamma)] = \mathscr{B}(\Gamma^l) \ ,$$

$$\text{(b)} \qquad \mathscr{D}^l[\mathscr{B}^\omega(\Gamma)] = \mathscr{B}^\omega(\Gamma^l) \ .$$

PROOF. (IA) We prove first that $\mathscr{D}[\mathscr{B}^\omega(\Gamma)] = \mathscr{B}^\omega(\Gamma^\diamond)$. Let $\Gamma^\diamond = \dot{\Gamma} = (\dot{S}, \dot{G}, \dot{R}, \dot{P}, \dot{Q})$. It follows from the definition of the unit shift operator that there is a one-one correspondence between the set of infinite Γ-sequences and the set of infinite $\dot{\Gamma}$-sequences with corresponding sequences being of the form

$$(1) \qquad s_0, s_1, s_2, s_3, \cdots \ ,$$

$$(2) \qquad \langle \Lambda, s_0\rangle, \langle s_0, s_1\rangle, \langle s_1, s_2\rangle, \langle s_2, s_3\rangle, \cdots \ .$$

When $P \times Q$ is applied to (1), we get

$$(3) \qquad \langle p_0, q_0\rangle, \langle p_1 q_1\rangle, \langle p_2, q_2\rangle, \langle p_3, q_3\rangle, \cdots$$

as an element of $\mathscr{B}^\omega(\Gamma)$, and when $\dot{P} \times \dot{Q}$ is applied to (2), we get

$$(4) \qquad \langle p_0, \Lambda\rangle, \langle p_1, q_0\rangle, \langle p_2, q_1\rangle, \langle p_3, q_2\rangle, \cdots$$

as an element of $\mathscr{B}^\omega(\Gamma)$. By definition of $\mathscr{D}$,

$$\mathscr{D}[(3)] = (4) \ .$$

Hence

$$\mathscr{D}[\mathscr{B}^\omega(\Gamma)] = \mathscr{B}^\omega(\Gamma^\diamond) \ .$$

(IB) Applying mathematical induction to result (IA), we get

$$\mathscr{D}^l[\mathscr{B}^\omega(\Gamma)] = \mathscr{B}^\omega(\Gamma^l)] \ .$$

(IIA) An argument similar to that of (IA) may be given for finite Γ-sequences and finite $\Gamma^\diamond$-sequences. When the result is combined with result (IA), we get

$$\mathscr{D}[\mathscr{B}(\Gamma)] = \mathscr{B}(\Gamma^\diamond) \ .$$

(IIB) Applying mathematical induction to (IIA), we get

$$\mathscr{D}^l[\mathscr{B}(\Gamma)] = \mathscr{B}(\Gamma^l) \ .$$

Corollary 3.2–2. *Let* $\Gamma = (S, G, R, P, Q)$ *and* $\Gamma^l = \dot{\Gamma} = (\dot{S}, \dot{G}, \dot{R}, \dot{P}, \dot{Q})$. [*$\Gamma$ is $(l+h)$-univalent*] {*(S, G, R, P) is solvable*} (*Γ is uniquely solvable*) *if and only if* [*Γ^l is h-univalent*] {*$(\dot{S}, \dot{G}, \dot{R}, \dot{P})$ is solvable*} (*Γ^l is uniquely solvable*).

This corollary is illustrated by Fig. 10. Consider Γ of this figure. It was shown in § 2.1 (in the paragraph preceding Lemma 2.1–4) that (S, G, R, P) is solvable. Also, Γ is unit-univalent. To see this, observe that every immediate successor (by the direct transition relation R) of a given complete state s has the same P-projection; e.g., $R(s_0) = \{s_0, s_1\}$ and $P(s_0) = P(s_1) = p_0$. Since (S, G, R, P) is solvable and Γ is unit-univalent, Γ is uniquely solvable. Turn now to Γ^1, the unit-shifted sequence generator of Γ. Since (S, G, R, P) is solvable and Γ is unit-univalent and uniquely solvable, by Corollary 3.2–2 Γ^1 (less its last projection) must be solvable, and Γ^1 must be zero-univalent and also uniquely solvable.

Lemma 3.2–3. *Let* $\Gamma = (S, G, R, P, Q)$ *and* $\Gamma^{h*\dagger} = \dot{\Gamma} = (\dot{S}, \dot{G}, \dot{R}, \dot{P}, \dot{Q})$. *Then* (a) *if Γ is h-univalent for some finite h and (S, G, R, P) is solvable, then $(\dot{S}, \dot{G}, \dot{R}, \dot{P})$ is deterministic,* (b) $\mathscr{D}^h[\mathscr{B}^\omega(\Gamma)] = \mathscr{B}^\omega(\dot{\Gamma})$.

Proof. (IA) We will prove first that $(\dot{S}, \dot{G}, \dot{R}, \dot{P})$ is solvable. It is given that (S, G, R, P) is solvable. By Corollary 3.2–2, Lemma 2.2–4, and Lemma 2.1–4a, $(\dot{S}, \dot{G}, \dot{R}, \dot{P})$ is solvable. (IB) We prove next that $(\dot{S}, \dot{G}, \dot{R}, \dot{P})$ is semi-deterministic. It is given that Γ is h-univalent. By Corollary 3.2–2, Γ^h is 0-univalent. By Lemma 3.1–4, $(\dot{S}, \dot{G}, \dot{R}, P)$ is semi-deterministic. (IC) It follows from Lemma 2.1–4b and the definition of solvable that if every complete state of any sequence generator $\Gamma = (S, G, R, P)$ is Γ-accessible, then Γ is deter-

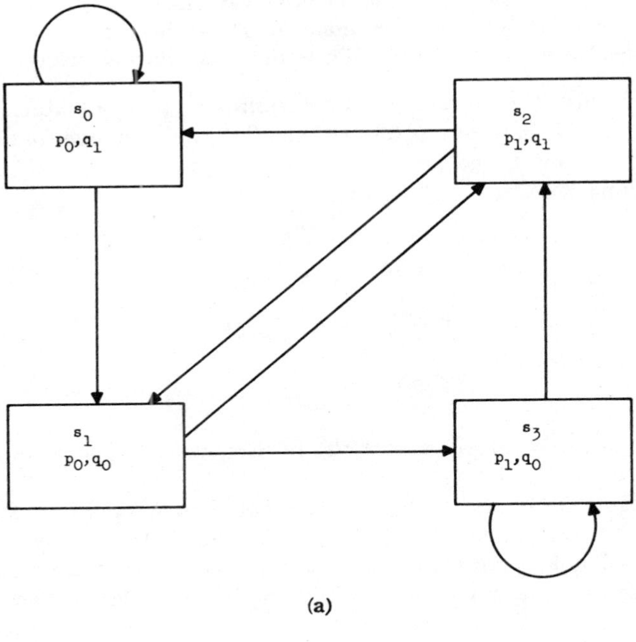

(a)

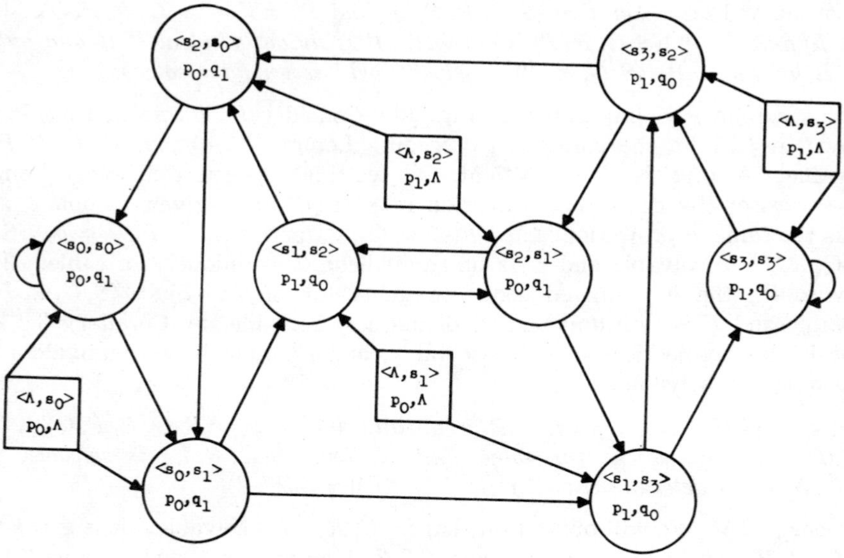

(b)

Figure 10. Illustration of Corollary 3.2–2. [Γ is 1-univalent] {Γ, less its last projection, is solvable} (Γ is uniquely solvable) if and only if [Γ^1 is 0-univalent] {Γ^1, less its last projection, is solvable} (Γ^1 is uniquely solvable).

(a) $\Gamma = (S, G, R, P, Q)$. (S, G, R, P) is solvable but not deterministic. Γ is unit-univalent and uniquely solvable.

(b) Γ^1, the unit-shifted sequence generator of Γ. Γ^1, less its last projection, is solvable, but not deterministic. Γ^1, is zero-univalent and uniquely solvable.

ministic if and only if Γ is both semi-deterministic and solvable. Applying this principle to $(\dot{S}, \dot{G}, \dot{R}, \dot{P})$ and using (IA) and (IB), we conclude that $(\dot{S}, \dot{G}, \dot{R}, \dot{P})$ is deterministic. This prove part (a) of the lemma.

(II) By Lemma 3.2–1b

(1) $$\mathscr{D}^h[\mathscr{B}^\omega(\Gamma)] = \mathscr{B}^\omega(\Gamma^h) .$$

By Theorem 2.2–3 $\mathscr{B}(\Gamma^h) = \mathscr{B}(\Gamma^{h*})$ and hence

(2) $$\mathscr{B}^\omega(\Gamma^h) = \mathscr{B}^\omega(\Gamma^{h*}) .$$

But by Corollary 2.1–1

(3) $$\mathscr{B}^\omega(\Gamma^{h*}) = \mathscr{B}^\omega(\Gamma^{h*\dagger}) .$$

Combining (1), (2), and (3) gives part (b) of Lemma 3.2–3 and completes the proof of the present lemma.

We may apply this lemma to Fig. 8. As noted in § 3.1, (S, G, R, P) is not semi-deterministic but (S, G, R, P, Q) is 0-univalent. It is easy to see that (S, G, R, P) is solvable. Applying Lemma 3.2–3 with $h = 0$, we conclude that $(\dot{S}, \dot{G}, \dot{R}, \dot{P})$ is deterministic and that $\mathscr{B}^\omega(\Gamma) = \mathscr{B}^\omega(\Gamma)$. These two facts may be confirmed by inspection of Fig. 8(b).

Actually $\mathscr{B}(\Gamma) = \mathscr{B}(\dot{\Gamma})$ in Fig. 8, i.e., the finite behaviors of Γ and $\dot{\Gamma}$ are equal as well as the infinite behaviors. There is a variant of Lemma 3.2–3 which covers this point. Since our main interest in the present section is in infinite behavior, we will merely state this result without proof. Let $\Gamma = (S, G, R, P, Q)$ be h-univalent and (S, G, R, P) solvable; then Γ^{h*} less its Q-projection is deterministic, $\mathscr{D}^h[\mathscr{B}(\Gamma)] = \mathscr{B}(\Gamma^{h*})$, and if Γ is in reduced form then $\mathscr{B}(\Gamma^{h*}) = \mathscr{B}(\Gamma^{h*\dagger})$.

Fig. 11 also illustrates Lemma 3.2–3. We begin with $\Gamma = (S, G, R, P, Q)$, where Γ is unit-univalent and (S, G, R, P) is solvable. Lemma 3.2–3 tells us that $\Gamma^{1*\dagger}$ (less its last projection) is deterministic, and also that $\mathscr{D}^1[\mathscr{B}^\omega(\Gamma)] = \mathscr{B}^\omega(\Gamma^{1*\dagger})$. [$\Gamma^1$ is shown in Fig. 10(b); it has 12 complete states. Γ^{1*} has 28 states, but only 6 of these are Γ^{1*}-admissible, so $\Gamma^{1*\dagger}$ has only 6 states.]

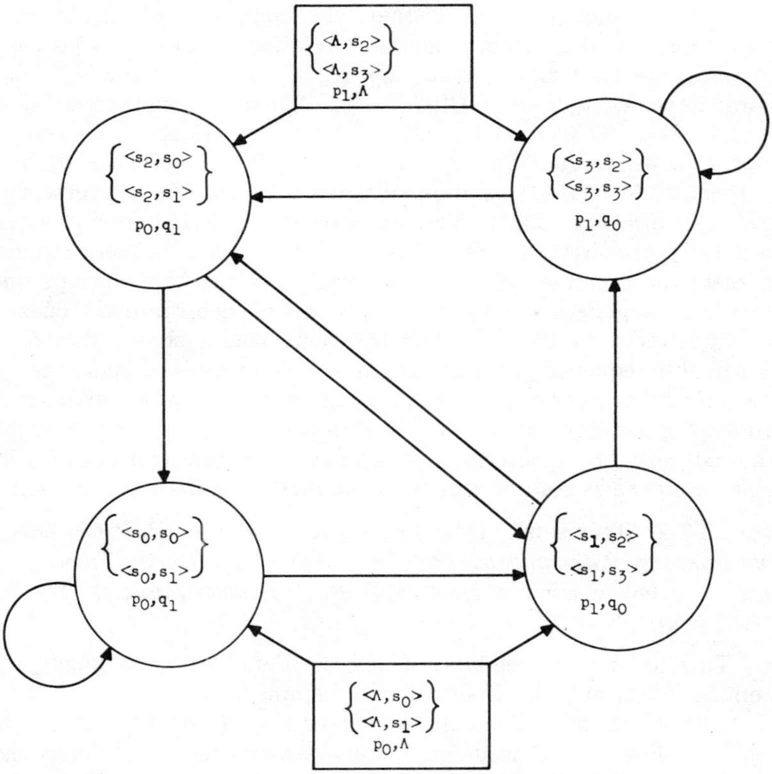

Figure 11. $\Gamma^{1*\dagger}$, where Γ is Figure 10(a). Γ and Γ^1 (less their last projections) are not deterministic, but $\Gamma^{1*\dagger}$ (less its last projection) is deterministic. $\mathscr{D}^1[\mathscr{B}^\omega(\Gamma)] = \mathscr{B}^\omega(\Gamma^{1*\dagger})$. This illustrates Lemma 3.2–3.

3.3. **Time-shift Theorem.** We will prove now a lemma which is used in proving one of our main theorems (the Time-shift Theorem) and in validating a procedure for h-univalence.

LEMMA 3.3–1 (FIXED BOUND LEMMA). *Let $\Gamma = (S, G, R, P, Q)$ be a sequence generator with k Γ-admissible complete states. Then Γ is ω-univalent if and only if it is k^2-univalent.*

PROOF. The proof in one direction is obvious. To prove that if Γ is ω-univalent it is k^2-univalent, we consider any two Γ-sequences $[s_1](0, \omega)$, $[s_2](0, \omega)$ and any time t such that $P([s_1](0, t + k^2)) = P([s_2](0, t + k^2))$. Since there are k^2 distinct pairs of complete states, there are two times t_1, t_2 such that $t \leqq t_1 < t_2 \leqq t + k^2$, $s_1(t_1) = s_1(t_2)$, and $s_2(t_1) = s_2(t_2)$. Form the sequences

$$[s_3](0, \omega) = [s_1](0, t_2 - 1), [s_1](t_1, t_2 - 1), [s_1](t_1, t_2 - 1), \cdots ,$$
$$[s_4](0, \omega) = [s_2](0, t_2 - 1), [s_2](t_1, t_2 - 1), [s_2](t_1, t_2 - 1), \cdots .$$

These are both Γ-sequences since they are composed of segments of Γ-sequences linked by the direct transition relation. Since $P([s_1](0, t + k^2)) = P([s_2](0, t + k^2))$ we have by construction $P([s_3](0, \omega)) = P([s_4](0, \omega))$. Because Γ is ω-univalent, $Q([s_3](0, \omega)) = Q([s_4](0, \omega))$. Then by construction $Q(s_1(t)) = Q(s_3(t))$ and $Q(s_2(t)) = Q(s_4(t))$, and so $Q(s_1(t)) = Q(s_2(t))$. Hence $\dot{\Gamma}$ is k^2-univalent.

Consider a sequence generator $\Gamma = (S, G, R, P, Q)$. If (S, G, R, P) is deterministic then (S, G, R, P, Q) is uniquely solvable, but the converse does not in general hold [see Fig. 8(a)]. We noted earlier (§ 3.1) that unique solvability is essentially a property of the infinite behavior of a sequence generator. This suggests the question: what is the relation of the behaviors of uniquely solvable sequence generators to the behaviors of deterministic ones? This question is answered by the following theorem, which shows that for every uniquely solvable sequence generator there is a deterministic sequence generator whose infinite behavior is a displacement of the infinite behavior of the given sequence generator. In § 4 we will introduce a concept of "computation." Using this concept, the result may by expressed: the behavior of every uniquely solvable sequence generator can be computed by a finite automaton.

THEOREM 3.3–2 (TIME-SHIFT THEOREM). *Let $\Gamma = (S, G, R, P, Q)$ have k Γ-admissible complete states and let $\Gamma^{k^2*\dagger} = \dot{\Gamma} = (\dot{S}, \dot{G}, \dot{R}, \dot{P}, \dot{Q})$. Then*

(a) *if Γ is uniquely solvable, then $(\dot{S}, \dot{G}, \dot{R}, \dot{P})$ is deterministic,*

(b) $\mathscr{D}^{k^2}[\mathscr{B}(\Gamma)] = \mathscr{B}^{\omega}(\dot{\Gamma})$.

PROOF. This follows immediately from the definition of uniquely solvable (§ 3.1) Lemma 3.2–3, and the Fixed Bound Lemma (3.3–1).

Consider the Time-shift Theorem in relation to Γ of Fig. 10(a) and the derived $\Gamma^{1*\dagger}$ of Fig. 11. Γ is uniquely solvable and has 4 Γ-admissible complete states. Then the Time-shift Theorem tells us that $\Gamma^{16*\dagger}$, less its last projection, is deterministic. This is clearly so, for $\Gamma^{1*\dagger}$, less its last projection, is deterministic, and further applications of the l-shift operation will obviously not destroy this property.

We pause to note an analogue of the Time-shift Theorem in which the shifting takes place in the opposite direction. The displacement operator $\mathscr{D}^{l}$

was defined to produce a *right-shift* of the Q-projections of a Γ-sequence; that is, it shifts the Q-projections l steps later in time, leaving the P-projections as they were. One could easily extend this operator to cover shifts in the opposite direction (i.e., with the Q-projections moved earlier in time); this could be symbolized by using the same operator $\mathscr{D}^l$, allowing negative as well as positive integer values for l. Similarly, the l-shift operator can be extended to produce shifts of the Q-projections to the left; again, we can use the same symbolism Γ^l and signify left-shifts by negative values of l. We then get the following partial analogue to the Time-shift Theorem. Let $\Gamma = (S, G, R, P, Q)$ be a sequence generator, with (S, G, R, P) deterministic. Let $\dot{\Gamma} = \Gamma^l$, where l is negative. Then $\dot{\Gamma}$ is uniquely solvable, and $\mathscr{D}^l[\mathscr{B}^\omega(\Gamma)] = \mathscr{B}^\omega(\dot{\Gamma})$. Combining this with the Time-shift Theorem, we obtain the following result: the set of infinite behaviors of uniquely solvable sequence generators is exactly the set of displaced infinite behaviors of deterministic sequence generators.

It is not obvious from the definition that the class of h-univalent sequence generators is decidable. However, this is in fact the case, as we will now show.

h-univalence Procedure (where h is any non-negative integer or ω): Let $\Gamma = (S, G, R, P, Q)$ be the given sequence generator. Find k, the number of admissible complete states, by the Reduced Form Algorithm. Let $l = \min(h, k^2)$. Form $\Gamma^{l*\dagger} = \dot{\Gamma} = (\dot{S}, \dot{G}, \dot{R}, \dot{P}, \dot{Q})$. Answer "yes" or "no" as $(\dot{S}, \dot{G}, \dot{R}, \dot{P})$ is semi-deterministic or not.

THEOREM 3.3–3. *The h-univalence procedure is a decision procedure for the class of h-univalent sequence generators.*

PROOF. We will use the notation of the algorithm. By the Fixed Bound Lemma Γ is h-univalent if and only if Γ is l-univalent. By Corollary 3.2–2 Γ is l-univalent if and only if Γ^l is 0-univalent. By Lemma 3.1–4 Γ^l is 0-univalent if and only if $(\dot{S}, \dot{G}, \dot{R}, \dot{P})$ is semi-deterministic. As noted in § 2.1, it is obvious from the definition of semi-determinism that there is a decision procedure for the class of semi-deterministic sequence generators. This completes the proof of the theorem.

It can be shown that the following is a characterization of h-univalence. Let $\Gamma = (S, G, R, P, Q)$, k the number of Γ-admissible complete states, and $l = \min(h, k^2)$. Then Γ is h-univalent, if and only if, for any two Γ-sequences $[s_1](0, \omega)$ and $[s_2](0, \omega)$ and any time $t \leq k^2$, if $P([s_1](0, t + l)) = P([s_2](0, t + l))$, then $Q(s_1(t)) = Q(s_2(t))$. This characterization can be made the basis of a decision procedure for h-univalence which is more efficient than the one we have given.

Since unique solvability is defined in terms of solvability and ω-univalence (§ 3.1), by combining the ω-univalence Procedure with the decision procedure for solvability of Theorem 2.3–2, we obtain a decision procedure for unique solvability.

4. Generalizations and applications

4.1. **Computation.** We will next define a concept of "computation" which corresponds more closely to the way a digital computer is used than does the concept of behavior; essentially the same concept is defined in Burks [**3**]. Computers are employed to produce answers to questions; the questions go in as inputs, the answers appear as outputs. Generally speaking, the answer is not produced immediately, but only after a time delay. Moreover, except in real-time computation, the answer will not appear at the same rate as the input information is received. In contrast, all outputs of a computer are part of its behavior, whether they contribute to the answer or not. For these reasons the concept of behavior does not fit the question-answer mode of using a computer as closely as the concept of computation to be defined. Computation differs from behavior in that in the case of computation not all "output states" are interpreted as part of the answer, but only those selected as the "computed output states" by the sequence generator itself. The concept of computation reflects the fact that in our theory the internal operations of a sequence generator are strictly correlated to the basic time scale, while the computed outputs are not.

In the definition of computation we will need the μ-operator (selection operator). Suppose ϕx expresses some condition on the natural numbers. "$(\mu x)\phi x$" designates the smallest number satisfying the condition ϕx if there is one; if no number satisfies ϕx, then "$(\mu x)\phi x$" is undefined. For example, $(\mu x)(x > 3^2) = 10$, while "$(\mu x)(x^2 = 2)$" is undefined.

Definition. Let $\Gamma = (S, G, R, P, Q, C)$ be a sequence generator with the projection C having only the two values 0, 1. Let $[s](0, \omega)$ be an arbitrary infinite Γ-sequence and define

$$\gamma(t) = Q\left\{s\left((\mu x)\left[\sum_{y=0}^{x} C\{s(y)\} = t + 1\right]\right)\right\} ;$$

note that if $\sum_{y=0}^{\infty} C\{s(y)\}$ is finite then $\gamma(t)$ is undefined for any $t \geqq \sum_{y=0}^{\infty} C\{s(y)\}$. If $\sum_{y=0}^{\infty} C\{s(y)\}$ is unbounded, then the sequence

$$\langle P\{s(0)\}, \gamma(0)\rangle, \langle P\{s(1)\}, \gamma(1)\rangle, \cdots$$

is an *infinite computation element* of Γ. If $\sum_{y=0}^{\infty} C\{s(y)\} = k$ (k being finite), then the sequence

$$\langle P\{s(0)\}, \gamma(0)\rangle, \cdots, \langle P\{s(k-1)\}, \gamma(k-1)\rangle, \langle P\{s(k)\}\rangle, \langle P\{s(k+1)\}\rangle, \cdots$$

is a *finite computation element* of Γ. The *computation* of Γ, denoted by $\mathscr{C}(\Gamma)$, is the set composed of all infinite computation elements and all finite computation elements of Γ.

An example will help make the concept of computation clear. Consider a sequence generator Γ which has only two infinite Γ-sequences. These are shown below, together with the projections of each and the derived sequences $[\gamma](0, k)$.

$$[s_1](0, \omega) = s_0, s_1, s_2, s_3, s_3, s_3, \cdots ,$$

$$P([s_1](0,\omega)) = p_0, p_1, p_1, p_3, p_3, p_3, \cdots,$$
$$Q([s_1](0,\omega)) = q_0, q_0, q_1, q_1, q_1, q_1, \cdots,$$
$$C([s_1](0,\omega)) = 0, 1, 1, 0, 0, 0, \cdots,$$
$$[\gamma_1](0,1) = q_0, q_1 ;$$
$$[s_2](0,\omega) = s_0, s_1, s_2, s_4, s_5, s_4, s_5, s_4, s_5, \cdots,$$
$$P([s_2](0,\omega)) = p_0, p_1, p_1, p_3, p_5, p_3, p_5, p_3, p_5, \cdots,$$
$$Q([s_2](0,\omega)) = q_0, q_0, q_1, q_2, q_3, q_2, q_3, q_2, q_3, \cdots,$$
$$C([s_1](0,\omega)) = 0, 1, 1, 0, 1, 0, 1, 0, 1, \cdots,$$
$$[\gamma_2](0,\omega) = q_0, q_1, q_3, q_3, q_3, \cdots .$$

Hence the computation $\mathscr{C}(\Gamma)$ is the set consisting of the two sequences

$$\langle p_0, p_0\rangle, \langle p_1, q_1\rangle, \langle p_1\rangle, \langle p_3\rangle, \langle p_3\rangle, \langle p_3\rangle, \cdots,$$
$$\langle p_0, q_0\rangle, \langle p_1, q_1\rangle, \langle p_1, q_3\rangle, \langle p_3, q_3\rangle, \langle p_5, q_3\rangle, \cdots .$$

Note that the first of these is a finite computation element of Γ while the second is an infinite computation element of Γ.

We will now show that the concept of computation is a bona fide generalization of the concept of behavior. Let α be the class of all two-projection sequence generators. Let β be the class of all three-projection sequence generators $\Gamma = (S, G, R, P, Q, C)$ such that C has the values 0, 1 and every element of $\mathscr{C}(\Gamma)$ is infinite. Then the class $\{\mathscr{B}^\omega(\Gamma) \mid \Gamma \in \alpha\}$ of infinite behaviors of elements of α is a proper subclass of the class $\{\mathscr{C}(\Gamma) \mid \Gamma \in \beta\}$ of the computations of elements of β. To see that

$$\{\mathscr{B}^\omega(\Gamma) \mid \Gamma \in \alpha\} \subseteq \{\mathscr{C}(\Gamma) \mid \Gamma \in \beta\},$$

note that for each element $\Gamma = (S, G, R, P, Q)$ of α there is an element $\dot{\Gamma} = (S, G, R, P, Q, C)$ of β in which $C(s) = 1$ for every $s \in S$, so that $\mathscr{B}^\omega(\Gamma) = \mathscr{C}(\dot{\Gamma})$. That the inclusion of $\{\mathscr{B}^\omega(\Gamma) \mid \Gamma \in \alpha\}$ in $\{\mathscr{C}(\Gamma) \mid \Gamma \in \beta\}$ is a proper one is shown by Fig. 12. In Fig. 12, $\dot{P}(\dot{s}_0) = \dot{P}(\dot{s}_2) = 0$, $\dot{P}(\dot{s}_1) = \dot{P}(\dot{s}_3) = 1$, $\dot{P}(\dot{s}) = \dot{Q}(\dot{s})$ for every $\dot{s} \in \dot{S}$, $C(\dot{s}_0) = C(\dot{s}_1) = 1$, and $C(\dot{s}_2) = C(\dot{s}_3) = 0$. The sequence generator $\dot{\Gamma}$ of this figure has the property that

$$\mathscr{C}(\dot{\Gamma}) \notin \{\mathscr{B}^\omega(\Gamma) \mid \Gamma \in \alpha\} .$$

To see this, note that since $\dot{P}(\dot{s}) = \dot{Q}(\dot{s})$ for every $\dot{s} \in \dot{S}$, the computation $\mathscr{C}(\dot{\Gamma})$ consists of all infinite sequences of the form

$$\langle \dot{P}(\dot{s}(0)), \dot{P}(\dot{s}(0))\rangle, \langle \dot{P}(\dot{s}(1)), \dot{P}(\dot{s}(2))\rangle, \cdots, \langle \dot{P}(\dot{s}(t)), \dot{P}(\dot{s}(2t))\rangle, \cdots,$$

where $\dot{P}(\dot{s})$ is either zero or one. By the Fixed Bound Lemma (3.3–1) no uniquely solvable sequence generator can have $\mathscr{C}(\dot{\Gamma})$ as its behavior, and since, as we remarked in § 3.1, unique solvability is essentially a property of the infinite behavior of a sequence generator, no sequence generator can have $\mathscr{C}(\dot{\Gamma})$ as its infinite behavior. Since ω-univalence and unique solvability apply to sequences of pairs (§ 3.1), these concepts may be extended to cover the computation of a sequence generator; the computation $\mathscr{C}(\dot{\Gamma})$ of the sequence generator of Fig. 12 is uniquely solvable.

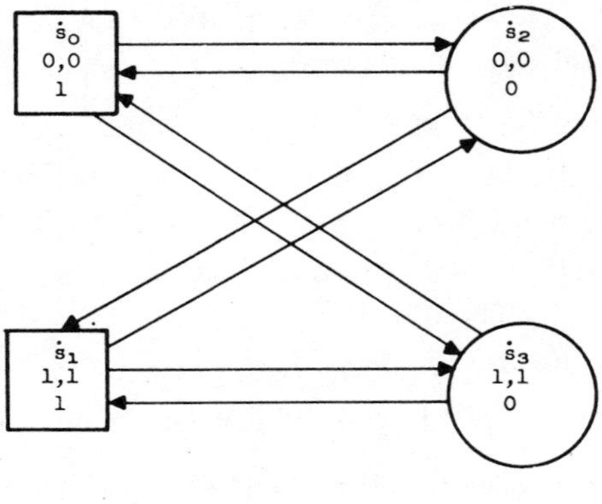

Figure 12. $\dot{\Gamma} = (\dot{S}, \dot{G}, \dot{R}, \dot{P}, \dot{Q}, \dot{C})$. No sequence generator can have $\mathscr{C}(\dot{\Gamma})$ as its infinite behavior.

Since computation is a bona fide generalization of behavior, the existence of certain algorithms for behavior (§ 2.3) does not guarantee the existence of corresponding algorithms for computation. It is not known whether either of these two algorithms exist: an algorithm to decide whether the computation of one sequence generator is included in the computation of another, a decision procedure for the computational equivalence of two sequence generators. We do have an algorithm to decide of any pair of sequence generators $\Gamma = (S, G, R, P, Q, C)$ and $\dot{\Gamma} = (\dot{S}, \dot{G}, \dot{R}, \dot{P}, \dot{Q}, \dot{C})$ such that $\mathscr{C}(\dot{\Gamma})$ is ω-univalent, whether the computation $\mathscr{C}(\Gamma)$ is included in the computation $\mathscr{C}(\dot{\Gamma})$ or not, but this algorithm and its justification are too long and involved to be included here.

There is an algorithm for deciding of a sequence generator $\Gamma=(S,G,R,P,Q,C)$, where C has only the values $0, 1$, whether all, some but not all, or none of the elements of $\mathscr{C}(\Gamma)$ are infinite. One first finds the reduced form $\Gamma^{\dagger}$. Note that $\mathscr{C}(\Gamma^{\dagger}) = \mathscr{C}(\Gamma)$ for any Γ. Then each non-repetitive cycle of complete states of $\Gamma^{\dagger}$ is examined to determine whether $C(s_i) = 1$ for any state s_i of this cycle. If $C(s_i) = 1$ for some state s_i of this cycle, there will be an infinite $\Gamma^{\dagger}$-sequence made of iterations of the cycle, which $\Gamma^{\dagger}$-sequence will give rise to an infinite computation element of $\mathscr{C}(\Gamma^{\dagger})$ and hence of $\mathscr{C}(\Gamma)$. On the other hand, if for every state s_i of a cycle $C(s_i) = 0$, then there is an infinite $\Gamma^{\dagger}$-sequence made of repetitions of this cycle which will give rise to a finite computation element of $\mathscr{C}(\Gamma)$.

Let us now return to the Time-shift Theorem (3.3–2) and extend it to cover the case of computation. Let $\Gamma = (S, G, R, P, Q, C)$ have k Γ-admissible complete states and let C have only the values $0, 1$. The l-shift operator was defined for two-projection sequence generators, but it may easily be extended to cover sequence generators with a computation projection C in the following way. Form $(S, G, R, P, Q \times C)$; apply the k^2-shift operator to it with the modification that $\langle \Lambda, 0\rangle$ is the $Q \times C$ projection of those admissible complete

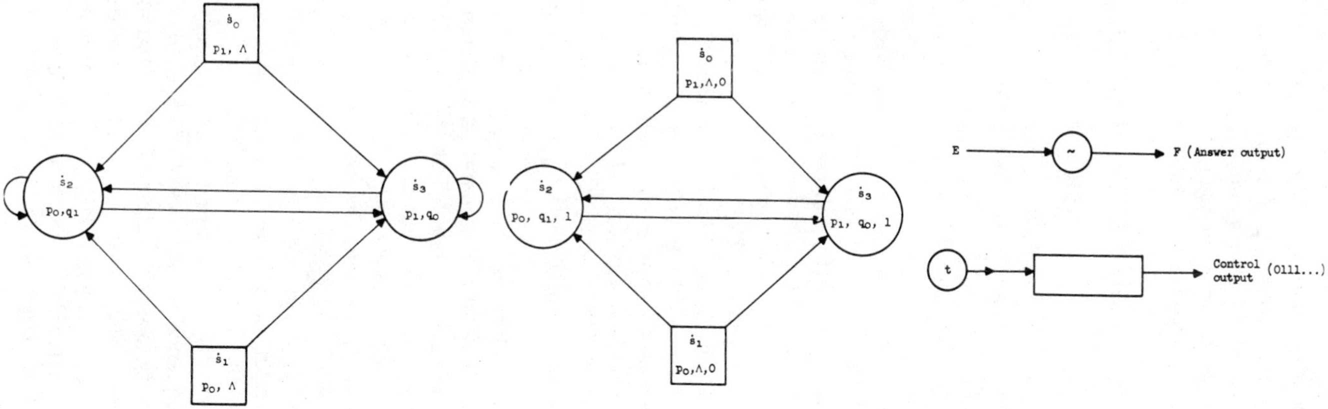

(a) (b) (c)

Figure 13.

(a) $\dot{\Gamma} = (\dot{S}, \dot{G}, \dot{R}, \dot{P}, \dot{Q})$, which results from $\Gamma^{1*\dagger}$ of Fig. 11 by identifying and re-naming behaviorally equivalent states. $\dot{\Gamma}$, less its last projection, is deterministic. $\mathscr{B}(\dot{\Gamma}) = \mathscr{B}(\Gamma^{1*\dagger})$.

(b) $\ddot{\Gamma} = (\dot{S}, \dot{G}, \dot{R}, \dot{P}, \dot{Q}, C)$, which results from $\dot{\Gamma}$ by adding a control projection $C(\dot{s}_0) = C(\dot{s}_1) = 0$, $C(\dot{s}_2) = C(\dot{s}_3) = 1$. $\mathscr{C}(\Gamma) = \mathscr{B}^{\omega}(\Gamma)$, where Γ is Fig. 10(a).

(c) A net which has the computation $\mathscr{C}(\ddot{\Gamma})$.

states that occur before time k^2 (cf. § 3.2). Call the result $(\dot{S}, \dot{G}, \dot{R}, \dot{P}, \dot{Q} \times \dot{C})$. Then form $(\dot{S}, \dot{G}, \dot{R}, \dot{P}, \dot{Q}, \dot{C})^{*\dagger} = \ddot{\Gamma} = (\ddot{S}, \ddot{G}, \ddot{R}, \ddot{P}, \ddot{Q}, \ddot{C})$. Now apply Lemmas 3.2–3 and 3.3–1, noting that a displacement of the infinite behavior of a sequence generator with a computation projection does not alter its computation, since the behavior includes the computed output. The net result of all this is as follows. Given a sequence generator $\Gamma = (S, G, R, P, Q, C)$ with C having the values 0, 1, there is an effective procedure for constructing $\ddot{\Gamma} = (\ddot{S}, \ddot{G}, \ddot{R}, \ddot{P}, \ddot{Q}, \ddot{C})$ such that:

(a) If $(S, G, R, P, Q \times C)$ is uniquely solvable, then $(\ddot{S}, \ddot{G}, \ddot{R}, \ddot{P})$ is deterministic,

(b) $\mathscr{C}(\Gamma) = \mathscr{C}(\ddot{\Gamma})$.

When the foregoing statement is applied to the case where C is always one, it implies that the infinite behavior of any uniquely solvable sequence generator is the computation of a deterministic sequence generator. This justifies the statement made prior to the Time-shift Theorem (3.3–2) that the infinite behavior of any uniquely solvable sequence generator can be computed by a finite automaton. We will illustrate this with an example (Fig. 13). We begin with Γ of Fig. 10(a), which is uniquely solvable (but not deterministic). In this case a unit-shift, followed by an application of the * and †, suffices to produce a deterministic sequence generator $\Gamma^{1*\dagger}$, shown in Fig. 11. $\dot{\Gamma}$ of Fig. 13(a) is a simplification of this. To $\dot{\Gamma}$ we now add a projection C which is 0 for the generators, 1 otherwise, obtaining $\ddot{\Gamma}$ of Fig. 13(b). $\ddot{\Gamma}$ is a deterministic sequence generator which computes the infinite behavior of the original Γ, i.e., $\mathscr{C}(\ddot{\Gamma}) = \mathscr{B}^{\omega}(\Gamma)$. By a slight generalization of the process described in § 2.1, we can pass from the deterministic sequence generator $\ddot{\Gamma}$ to a finite automaton and to a w.f.n. The w.f.n. which produces the computation $\mathscr{C}(\ddot{\Gamma})$ is shown in Fig. 13(c). It is perhaps worth noting that the computation of this net [Fig. 13(c)] is the infinite behavior of the net of Fig. 2(a).

The concept of unique solvability may be generalized to cover computation. Let $\Gamma = (S, G, R, P, Q, C)$ be a sequence generator, C having the values 0, 1. The computation $\mathscr{C}(\Gamma)$ is "uniquely solvable" if for each infinite sequence of P-states $[p](0, \omega)$ there is exactly one element of $\mathscr{C}(\Gamma)$ which contains $[p](0, \omega)$ (i.e., which is composed of this sequence of P-states together with zero or more Q-states). The motivation behind the concept of a computation being uniquely solvable is this. We may think of a sequence generator with an input projection P, an output projection Q, and a control projection C as specifying a computational relation between infinite input sequences and *computed* output sequences, and the sequence generator does it uniquely if each input sequence determines exactly one computed output sequence. It is easy to construct examples of sequence generators $\Gamma = (S, G, R, P, Q, C)$ such that (S, G, R, P, Q) is not uniquely solvable, though the computation $\mathscr{C}(\Gamma)$ is. Thus we are led naturally to the question: Is there a decision procedure which will tell whether the computation $\mathscr{C}(\Gamma)$ of a sequence generator $\Gamma = (S, G, R, P, Q, C)$ is uniquely solvable? It is also of interest to know whether the analogue of the Time-shift Theorem (3.3–2) holds for uniquely solvable computations, i.e., whether or not the following is true of a sequence generator $\Gamma = (S, G, R, P, Q, C)$, with C having the values 0, 1: if $\mathscr{C}(\Gamma)$ is uniquely solvable, then there is a sequence generator $\dot{\Gamma} = (\dot{S}, \dot{G}, \dot{R}, \dot{P}, \dot{Q}, \dot{C})$ such that $(\dot{S}, \dot{G}, \dot{R}, \dot{P})$ is deterministic and $\mathscr{C}(\Gamma) = \mathscr{C}(\dot{\Gamma})$. We do not know the answer to either of these questions.

4.2. **Formulas and sequence generators.** We will now discuss the relation of some formulas of symbolic logic to sequence generators. Because of limitations of space we will not give a detailed or rigorous treatment of the subject but will rely heavily on examples and will present theorems without proofs.

The language L being considered here is a first-order monadic predicate calculus with a successor function and zero. The symbols of L are: an infinite list of monadic predicate variables $A, B, C, \cdots$; an infinite list of individual variables $t, t_1, t_2, \cdots$; the successor function $'$; the individual constant 0; all truth-functional connectives; and parentheses. The individual variables range over natural numbers $0, 0', 0'', \cdots$. The predicate variables range over predicates of natural numbers, i.e., over sets of natural numbers.

Consider an arbitrary w.f.f. of L; the result of universally quantifying all the individual variables of L is called an *A-formula* (arbitrary-formula). $B(t_1) \equiv B(t_2''')$ is a w.f.f. of L, and so $(t_1)(t_2)[B(t_1) \equiv B(t_2''')]$ is an A-formula. $(t_1)(t_2)[B(t_1) \equiv [B(t_2''')]$ means that for all times t_1, t_2, B is true at time t_1 if and only if B is true at time $t_2 + 3$. Note that since language L does not contain quantifiers, an A-formula is not a w.f.f. of L; this does not matter for our purposes.

The *extension* of an A-formula with predicate variables $B_1, B_2, \cdots, B_k$ is the set of all k-tuples of predicates which satisfy the formula, i.e., for which the formula is true of the natural numbers when B_i is interpreted as the ith predicate of the k-tuple. We will regard a predicate, e.g., B_i, as an infinite binary sequence $[s](0, \omega)$ in which $s(t)$ is 1 or 0 according to whether $B_i(t)$ is true or false. When predicates are viewed in this way, a k-tuple of predicates becomes an ω-sequence of k-tuples or column vectors, i.e., a k by ω binary matrix.

We will give some examples. $(t_1)(t_2)[B(t_1) \equiv B(t_2''')]$ is satisfied by $1111\cdots$ but not by $1010\cdots$, for in the later case $B(0) \not\equiv B(0''')$. The extension of $(t)[B(0'') \;\&\; (B(t) \equiv B(t''))]$ consists of all sequences of the form

$$1, x, 1, x, 1, x, 1, x, \cdots .$$

The extension of $(t)\{\overline{B_2(0)} \;\&\; [B_1(t) \equiv B_2(t')]\}$ consists of all two by omega matrices of the form

$$\begin{matrix} x_0 & x_1 & x_2 & x_3 & \cdots, \\ 0 & x_1 & x_2 & x_3 & \cdots. \end{matrix}$$

Sequence generators without projections may be translated into a particular kind of A-formula, called an M-formula. For example, the sequence generator of Fig. 14 may be translated into

$$(t)[\overline{\{(B_1(0)} \ \& \ \overline{B_2(0))} \vee (B_1(0) \ \& \ B_2(0))\} \ \& \ \overline{\{(B_1(t)} \ \& \ \overline{B_2(t)} \ \& \ \overline{B_1(t'}) \ \& \ B_2(t'))$$
$$\vee \ \overline{(B_1(t)} \ \& \ B_2(t) \ \& \ \overline{B_1(}t') \ \& \ \overline{B_2(t')})\}] \, .$$

The first braced conjunct of this A-formula contains no individual symbol other than the constant zero; it tells us what the generators are. The second braced conjunct contains no individual symbols other than t and t'; it tells what direct transitions are possible. Clearly, the set of infinite Γ-sequences of Fig. 14 equals the extension of the formula given above. Any A-formula consisting of a universal individual quantifier operating on a conjunction, the first conjunct having only zero as argument, the second conjunct having only

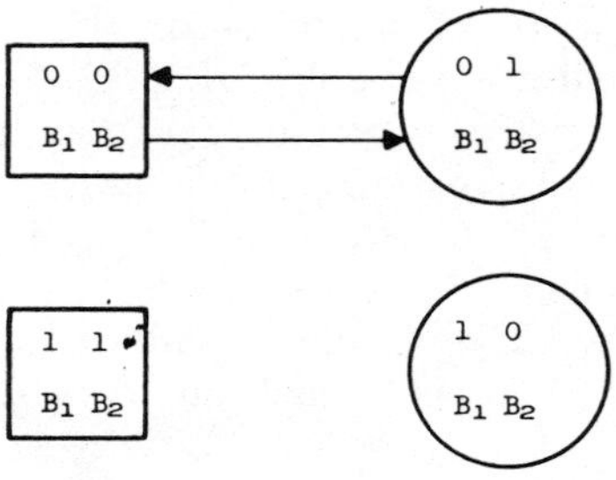

Figure 14. $\Gamma = (S, G, R)$.

one variable with at most one prime as argument, is called an *M-formula* (minimal formula). Our previous example illustrates the fact that any sequence generator $\Gamma = (S, G, R)$ can be translated into an M-formula $\mathscr{F}$ with k predicates by coding its states into binary k-tuples, the extension of $\mathscr{F}$ being the set of Γ-sequences. Conversely, any M-formula $\mathscr{F}$ may be translated into a sequence generator $\Gamma = (S, G, R)$ such that the set of infinite Γ-sequences is the extension of $\mathscr{F}$.

We will consider next two types of A-formulas more general than M-formulas. Any A-formula of the form $(t)\mathscr{F}_1(0, 0^1, \cdots, t^{j_1}) \ \& \ \mathscr{F}_2(t, t^1, \cdots, t^{j_2})$, where j_1, j_2, indicate the maximum number of primes of arguments of $\mathscr{F}_1$ and $\mathscr{F}_2$, respectively, is a *D-formula* (decomposable-formula). "$[B(0) \ \& \ B(1) \ \& \ \overline{\{(B(t)} \ \& \ \overline{B(t'))} \equiv B(t'')\}]$" is a D-formula whose extension is the single infinite sequence $001001 \cdots$. In § 2.1 (immediately after Corollary 2.1–1) we proved that there is no sequence generator $\dot{\Gamma} = (\dot{S}, \dot{G}, \dot{R})$ whose set of infinite Γ-sequences consists of the single sequence $001001 \cdots$. It follows by the results described in the preceding paragraph that there is no M-formula whose extension consists of this sequence, i.e., there is no M-formula logically equivalent to the D-formula given above.

Any A-formula with at most one individual variable is an *O-formula* (one-variable formula). $(t)[B(0) \supset (B(t) \equiv B(t'))]$ is an O-formula whose extension consists of $111 \cdots$ together with all binary sequences beginning with zero. It can be proved that there is no D-formula with this extension, i.e., $(t)[B(0) \supset (B(t) \equiv (B(t'))]$ cannot be decomposed into a D-formula $(t)(\mathscr{F}_1 \ \& \ \mathscr{F}_2)$ logically equivalent to it. Thus O-formulas are stronger than D-formulas, just as D-formulas are stronger than M-formulas. A-formulas are stronger than all these, for it can be proved that there is no O-formula with the extension of the A-formula $[\overline{B(t_1)} \vee \overline{B(t_1')}] \vee [B(t_2) \vee B(t_2')]$. This formula is equivalent to the condition that a binary sequence cannot have both consecutive ones and consecutive zeros.

There is thus a hierarchy of A-formulas, with sequence generators corresponding to the formulas of lowest level, i.e., the M-formulas. This suggests generalizing the concept of sequence generator so there is a kind of sequence generator corresponding to each level of the hierarchy. This could be done, as an example will make clear. The generalized sequence generator $\Gamma = (S, R_G^3)$ is defined as follows. S is the set $\{0, 1\}$. R_G^3 is the triadic relation $\{\langle 100\rangle, \langle 111\rangle, \langle 000\rangle, \langle 001\rangle, \langle 010\rangle, \langle 011\rangle\}$. An infinite Γ-sequence of this generalized sequence generator is any binary sequence $[s](0, \omega)$ such that all t, $R_G^3(s(0), s(t), s(t'))$. Note that generators and the direct transition relation are not defined separately in $\Gamma = (S, R_G^3)$. The set of Γ-sequences of (S, R_G^3) is equivalent to the extension of the O-formula $(t)[B(0) \supset (B(t) \equiv B(t'))]$. In this example the states of the sequence generator are zero and one, with the associated A-formula having one predicate (B). In general, of course, the states will be k-tuples of zeros and ones and the associated A-formula will have k predicates. Proceeding in a similar manner one can define generalized sequence generators corresponding to D-formulas and to A-formulas, with the result that for each formula $\mathscr{F}$ of a given type there is a corresponding generalized sequence generator Γ such that the extension of $\mathscr{F}$ equals the set of infinite Γ-sequences of Γ.

While these generalizations are of interest in showing the relations of formulas to sequence generators, they are not as easy to work with as the corresponding D-formulas, O-formulas, and A-formulas. In contrast, sequence generators are easier to work with than the corresponding M-formulas. Moreover, by employing projections it is possible to reduce any A-formula $\mathscr{F}$ to a sequence generator $\Gamma = (S, G, R, P)$ so that the infinite behavior $\mathscr{B}^\omega(\Gamma)$ equals the extension of $\mathscr{F}$, and in this way to investigate A-formulas of all kinds by means of sequence generators with projections. After defining some terms we will explain this process in more detail.

A *ΣA-formula* is the result of prefixing a sequence of existential predicate quantifiers to an A-formula; "ΣO-formula," "ΣD-formula," and "ΣM-formula" are similarly defined. We will call the set of k-tuples of predicates which satisfy a ΣA-formula the *behavior* of the formula. For example, the binary sequence 00110011 $\cdots$ is the behavior of the M-formula

$$(\Sigma C)(t)\overline{\{[B(0)}\ \&\ \overline{C(0)]}\ \&\ [C(t) \not\equiv C(t')]\ \&\ (B(t') \not\equiv (B(t) \equiv C(t)))]\}\ .$$

Our preceding examples show that the sets of extensions of M-formulas, D-formulas, O-formulas, and A-formulas get progressively larger. In contrast, the set of behaviors of ΣA-formulas equals the set of behaviors of ΣM-formulas. Given any ΣA-formula, one can, by a procedure of Church [8, p.

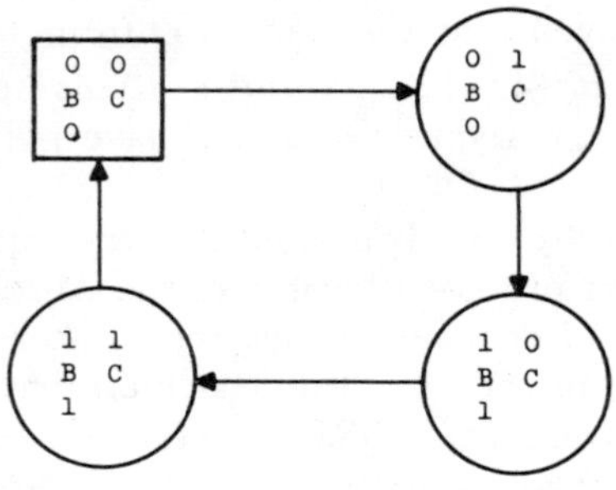

Figure 15. $\Gamma = (S, G, R, P)$.

36 ff.], construct a behaviorally equivalent ΣO-formula. There is, moreover, a process for reducing a ΣO-formula to a behaviorally equivalent ΣM-formula. A description of the process is too long to include here, but the essential steps are illustrated in the following example. Consider the O-formula

$$(t)[B(0') \supset \{B(t'') \equiv B(t)\}] .$$

We introduce a predicate C_1 which is defined by the conditions $C_1(t') \equiv C_1(t)$ and $B(0') \equiv C_1(0)$, so that $C_1(t) \equiv B(0')$ for all t. Since these conditions imply that $C_1(t) \equiv B(0')$, we may substitute $C_1(t)$ for $B(0')$ in the original O-formula and conjoin the two conditions to obtain a ΣD-formula $(\Sigma C_1)(t)[B(0') \equiv C_1(0)] \;\&\; [\{C_1(t) \supset (B(t'') \equiv B(t))\} \;\&\; \{C_1(t') \equiv C_1(t)\}]$ which has the same behavior as the given O-formula. We next introduce a predicate C_2 defined by the condition $B(t') \equiv C_2(t)$, which implies that $B(t'') \equiv C_2(t')$ and $B(0') \equiv C_2(0)$. Finally, we substitute $C_2(t')$ for $B(t'')$ and $C_2(0)$ for $B(0')$ in the ΣD-formula just obtained and conjoin the condition $B(t') \equiv C_2(t)$, thereby obtaining a ΣM-formula

$$(\Sigma C_2)(\Sigma C_1)(t)[C_2(0) \equiv C_1(0)] \;\&\; [\{C_1(t) \supset (C_2(t') \equiv B(t))\} \;\&\; \{C_1(t') \equiv C_1(t)\} \;\&\; \{B(t') \equiv C_2(t)\}]$$

which has the same behavior as the ΣD-formula and hence as the original O-formula.

Each ΣM-formula $\mathscr{F}$ may be converted into a one-projection sequence generator $\Gamma = (S, G, R, P)$ such that $\mathscr{B}^{\omega}(\Gamma)$ is the behavior of $\mathscr{F}$, and conversely. Again we will not state the algorithm for this conversion but will illustrate it. Consider the ΣM-formula

$$(\Sigma C)(t)\{[\overline{B(0)} \;\&\; \overline{C(0)}] \;\&\; [(C(t) \not\equiv C(t')) \;\&\; (B(t') \not\equiv (B(t) \equiv C(t)))]\} .$$

Drop the existential quantifier and convert the resultant M-formula into a sequence generator (S, G, R) in the way indicated before; see Fig. 15. Next, add a projection P defined so that $P(B(t) \;\&\; C(t)) = B(t)$, obtaining $\Gamma = (S, G, R, P)$. $\mathscr{B}^{\omega}(\Gamma)$ is the behavior of the original M-formula. The reverse procedure of going from an arbitrary one-projection sequence generator to a ΣM-formula with the same behavior is only slightly more difficult.

To sum up: given a ΣA-formula, one can effectively construct a ΣM-formula which has the same behavior. Given a ΣM-formula $\mathscr{F}$, one can effectively construct $\Gamma = (S, G, R, P)$ such that $\mathscr{B}^{\omega}(\Gamma)$ equals the behavior of $\mathscr{F}$. Conversely, given $\Gamma = (S, G, R, P)$ one can effectively find a ΣM-formula $\mathscr{F}$ whose behavior is $\mathscr{B}^{\omega}(\Gamma)$. Hence the class of infinite behaviors of one-projection sequence generators equals the class of behaviors of ΣA-formulas, and so ΣA-formulas may be investigated by means of sequence generators with one projection.

One can go further than this by classifying the free predicate variables of a ΣA-formula so as to correspond to several projections. For example, one can divide the free predicate variables of a ΣA-formula into two categories, the first corresponding to a projection P, the second to a projection Q. When this is done, the two-projection concepts and theorems of § 3 apply to ΣA-formulas. It is worth noting what the Time-shift Theorem (3.3–2) becomes when looked at in this way. A deterministic sequence generator corresponds to a ΣM-formula in which the Q-predicates (outputs) and bound predicates at time $t + 1$ are defined recursively in terms of themselves at time t and of the P-predicates (inputs) at times t and $t + 1$; this is essentially what Church [8, p. 11] calls a system of "restricted recursions." One may also consider recursive definitions in which the values of the Q-predicates and bound predicates at time t depend on the values of the P-predicates at later times; these recursions are essentially what Church [8, p. 12] calls "unrestricted singulary recursions," and can be expressed by a special type of ΣD-formula.

Put in these terms, the Time-shift Theorem becomes: for every uniquely solvable ΣA-formula there is a logically equivalent ΣD-formula which is

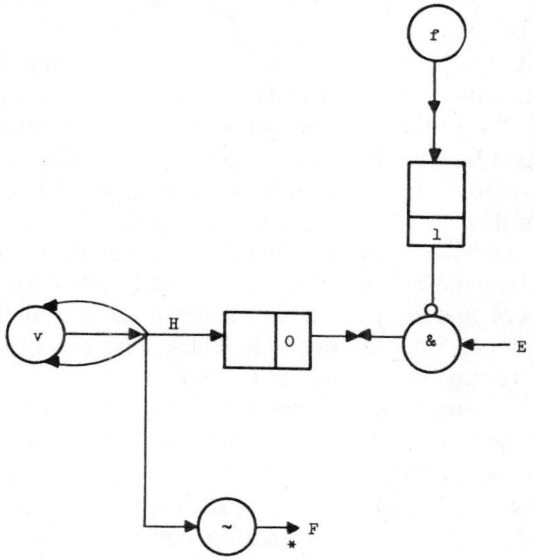

Figure 16. Normal form of the well-behaved net of Figure 2(a). $F(t) \equiv \sim E(t')$.

composed of unrestricted singulary recursions. This ΣD-formula is of a special form which may be thought of as a normal form for uniquely solvable ΣA-formulas. A normal form ΣD-formula translates into a well-behaved net (see the last paragraph of § 3.1) which consists of a well-formed part (corresponding to a deterministic sequence generator) together with a sequence of delays (which shifts the inputs earlier in time), and so the Time-shift Theorem can also be interpreted as saying that every well-behaved net has a normal form. Fig. 16 gives the normal form for the well-behaved net of Fig. 2(a). The upper part produces the time-shift $H(t) \equiv E(t')$ while the lower part, which is well-formed, produces $F(t) \equiv \sim H(t)$; hence $F(t) \equiv \sim E(t')$, as in Fig. 2(a).

One can think of a uniquely solvable ΣA-formula whose free predicate variables are divided into P-predicates and Q-predicates as giving an implicit definition of the Q-predicates in terms of the P-predicates. Looked at in this way the Time-shift Theorem tells us that for every ΣA-formula which implicitly defines the Q-predicates in terms of the P-predicates, there is a ΣD-formula which recursively defines the Q-predicates in terms of the P-predicates.

4.3. **Sequence generators and conditions.** Suppose one wishes to design an automaton or other deterministic information-processing system. Using a formula, a diagram, or a set of tables, one may specify the output of the system as a deterministic function of the inputs of the system. Sometimes, however, the designer has in mind only a condition or requirement which he wishes the behavior of the device to satisfy, there being many different behaviors which satisfy this condition. The designer may wish to consider all systems whose behaviors satisfy the given condition and select from among these by some criterion, such as minimality of components. A well-known example of this is the use of "don't care" cases in formulating a switch requirement. Because the requirement imposes no restrictions on the switch behavior for the "don't care" inputs, it may be satisfied by different switching functions. The designer wishes to select from all switches which satisfy this requirement one with a minimal number of switching elements.

Many different languages may be used for expressing conditions on information-processing systems and for describing such systems. Consider first the language of ΣA-formulas (§ 4.2). Suppose $\mathscr{F}$ and $\dot{\mathscr{F}}$ are ΣA-formulas whose free predicate variables are divided into input variables and output variables. $\mathscr{F}$ may describe a computer system and $\dot{\mathscr{F}}$ may express some relation between inputs and outputs which a designer would like the digital system to satisfy. The idea of a system $\mathscr{F}$ satisfying condition $\dot{\mathscr{F}}$ can be formulated in logical terms by saying, first, that $\mathscr{F}$ and $\dot{\mathscr{F}}$ have the same input variables and the same output variables, and second, that $\mathscr{F} \supset \dot{\mathscr{F}}$ is valid. Whenever the pair $\mathscr{F}, \dot{\mathscr{F}}$ satisfies these two conditions we will speak of formula $\mathscr{F}$ being "a solution of" formula $\dot{\mathscr{F}}$.

Let us see next how to formulate these ideas in sequence-generator terms. In passing between formulas and sequence generators, one must do some coding or decoding, since the predicates of the formulas are two-valued, whereas a sequence generator has, in general, more than two states. If this coding is handled properly, the following definition of "Γ is a solution of $\dot{\Gamma}$" is essentially the same as the definition of "$\mathscr{F}$ is a solution of $\dot{\mathscr{F}}$" just given. Let $\Gamma = (S, G, R, P, Q)$ describe some digital system and let $\dot{\Gamma} = (\dot{S}, \dot{G}, \dot{R}, \dot{P}, \dot{Q})$ express a condition. P and $\dot{P}$ are interpreted as input projections and Q and $\dot{Q}$ as output projections. *Γ is a solution of $\dot{\Gamma}$* whenever, first, $\mathscr{B}^{\omega}(\Gamma) \subseteqq \mathscr{B}^{\omega}(\dot{\Gamma})$, and second, every $\dot{P}$-state occurring in $\mathscr{B}^{\omega}(\dot{\Gamma})$ occurs in $\mathscr{B}^{\omega}(\Gamma)$.

There are several kinds of design algorithms which make use of conditions. Büchi, Elgot, and Wright [2] define three kinds of algorithms for sets of formulas whose free predicate variables are divided into input variables and output variables. We will define analogous kinds of algorithms for sequence generators. Let α and β be two classes of two-projection sequence generators. A solution algorithm for $\langle \alpha, \beta \rangle$ is a decision procedure which applies to any pair $\langle \Gamma, \dot{\Gamma} \rangle$ such that $\Gamma \in \alpha$ and $\dot{\Gamma} \in \beta$ and answers the question: is Γ a solution of $\dot{\Gamma}$? A solvability algorithm for $\langle \alpha, \beta \rangle$ applies to any $\dot{\Gamma} \in \beta$ and is a decision procedure for the question: does there exist a $\Gamma \in \alpha$ such that Γ is a solution of $\dot{\Gamma}$? A synthesis algorithm for $\langle \alpha, \beta \rangle$ applies to any $\dot{\Gamma} \in \beta$ and produces a $\Gamma \in \alpha$ such that Γ is a solution of $\dot{\Gamma}$, if there is one.

Note that as a synthesis algorithm is here defined it may be non-terminating, but that a synthesis algorithm combined with a solvability algorithm is terminating, producing a sequence generator of the desired kind if one exists, terminating in a "no" otherwise. Note also that, if α can be recursively enumerated, the existence of a synthesis algorithm for $\langle \alpha, \beta \rangle$ follows from the existence of a solution algorithm for $\langle \alpha, \beta \rangle$, since the elements of α can be enumerated and each compared with the given element of β by the solution algorithm.

Clearly the Behavior Inclusion Procedure of § 2.3 constitutes the basis of a solution algorithm for any classes of sequence generators α, β. Now there are decision procedures for the sets of deterministic and semi-deterministic (§ 2.2), solvable (§ 2.3), h-univalent, and uniquely solvable (§ 3.3) sequence generators. Consequently, if α is any one of these sets it can be recursively enumerated, and so there is a synthesis algorithm for α and any set of sequence generators β.

For our next result we need the concept of one sequence generator being "a part of" another. $\Gamma = (S, G, R, P^1, \cdots, P^n)$ is a part of $\dot{\Gamma} = (\dot{S}, \dot{G}, \dot{R}, \dot{P}^1, \cdots, \dot{P}^n)$ whenever $S \subseteqq \dot{S}$, $G \subseteqq \dot{G}$, $R \subseteqq \dot{R}$ and each P^i is $\dot{P}^i$ cut down to S. As an example of this concept we mention that the reduced form $\Gamma^{\dagger}$ is a part of Γ. Clearly if Γ is a part of $\dot{\Gamma}$, $\mathscr{B}(\Gamma) \subseteqq \mathscr{B}(\dot{\Gamma})$. The following is a theorem: let $\dot{\Gamma} = (\dot{S}, \dot{G}, \dot{R}, \dot{P}, \dot{Q})$ be semi-deterministic with respect to $\dot{P} \times \dot{Q}$; there is a P-deterministic $\Gamma = (S, G, R, P, Q)$ which is a solution of $\dot{\Gamma}$ if and only if there is a P-deterministic $\Gamma = (S, G, R, P, Q)$ which is a part of $\dot{\Gamma}$ and such that every $\dot{P}$-state occurring in $\mathscr{B}^{\omega}(\dot{\Gamma})$ occurs in $\mathscr{B}^{\omega}(\Gamma)$.

There is a proof of this theorem which consists of two steps. In the first step the theorem is established for the special case where no two complete states of the given condition $\dot{\Gamma}$ agree on both of their projections. In terms of formulas (§ 4.2), this means that the theorem is proved for conditions which are M-formulas. The second step is to extend this justification to ΣM-formulas, that is, to extend the proof of the theorem to the general case, where the given condition $\dot{\Gamma}$ is an arbitrary two-projection sequence generator. The second step of the proof involves a construction which is of some interest in its own right. Let $\Gamma = (S, G, R, P)$ and $\dot{\Gamma} = (\dot{S}, \dot{G}, \dot{R}, \dot{P})$. The cross product of these two sequence generators $\dot{\Gamma} \times \Gamma = \ddot{\Gamma}(\ddot{S}, \ddot{G}, \ddot{R}, \ddot{P})$ is defined by

$$\ddot{S} = \{\langle s, \dot{s}\rangle \mid s \in S \ \& \ \dot{s} \in \dot{S} \ \& \ P(s) = \dot{P}(\dot{s})\} ,$$
$$\ddot{G} = (G \times \dot{G}) \cap \ddot{S} ,$$
$$\ddot{R}(\langle s_1, \dot{s}_1\rangle, \langle s_2, \dot{s}_2\rangle) = [R(s_1, s_2) \ \& \ \dot{R}(\dot{s}_1, \dot{s}_2)] ,$$
$$\ddot{P}(\langle s, \dot{s}\rangle) = P(s) .$$

The preceding construction can be used to establish the following lemma: Let $\dot{\Gamma} = (\dot{S}, \dot{G}, \dot{R}, \dot{P}, \dot{Q})$ be a sequence generator which is semi-deterministic with regard to $\dot{P} \times \dot{Q}$ and $\ddot{\Gamma} = (\dot{S}, \dot{G}, \dot{R}, \dot{P}, \dot{I})$, where $\dot{I}(\dot{s}) = \dot{s}$. There is a P-deterministic sequence generator $\Gamma = (S, G, R, P, Q)$ which is a solution of $\dot{\Gamma}$ if and only if there is a P-deterministic $\ddot{\Gamma}$ which is a solution of $\ddot{\Gamma}$.

This theorem leads directly to a combined solvability-synthesis algorithm for $\langle \alpha, \beta\rangle$, where α is the class of all two-projection sequence generators which are deterministic with respect to the first projection, and β is any class of two-projection sequence generators; the essential steps of the algorithm are to apply the subset sequence generator operation to the given condition and to examine each part of the result for determinism. By the results of § 4.2, this gives us a combined solvability-synthesis algorithm for $\langle \alpha, \beta\rangle$, where α is the class of systems of restricted recursions (i.e., representations of deterministic information processing systems) and β is any class of ΣA-formulas, the free predicate variables of all formulas being divided into input variables and output variables. This extends a result of Church [**8**, pp. 33a ff., 36 ff.], which is a solvability-synthesis algorithm for $\langle \alpha, \beta\rangle$, where α is the class of systems of restricted recursions but β is any class of A-formulas.

Wang [**33**, p. 312, Theorem 6] has a result which is (in our terms) a solvability-synthesis algorithm for $\langle \alpha, \beta\rangle$, where α consists of systems of unrestricted singulary recursions and β is any set of O-formulas. Our last stated lemma holds with "uniquely solvable" replacing "P-deterministic." By applying this lemma, Wang's result can be extended to provide a solvability-synthesis algorithm for $\langle \alpha, \beta\rangle$, where α consists of systems of unrestricted singulary recursions and β is any set of ΣA-formulas, the free predicate variables of all formulas being divided into input variables and output variables, as before.

We give next an example (already to be found in the literature) of the use of conditions. Sometimes an automaton is given with "input restrictions" by drawing an internal state diagram which does not provide transitions for all inputs (Aufenkamp and Hohn [**1**, § IV]). One is then interested in an automaton which has the same behavior as this diagram with respect to all input sequences which are provided for. This situation may be described in sequence-generator terms. The internal state diagram with input restrictions may be converted into a $\Gamma = (S, G, R, P, Q)$ which is semi-deterministic with respect to P (see § 1.2); in fact, the semi-determinism is not required for what we are going to do. We define a $\dot{\Gamma} = (\dot{S}, \dot{G}, \dot{R}, \dot{P}, \dot{Q})$ which will have Γ as a part. Let B consist of complete states $\langle p, q \rangle$ for all P-states p and Q-states q. $\dot{S} = S \cup B$. $\dot{R}$ is R extended so that (1) $\dot{R}(\dot{s}_1, \dot{s}_2)$ for all $\dot{s}_1, \dot{s}_2 \in B$ and (2) for any $s \in S$ and any P-state p, if there is no $s_1 \in S$ such that $R(s, s_1)$ and $P(s_1) = p$, then $\dot{R}(\dot{s}, \dot{s})$ for every $\dot{s} \in B$. $\dot{P}$ and $\dot{Q}$ are P and Q extended to cover the elements of B so that $\dot{P}(\langle p, q \rangle) = p$ and $\dot{Q}(\langle p, q \rangle) = q$. The original problem now becomes that of finding a $\ddot{P}$-deterministic $\ddot{\Gamma} = (\ddot{S}, \ddot{G}, \ddot{R}, \ddot{P}, \ddot{Q})$ which is a solution of $\dot{\Gamma}$ (cf. the penultimate paragraph of § 2.1). An answer will always be given by the solvability-synthesis algorithm for deterministic sequence generators described two paragraphs back.

The solution, solvability, and synthesis algorithms we have been discussing are defined in terms of behavior. It is worth noting that other solution, solvability, and synthesis algorithms may be defined by using different concepts in place of behavior in the definition of "Γ is a solution of $\dot{\Gamma}$;" for example, the set of Γ-sequences of a sequence generator, the computation of a sequence generator (§ 4.1), or the behavior of the final state sequence generators to be introduced next. We do not have space to discuss these algorithms.

4.4. **Sequence generators and regularity.** We now extend the concept of sequence generator to include a set F of final states. $\Gamma = (S, G, F, R, P)$ is defined as in § 1.1, with the additional proviso that $F \subseteq S$. The definition of Γ-sequence is altered to require that the last complete state of a finite Γ-sequence belong to F (be a final state) and to require that an infinite Γ-sequence contain infinitely many occurrences of members of F. (Contrast the concept of an infinite computation element of § 4.1 and note that the functions of the sets G and F could be performed by two two-valued projections; for an illustration of this, see the last example of § 1.3.) Behavior is defined as in § 2.1, using the new concept of Γ-sequence. It will be convenient to have a concept of finite behavior $\mathscr{B}^f$ for both ordinary and final state sequence generators: $\mathscr{B}^f(\Gamma) = \mathscr{B}(\Gamma) - \mathscr{B}^\omega(\Gamma)$. We will assume that those earlier concepts (e.g., determinism) which are easily extended to cover sequence generators with final states are in fact so extended.

Final-state sequence generators are of interest in connection with "regular sets." A regular set is a set of finite sequences defined in terms of certain algebraic operations on a given finite set of finite sequences. It has been shown that the following three sets are equivalent:

(1) The set of regular sets.

(2) The set of finite behaviors $\mathscr{B}^f(\Gamma)$ of final-state sequence generators $\Gamma = (S, G, F, R, P)$.

(3) The set of finite behaviors $\mathscr{B}^f(\Gamma)$ of final-state sequence generators $\Gamma = (S, G, F, R, P)$ which are deterministic.

See Kleene [**17**], Copi, Elgot, and Wright [**10**], and Myhill [**24**].

The concept of a final-state sequence generator is a bona fide generalization of the concept of a sequence generator without final states. For every sequence generator (S, G, R, P) of the latter kind there is a behaviorally equivalent sequence generator of the former variety, namely, (S, G, S, R, P). But not every behavior of a final-state sequence generator is the behavior of some sequence generator without final states. This may be shown by means of the concept of "open behavior." A behavior (finite or unrestricted) of a sequence generator is said to be "open" if every initial segment of an element of the behavior belongs to the behavior. The Infinity Theorem (2.1–3) implies that the behavior of a sequence generator without final states is open. The corresponding theorem does not hold for final-state sequence generators, as the simple example of Fig. 17 shows; $\mathscr{B}(\Gamma)$ consists of the single sequence p_0, p_0 and hence is not open. Clearly, then, no sequence generator without final states has the behavior of Fig. 17.

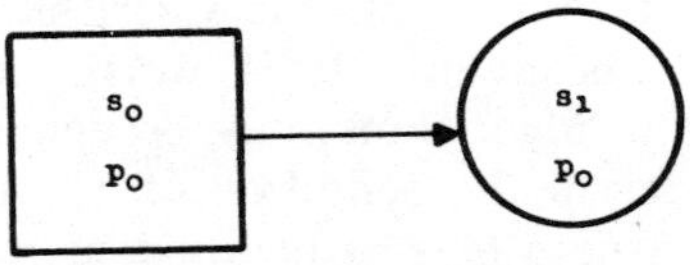

Figure 17. Final-state sequence generator $\Gamma = (S, G, F, R, P)$. $F = \{s_1\}$. $\mathscr{B}(\Gamma)$ is not open.

Though the behavior of a final-state sequence generator is not in general the behavior of a sequence generator without final states, any open finite behavior of a final-state sequence generator is the finite behavior of some sequence generator without final states. That is, for any $\Gamma = (S, G, F, R, P)$, if $\mathscr{B}^f(\Gamma)$ is open, then there is a $\dot{\Gamma} = (\dot{S}, \dot{G}, \dot{R}, \dot{P})$ such that $\mathscr{B}^f(\dot{\Gamma}) = \mathscr{B}^f(\Gamma)$. Given any $\Gamma = (S, G, F, R, P)$, whose $\mathscr{B}^f(\Gamma)$ is open, the desired $\dot{\Gamma}$ may be constructed as follows. By the results of paragraph two of this subsection, there exists a $\ddot{P}$-deterministic sequence generator $\ddot{\Gamma} = (\ddot{S}, \ddot{G}, \ddot{F}, \ddot{R}, \ddot{P})$ whose $\mathscr{B}^f(\ddot{\Gamma}) = \mathscr{B}^f(\Gamma)$. Since $\mathscr{B}^f(\Gamma)$ is open, $\mathscr{B}^f(\ddot{\Gamma})$ is open. Now construct $\dot{\Gamma} = (\ddot{S}, \dot{G}, \dot{R}, \ddot{P})$, where $\dot{G} = \ddot{G} \cap \ddot{F}$ and $\dot{R}$ is $\ddot{R}$ cut down to $\ddot{F} \times \ddot{F}$. It can be shown from the openness of $\mathscr{B}^f(\ddot{\Gamma})$ and the determinism of $\ddot{\Gamma}$ that every complete state which belongs to a $\ddot{\Gamma}$-sequence is an $\ddot{F}$. Consequently, $\mathscr{B}^f(\dot{\Gamma}) = \mathscr{B}^f(\ddot{\Gamma})$. Hence $\mathscr{B}^f(\dot{\Gamma}) = \mathscr{B}^f(\Gamma)$, and $\dot{\Gamma}$ is the desired sequence generator without final states.

The result just established for open finite behaviors does not hold for open behaviors in general. That is, there is a $\Gamma = (S, G, F, R, P)$ with open $\mathscr{B}(\Gamma)$ for which there is no behaviorally equivalent sequence generator without final states. An example is given in Fig. 18. $\mathscr{B}(\Gamma)$ consists of all finite sequences of p_0 and is thus open. But if the behavior of a sequence generator without final states contains all finite sequences of p_0, by the Infinity Theorem (2.1–3) it must also contain the infinite sequence $p_0, p_0, p_0, \cdots$. Hence $\mathscr{B}(\Gamma)$ is not the behavior of any sequence generator without final states.

We will conclude this subsection with a discussion of Boolean operations on sequence generator behaviors. Let $\mathscr{P}(\Gamma)$, $\mathscr{P}^f(\Gamma)$, $\mathscr{P}^\omega(\Gamma)$ be the sets of all,

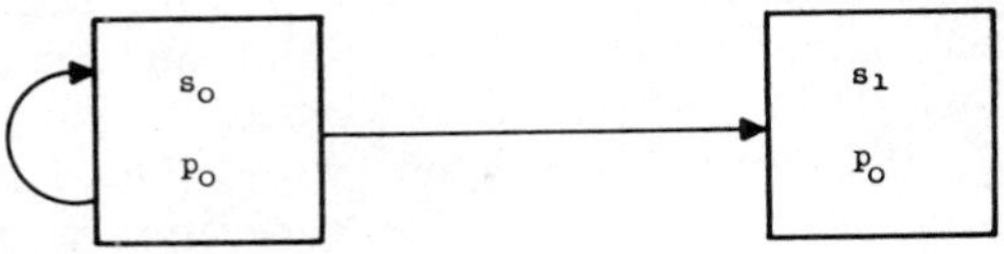

Figure 18. Final-state sequence generator $\Gamma = (S, G, F, R, P)$. $F = \{s_1\}$. $\mathscr{B}(\Gamma)$ is open, but it is not the behavior of any sequence generator without final states.

all finite, all infinite sequences of P-states, respectively. The complement of a behavior is defined with respect to the appropriate one of these: $\sim\mathscr{B}(\Gamma) = \mathscr{P}(\Gamma) - \mathscr{B}(\Gamma)$, $\sim\mathscr{B}^f(\Gamma) = \mathscr{P}^f(\Gamma) - \mathscr{B}^f(\Gamma)$, and $\sim\mathscr{B}^\omega(\Gamma) = \mathscr{P}^\omega(\Gamma) - \mathscr{B}^\omega(\Gamma)$.

Consider first sequence generators without final states. Let $\ddot{\Gamma}$ be the sequence generator represented by the state diagram obtained by juxtaposing the state diagrams for Γ and $\dot{\Gamma}$, treating the complete states of Γ and $\dot{\Gamma}$ as distinct. Then clearly

$$\mathscr{B}(\ddot{\Gamma}) = \mathscr{B}(\Gamma) \cup \mathscr{B}(\dot{\Gamma}),$$
$$\mathscr{B}^f(\ddot{\Gamma}) = \mathscr{B}^f(\Gamma) \cup \mathscr{B}^f(\dot{\Gamma}),$$
$$\mathscr{B}^\omega(\ddot{\Gamma}) = \mathscr{B}^\omega(\Gamma) \cup \mathscr{B}^\omega(\dot{\Gamma}).$$

Let $\ddot{\Gamma} = \Gamma \times \dot{\Gamma}$, where "$\times$" is the cross product operation defined in § 4.3. It may be proved that

$$\mathscr{B}(\ddot{\Gamma}) = \mathscr{B}(\Gamma) \cap \mathscr{B}(\dot{\Gamma}),$$
$$\mathscr{B}^f(\ddot{\Gamma}) = \mathscr{B}^f(\Gamma) \cap \mathscr{B}^f(\dot{\Gamma}),$$
$$\mathscr{B}^\omega(\ddot{\Gamma}) = \mathscr{B}^\omega(\Gamma) \cap \mathscr{B}^\omega(\dot{\Gamma}).$$

Thus the union and intersection of sequence-generator behaviors are always sequence-generator behaviors. But the complement of a sequence-generator behavior is not generally a sequence-generator behavior. This is shown by Fig. 19; it can be proved by means of the openness of the behavior of a sequence generator without final states and the Infinity Theorem (2.1–3) that no sequence generator without final states has $\sim\mathscr{B}(\Gamma)$, $\sim\mathscr{B}^f(\Gamma)$, or $\sim\mathscr{B}^\omega(\Gamma)$ as its behavior.

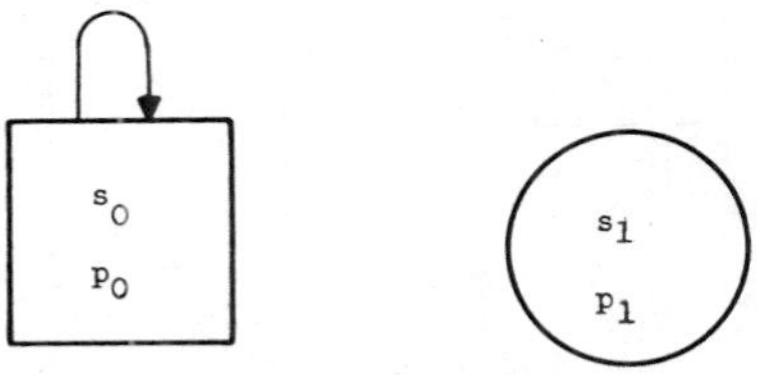

Figure 19. $\Gamma = (S, G, R, P)$. No sequence generator without final states has $\sim\mathscr{B}(\Gamma)$, $\sim\mathscr{B}^f(\Gamma)$, or $\sim\mathscr{B}^\omega(\Gamma)$ as its behavior.

The situation is very different with final-state sequence generators. The class of finite behaviors of sequence generators with final states is closed under union, intersection, and complement (Rabin and Scott [**27**]). Consider next infinite behaviors. In § 4.2 we showed how to pass between various kinds of formulas and sequence generators without final states. There are also formulas which correspond to sequence generators with final states. J. Richard Büchi (unpublished) discusses such formulas which he calls "quasi-Σ_1-formulas" and establishes that the complement of a set represented by a quasi-Σ_1-formula may be represented by a quasi-Σ_1-formula. The same result is, in sequence-generator terms, that for every $\Gamma = (S, G, F, R, P)$ there is a $\dot{\Gamma} = (\dot{S}, \dot{G}, \dot{F}, \dot{R}, \dot{P})$ such that $\mathscr{B}^{\omega}(\dot{\Gamma}) = \sim \mathscr{B}^{\omega}(\Gamma)$. As before, if $\ddot{\Gamma}$ is the result of juxtaposing Γ and $\dot{\Gamma}$, $\mathscr{B}^{\omega}(\ddot{\Gamma}) = \mathscr{B}^{\omega}(\Gamma) \cup \mathscr{B}^{\omega}(\dot{\Gamma})$. Hence by De Morgan's theorem, the class of infinite behaviors is closed under union, intersection, and complement. Consider finally (complete) behaviors. As before, if $\ddot{\Gamma}$ is the result of juxtaposing Γ and $\dot{\Gamma}$, $\mathscr{B}(\ddot{\Gamma}) = \mathscr{B}(\Gamma) \cup \mathscr{B}(\dot{\Gamma})$. However, the intersection of the behaviors of sequence generators with final states is not always a sequence-generator behavior, and hence by De Morgan's theorem the complement of the behavior of a sequence generator with final states is not always a sequence-generator behavior. This is shown by Fig. 20. $\mathscr{B}(\Gamma)$ of Fig. 20(a) consists of all finite sequences terminating in p_1 and all infinite sequences with infinitely many occurrences of p_1. $\mathscr{B}(\dot{\Gamma})$ of Fig. 20(b) consists of all finite sequences terminating in p_0 and all infinite sequences with infinitely many occurrences of p_0. Hence $\mathscr{B}(\Gamma) \cap \mathscr{B}(\dot{\Gamma})$ contains only infinite sequences. But no sequence-generator behavior contains only infinite sequences, and therefore no sequence generator has the behavior $\mathscr{B}(\Gamma) \cap \mathscr{B}(\dot{\Gamma})$.

§ 2.3 contains a decision procedure for the inclusion of the behavior of one sequence generator without final states in the behavior of another. This is also a procedure for the inclusion of finite behaviors and a simple modification

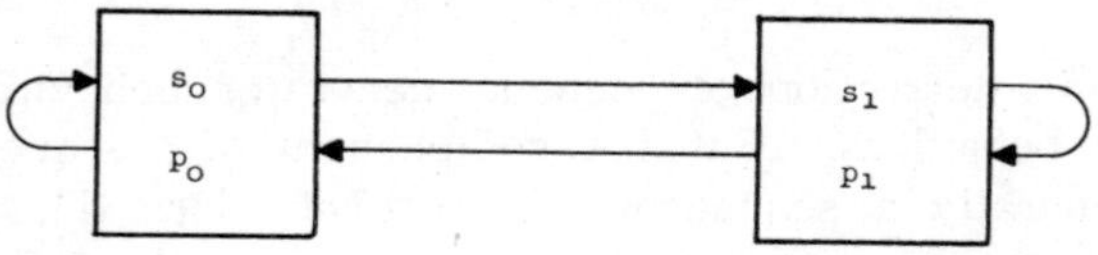

(a)

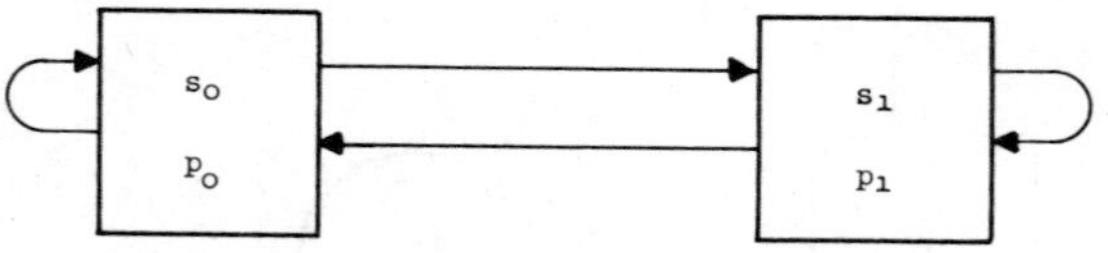

(b)

Figure 20.

(a) $\Gamma = (S, G, F, R, P)$. $F = \{s_1\}$.

(b) $\dot{\Gamma} = (S, G, \dot{F}, R, P)$. $\dot{F} = \{s_0\}$. No sequence generator has the behavior $\mathscr{B}(\Gamma) \cap \mathscr{B}(\dot{\Gamma})$.

of it gives a procedure for the inclusion of infinite behaviors. It follows from the fact that both the class of finite behaviors and the class of infinite behaviors of final state sequence generators are effectively closed under union and complement that there are decision procedures for "Is $\mathscr{B}^{\omega}(\Gamma) \subseteq \mathscr{B}^{\omega}(\dot{\Gamma})$?," "Is $\mathscr{B}^{f}(\Gamma) \subseteq \mathscr{B}^{f}(\dot{\Gamma})$?," "Is $\mathscr{B}(\Gamma) \subseteq \mathscr{B}(\dot{\Gamma})$?," where Γ and $\dot{\Gamma}$ are final-state sequence generators.

4.5. **Infinite-sequence generators.** In this subsection we will discuss briefly a generalization of the original concept of sequence generator obtained by dropping the requirement that the set of complete states S be finite (§ 1.1). The resultant generalization is called an "infinite-sequence generator."

Actually much of the content of this paper applies to infinite-sequence generators as well as finite ones. All our concepts apply to infinite-sequence generators except for the various decision procedures and the reduced form $\Gamma^{\dagger}$, and it is easy to define the reduced form of an infinite-sequence generator. Many of our theorems apply to infinite-sequence generators too, with only minor modifications being needed in the proofs we have given to cover this extension. Of course the various decision procedures we have given make essential use of the finitude of the number of complete states of sequence generator, and the Time-shift Theorem (3.3–2) applies only to finite-sequence generators.

In the first part of this paper we have made implicit use of infinite sequence generators. It will be seen on examination that Lemma 1.3–2 is in fact about an infinite-sequence generator (S, δ_0, ρ), where S is any set (finite or infinite) on which ρ is defined. The following theorem is very close to Lemma 2.1–2. Let $\Gamma = (S, G, R)$ be an infinite-sequence generator with G finite and $R(s)$ finite for every $s \in S$; if every set in the ω-sequence $G, R(G), R^2(G), R^3(G), \cdots$ is nonempty, then there is an infinite Γ-sequence. By means of this lemma the proof of the Infinity Theorem (2.1–3) can be rewritten, with minor modifications, to yield the following extension of the Infinity Theorem: Let $\Gamma = (S, G, R, P)$ be an infinite sequence generator with G finite and $R(s)$ finite for every $s \in S$; then any sequence of P-states belongs to $\mathscr{B}(\Gamma)$ if and only if every finite initial segment of it belongs to $\mathscr{B}(\Gamma)$.

The tree operator to be defined next was used implicitly in the proof of the Infinity Theorem as well as in step (2) of the Reduced Form Algorithm. The tree operator "t" applies to any (infinite or finite) sequence generator $\Gamma = (S, G, R, P^1, \cdots, P^n)$ and produces its "tree generator" $\Gamma^t = \dot{\Gamma} = (\dot{S}, \dot{G}, \dot{R}, \dot{P}^1, \cdots, \dot{P}^n)$:

$\dot{S} =$ the set of finite Γ-sequences,

$\dot{G} = G$,

$\dot{R}(\dot{s}, [s](0, j+1)) \equiv [([s](0, j+1) \in \dot{S}) \;\&\; (\dot{s} = [s](0, j))]$,

$P^i([s](0, k)) = P^i(s(k))$, for $i = 1, \cdots, n$.

Note that if $R(s)$ is always finite, then $R(\dot{s})$ is always finite. Moreover, if S is finite, then $\dot{S}$ is infinite if and only if Γ has a Γ-admissible complete state.

We conclude with some examples of infinite-sequence generators which have already been discussed in the literature, though not under this name. A Turing machine as defined by Turing [30] consists of a finite automaton to which is attached an infinite tape; odd-numbered squares are used for writing the digits of a real number. Such a machine corresponds to a sequence generator (S, G, R, Q), where S is the set of states of the machine (including its tape), G is the set of initial states of the machine, R is given by the transition rule, and the output projection Q when applied to a complete state s gives the number written on the odd-numbered squares of the tape when the machine is in that state. $R(s)$ is always a unit set. For the special-purpose machine (whose tape is initially blank), G is a unit set. For the universal machine, G is an infinite set consisting of those states whose tape part represents a program for a special purpose machine. The concept of computation (§ 4.1) can be extended to infinite-sequence generators. In particular, this can be done for Turing machines in such a way that two machines (one of these may be universal) are computationally equivalent if and only if they compute the same number in Turing's sense. This fact was part of the motivation for defining the concept of computation inasmuch as a similar statement cannot be made in terms of behavior.

The basis von Neumann uses for his construction of a self-reproducing automaton consists of an infinite number of cells each capable of 29 states, with the state of each cell at time $t+1$ determined by the states of itself and its neighbors at time t (Shannon [**28**]; Burks [**3**, § 4]). This basis corresponds to a sequence generator (S, G, R), with S and G both infinite, and $R(s)$ finite but unbounded. Burks [**4**, § 6], Church [**8**, p. 21 ff.], and Holland [**16**] have given definitions of infinite automata with inputs and outputs. These correspond to infinite-sequence generators with two projections; in the latter two formulations, infinitely many input states are allowed, with the consequence that $R(s)$ is infinite. Any Post canonical language (Post [**25**]) in which each production rule has only one premise may be represented by an infinite-sequence generator (S, G, R). S is the set of strings, G is the set of axioms, and R is given by the production rules.

Although we will not attempt to give a definition of "effective sequence generator," it should be noted that all the examples of infinite-sequence generators given above are effective in the sense that in each case integers may be assigned to the states in such a way that S, G, R, and the P^i's are all recursive.

4.6. **Probabilistic sequence generators.** For the sake of completeness we will discuss the relation of probabilistic to nonprobabilistic sequence generators before concluding this paper.

Let $\Gamma = (S, G, R, P^1, \cdots, P^n)$, $n = 0, 1, 2, \cdots$, be an infinite-sequence generator with non-null G. Let W be a weight function which assigns positive initial probabilities summing to one to the elements of G and, for each $s \in S$, assigns positive transition (conditional) probabilities summing to one to the elements of $\{\langle s, s_1\rangle \mid R(s, s_1)\}$, provided this set is non-null.[7] $(S, G, R, W, P^1, \cdots, P^n)$

[7] There is an alternative concept which is more difficult to define but easier to use in some applications: for each P-state p the initial probabilities sum to one over $\{s \mid s \in G \ \& \ P(s) = p\}$, and for each p and $s \in S$ the transition probabilities sum to one over $\{\langle s, s_1\rangle \mid R(s, s_1) \ \& \ P(s_1) = p\}$.

is a "probabilistic sequence generator." Note that, as we use the terms, "probabilistic" and "deterministic" are not contradictories. "Deterministic" was defined in § 2.1 for nonprobabilistic sequence generators; its opposite is "indeterministic," not "probabilistic." Moreover, given a deterministic sequence generator (S, G, R, P) one can form a probabilistic sequence generator (S, G, R, W, P) from it by adding a weight function W.

The weight function W induces a probability distribution on the finite Γ-sequences of $\Gamma = (S, G, R, P)$ and hence on the elements of $\mathscr{B}^f(\Gamma)$. The probability of a finite Γ-sequence $[s](0, k)$ is the initial probability of $s(0)$ multiplied by the probabilities of the transitions $\langle s(0), s(1)\rangle, \langle s(1), s(2)\rangle, \cdots, \langle s(k-1), s(k)\rangle$. The probability of any element of $\mathscr{B}^f(\Gamma)$ is the sum of the probabilities of those Γ-sequences which produce that behavior element.

A stochastic process in which the probability of a state occurring at time t depends only on the preceding states of the sequence can be represented by a sequence generator whose states are finite sequences of states of the stochastic process; compare the tree sequence generator Γ^t of § 4.5. A Markov chain with constant transition probabilities is a probabilistic sequence generator without projections (S, G, R, W) such that (S, G, R) has no terminal states (see, for example, Feller [**12**, p. 340]). If S is finite, then (S, G, R, W) is a finite Markov chain.

It is worth noting that many of the concepts employed in analyzing a finite Markov chain (S, G, R, W) depend only on the underlying nonprobabilistic sequence generator (S, G, R) and not on the probability function W. We will give some examples. s is an absorbing state if and only if $R(s) = \{s\}$. (It should be recalled in this connection that every probability assigned by W is positive; hence if the probability of s succeeding s is one, then $R(s)$ contains only s.) Let $R^\omega(s_1, s_2) \equiv s_2 \in \bigcup_{i=1}^{\omega} R^i(s_1)$. s is a transient state if and only if there is a state $s_1 \in S$ such that $R^\omega(s, s_1)$ but not $R^\omega(s_1, s)$; s is a persistent state if s is not transient. A sequence generator (S, G, R) which satisfies these two conditions is ergodic: first, for any two complete states $s_1, s_2 \in S$, $R^\omega(s_1, s_2)$ and $R^\omega(s_2, s_1)$, and second, the greatest common divisor of the lengths of all cycles of complete states is one (cf. Shannon [**29** p. 435]).

It was remarked immediately after Corollary 2.1-1 that the concept of a sequence generator with projections is a bona fide generalization of the concept of a sequence generator without projections; for example, the sequence $001001001\cdots$ belongs to the behavior of a sequence generator but it is not a Γ-sequence of any sequence generator (cf. the discussion of various ways of generalizing sequence generators in § 4.2). Similarly, probabilistic sequence generators with projections are bona fide extensions of probabilistic sequence generators without projections. In any probabilistic sequence generator (S, G, R, W, P) the probability of a given complete state occurring at any time depends only on what complete state occurred at the preceding time. This is not so for the P-states, for since the same P-state may be assigned to different complete states the occurrence of a P-state at t can be made to depend on what P-states occurred at times prior to $t - 1$. An example is given in Fig. 21; the probability that p_0 will occur, given that the three preceding P-states are p_1, p_0, p_0, in that order, is 0.8, while the probability that p_0 will occur, given that the three preceding states are p_1, p_1, p_0, in that order, is 0.5.

We give some examples of probabilistic sequence generators. Shannon, [**29**, p. 384 ff.], defines a "discrete information source." A discrete information source is a finite Markov chain $\Gamma = (S, G, R, W)$ to which has been added a set of symbols, each transition of Γ producing one of these symbols as outputs. One can construct another sequence generator $\dot{\Gamma} = (\dot{S}, \dot{G}, \dot{R}, \dot{W}, \dot{P})$, where $\dot{S}$ consists of pairs of elements of S and $\dot{P}$ is an output projection, such that $\dot{\Gamma}$ will produce the same output sequences with the same probabilities as the original discrete information source.

Von Neumann [**31**, p. 61 ff.] investigates finite probabilistic nets (cf. Moore

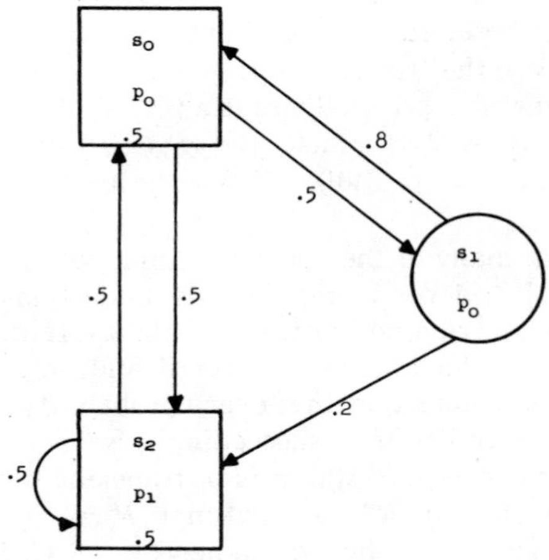

Figure 21. Probabilistic sequence generator (S, G, R, W, P).

and Shannon [**23**]). These are well-formed nets composed of combined switch-delay elements. An element produces the correct (desired) output at each time with probability $1 - \varepsilon$ and the complement (incorrect) output with probability ε. Thus in Fig. 22(a), the output B is defined probabilistically as

$$B(0) \text{ with probability } 1 - \varepsilon, \quad \overline{B(0)} \text{ with probability } \varepsilon,$$
$$B(t + 1) \equiv [A(t) \not\equiv B(t)] \text{ with probability } 1 - \varepsilon,$$
$$B(t + 1) \not\equiv [A(t) \not\equiv B(t)] \text{ with probability } \varepsilon.$$

Given a probabilistic net, one can derive a probabilistic sequence generator from it by methods similar to those of § 1.2. Figure 22(b) is the probabilistic sequence generator (S, G, R, W, I, θ) for the binary counter of Fig. 22(a); I is the input projection and θ is the output projection. The solid lines represent

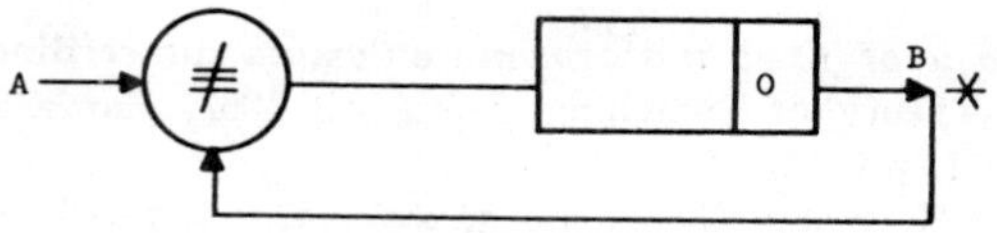

(a)

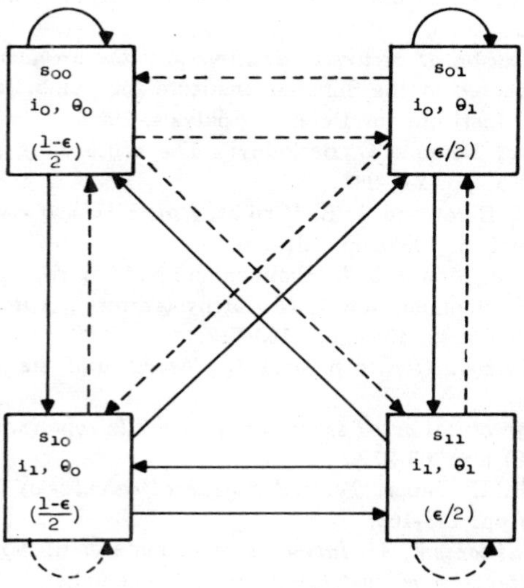

(b)

Figure 22.

(a) Binary counter (probabilistic).

(b) Probabilistic sequence generator (S, G, R, W, I, θ) for binary counter (a). Solid lines represent transitions with probability $(1 - \varepsilon)/2$. Dotted lines represent transitions with probability $\varepsilon/2$.

the desired (correct) transitions, the dotted lines the erroneous transitions; cf. Fig. 1. Note that (S, G, R, I, θ) of Fig. 22(b) is not deterministic. In a similar way, a probabilistic sequence generator (S, G, R, W, Q), may be obtained from a probabilistic Turing machine; here Q is the output projection, S is infinite; see de Leeuw et al. [11].

Bibliography

1. D. D. Aufenkamp and F. E. Hohn, *Analysis of sequential machines*, Institute of Radio Engineers, Transactions on Electronic Computers, 1957, EC-6, pp. 276-285.

2. J. R. Büchi, C. C. Elgot, and J. B. Wright, *Non-existence of certain algorithms of finite automata theory*, Abstract 543-12, Notices Amer. Math. Soc. vol. 5 (1958) p. 98.

3. A.W. Burks, *Computation, behavior, and structure in fixed and growing automata*, in M. Yovits and S. Cameron (editors), *Self-Organizing Systems*, New York, Pergamon Press, 1960, pp. 282–311.

4. ———, *The logic of fixed and growing automata*, Proceedings of an International Symposium on the Theory of Switching, 2-5 April 1957, Harvard University Press, Cambridge, 1959, Part I, pp. 147–188.

5. A.W. Burks and H. Wang, *The logic of automata*, J. Assoc. Comput. Mach. vol. 4 (1957) pp. 193–218, 279–297.

6. A.W. Burks and J. B. Wright, *Theory of logical nets*, Proc. I.R.E. vol. 41 (1953) pp. 1357–1365.

7. N. Chomsky and G. A. Miller, *Finite state languages*, Information and Control vol. 1 (1958) pp. 91–112.

8. A. Church, *Application of recursive arithmetic to the problem of circuit synthesis*, Summaries of talks presented at the Summer Institute for Symbolic Logic, Cornell University, 1957; Princeton, Institute for Defense Analysis, 1960.

9. ———, Review of Edmund C. Berkeley: *The algebra of states and events*, J. Symb. Logic vol. 20 (1955) pp. 286–287.

10. I. M. Copi, C. C. Elgot and J. B. Wright, *Realization of events by logical nets*, J. Assoc. Comput. Mach. vol. 5 (1958) pp. 181–196.

11. K. de Leeuw, E. F. Moore, C. E. Shannon and N. Shapiro, *Computability by probabilistic machines*, in C.E. Shannon and J. McCarthy (editors), *Automata Studies*, Princeton, Princeton University Press, 1956, pp. 183–212.

12. W. Feller, *An introduction to probability theory and its applications*, 2d ed., New York, Wiley, 1957.

13. F. B. Fitch, *Representation of sequential circuits in combinatory logic*, Philosophy of Science, vol. 25 (1958) pp. 263–279.

14. F. Harary and H. H. Paper, *Toward a general calculus of phonemic distribution*, Language vol. 33 (1957) pp. 143–169.

15. A. Heyting, *Intuitionism, an introduction*, Amsterdam, North-Holland, 1956.

16. J. H. Holland, *Iterative circuit computers*, Proceedings of the 1960 Western Joint Computer Conference, 1960.

17. S. C. Kleene, *Representation of events in nerve nets and finite automata*, in C.E. Shannon and J. McCarthy (editors), *Automata Studies*, Princeton, Princeton University Press, 1956, pp. 3–41.

18. D. König, *Theorie der endlichen und unendlichen Graphen*, Leipzig, Akademische Verlagsgesellschaft M.B.H., 1936.

19. G. H. Mealy, *A method for synthesizing sequential circuits*, Bell System Tech. J. vol. 34 (1955) pp. 1045–1079.

20. J. C. C. McKinsey, *Introduction to the theory of games*, New York, McGraw-Hill, 1952.

21. I. T. Medvedev, *On a class of events representable in a finite automaton*, translated by J. J. Schorr-Kon from a supplement to the Russian translation of *Automata Studies*, C. E. Shannon and J. McCarthy (editors), Group Report 34–73, Lexington, Massachusetts, Lincoln Laboratory, 1958.

22. E. F. Moore, *Gedanken experiments on sequential machines*, in C. E. Shannon and J. McCarthy (editors), *Automata Studies*, Princeton, Princeton University Press, 1956, pp. 129–153.

23. E. F. Moore and C. E. Shannon, *Reliable circuits using less reliable relays*, J. Franklin Inst. vol. 262 (1956) pp. 191–208, 281–287.

24. J. Myhill, *Finite automata and representation of events*, in *Fundamental Concepts in the Theory of Systems*, WADC Technical Report 57–624, ASTIA Document No.

AD 1557 41, 1957.

25. E. L. Post, *Formal reductions of the general combinatorial decision problem*, Amer. J. Math. vol. 65 (1943) pp. 197–215.

26. H. Putnam, *Decidability and essential undecidability*, J. Symb. Logic vol. 22 (1957) pp. 39–54.

27. M. O. Rabin, and D. Scott, *Finite automata and their decision problems*, IBM J. Res. Develop. vol. 3 (1959) pp. 114–125.

28. C. E. Shannon, *Computers and automata*, Proc. I.R.E. vol. 41 (1953) pp. 1235–1241.

29. ———, *A mathematical theory of communication*, Bell System Tech. J. vol. 27 (1948) pp. 379–423, 623–656.

30. A. M. Turing, *On computable numbers, with an application to the entscheidungs-problem*, Proc. London Math. Soc. series 2, vol. 42 (1936) pp. 230–265, and vol. 43 (1937) pp. 544–546.

31. J. von Neumann, *Probabilistic logics and the synthesis of reliable organisms from unreliable components*, in C. E. Shannon and J. McCarthy (editors), *Automata Studies*, Princeton, Princeton University Press, 1956 pp. 43–98.

32. ———, *The general and logical theory of automata*, in *Cerebral mechanisms in behavior*, New York, Wiley, 1951, pp.1–41.

33. H. Wang, *Circuit synthesis by solving sequential Boolean equations*, Z. Math. Logik Grundlagen Math. vol. 5 (1959) pp. 291–322.

University of Michigan,
Ann Arbor, Michigan

THE TREATMENT OF AMBIGUITY AND PARADOX IN MECHANICAL LANGUAGES[1]

BY

SAUL GORN

1. Introduction. As a representative of the newly emerging *Computer and Information Sciences,* I am happy to have this opportunity to address an audience which includes members of the pure sciences of mathematics and mathematical logic. In spite of the fact that it is hardly fair to demand the attention of the pure scientist to all possible applications of his subject, it is nonetheless pleasant for the applier when he does receive such attention. It *is* most relevant to our subject to begin by examining in what way this new discipline is an application of mathematics and mathematical logic. Indeed, the whole question of how ambiguity is resolved or remains unresolved in a machine, a mechanical language, or a recursive algorithm depends on the method by which context is determined in the application, and, in fact, on the meaning of the word 'context' itself.

If we hark back to the definition of *Semiotics* given by Charles Morris, as quoted by Carnap [**2**], Curry and Feys [**3**], and Martin [**5**; **6**], we have a subdivision into *syntax*, *semantics*, and *pragmatics*. Pragmatics is concerned with the relationship between symbols and their *users* or *interpreters*. Semantics is concerned with the relationship between symbols and their *meanings*. Syntax is concerned with the relationships of the symbols among themselves independent of their semantic content or pragmatic context.

Pure sciences are likely to reduce to purely formal systems in which all semantic content is abstracted from the *object* language and resides only in the syntax language over this object language. Pure sciences therefore tend to develop purely syntactic languages. The applied science, on the other hand, is directly concerned with the semantics of the object language, since *application* means returning semantic content to a formal system. The study of mechanical languages is among those studies, however, which are fully semiotic; it must concern itself with pragmatic matters, if we are willing to interpret the words *user* and *interpreter* in the mechanical sense of *processor*. The *context* of an expression composed of symbols then means, in the broadest

Received by the editor April 7, 1961.

[1] The background work on mechanical languages (see Gorn [**4**]) in this paper was originally supported by the U. S. Army Signal Corps from 1958 to 1960 under Contract DA-36-039-SC-74047 to the University of Pennsylvania's Institute for Cooperative Research and Moore School of Electrical Engineering. The developments in this paper were made possible by joint support from the National Science Foundation and the Air Force Office of Scientific Research to the University of Pennsylvania's Office of Computer Research and Education. (AF-49(638)-951 and NSF-G-14096).

sense, the processor under whose control that expression happens to be at the moment. Such an interpretation means that we have also assigned a mechanical meaning to the word 'intensional', and have equated 'intension' with *controlling processor* to mean the same as *control context*.

2. The semiotics of mechanical languages. The definition of *one-dimensional, digital mechanical languages* is, then, *not* merely the extensional one of 'a recursively enumerable set of finite, ordered strings of characters chosen from an alphabet'; it must include the intentional element of a set of processors such as *generators, recognizers, translators, etc.*, and even possibly such *control processors* as *applicability recognizers, priority controllers, context selectors*, and the like. If there are no processors, we might call what we have a code, but we would not call it a language.

Beginning with a 'base alphabet', we define a *one-dimensional, digital mechanical language* recursively as a set of processors whose data, the extent of the language, is a recursively enumerable set of finite ordered strings of characters from an alphabet; and we define an alphabet as a language possessing a processor called a *recognizer* which solves mechanically the decision problem of determining whether a 'word' does or does not 'belong' to the language. Thus the extension of an alphabet is not merely recursively enumerable, it is recursive, and it is not unusual to have alphabets of infinite extent, and with idealized processors with a potentially infinite memory, i.e. growing automata.

A central function, then, of the Computer and Information Sciences is the analysis and synthesis of mechanical languages and their processors. The word *processor* is just as deliberately ambiguous as the word 'word', and may refer to mechanisms, algorithms, instructions to a mechanism, programs for a machine, or systems, and organizations composed of machines, programs, and algorithmized people, all depending on the context determined by a conrol sub-processor. The language in which the whole theory is couched, as yet not completely mechanized, is self-referencing and includes both *descriptive syntax* of declarative statements for enunciation of theorems and proofs and specification of processors by logical design, and *command syntax* of imperative sentences for specification of processors by programming, with a strong interconnection between the two permitting the translation of statements into actions on call by appropriate control processors. Not only is this 'universal' mechanical language what we have called a *mixed* descriptive and command language, but it is also a language with 'unstratified control'. To recognize this phenomenon, let us note, for example, that a theorem about a class of programs for processing a fixed class of data is expressed in the descriptive syntax of the command syntax of an object language. In order to avoid an infinite ramified hierarchy of syntax languages, we envision a *control syntax* which can shift the context of an object expression to give it a syntactical or even a control interpretation. This is surely done in natural languages with their quotation marks and other punctuation marks, and even such words as 'mean' and 'define', whose function is to resolve momentary ambiguity by controlling sequencing, scope, and context. It is also done in general purpose digital computers, possessing "loop control", by the control instructions. The introduction of such a control syntax is therefore necessary to make a language complete in the Turing or Gödel sense, but makes us pay the price described in the various "incompleteness" and "undecidability" theorems, namely that the language is undecidable, can express inconsistencies and paradoxes in its descriptive side, and can possess programs which cycle indefinitely in its command side.

In a language with unstratified control there is not merely the well known double ambiguity of *use* and *mention*, but a triple or quadruple ambiguity at least. An expression may have a purely objective interpretation, a descriptive syntax interpretation, a command syntax interpretation, or a control syntax interpretation at different times, depending on the control interpreter's sequencing of such interpretations. Objects, commands, and controls are dealt with syntactically, semantically, and pragmatically. Ambiguities themselves may be classified at these levels. In purely object context an ambiguity is a *syntactic pun*, concerned with the question of multiple meaningful deconcatenation;[2] for example 'nowhere' might separate into 'no' and 'where' or into 'now' and 'here'. In syntactic context, command or descriptive, an ambiguity is a *semantic pun*, a pun in the usual sense of different meanings for the same word; for example, 'add' might refer to numbers or to matrices. In control context ambiguities may be across syntactic lines, so that the same expression in one context may be processed as an object, in another as a description, in another as a command; an example of such a *pragmatic pun* is a string of symbols which processor A decides is a good word (well formed formula) in an object language, processor B recognizes as the name of a proof, processor C uses to give the structure of the proof, and processor D causes to derive a theorem by going through the syntactic steps designated by the proof. We will quote an extended example of this type below, but a primitive example is a symbol for a variable, say x. At object level it refers to the symbol 'x'; in descriptive syntax it is a symbol for which substitutions may be made from a given domain, i.e., an indeterminate; in command syntax it means the contents of a certain storage cell, as in 'add x'; in control syntax it is the name of a storage cell, as in 'jump to x'. A similar primitive example of control ambiguity is the pair of words 'yes' and 'no'; in object selection they mean *recognized* or *not recognized* as belonging to some language or list; in statement selection they might mean *true* and *false*; in action selection they might mean *permitted* and *forbidden*.

As far as I have been able to see, the well known paradoxes are all pragmatic puns in which there is a cycle of shifting interpretations made possible by the same unstratified control which permits us to achieve universality and self-referencing.

3. Specification and processing of mechanical languages. In our pervasive universal syntax with unstratified control, the word 'specify' is as deliberately ambiguous as the word 'processor'. The *specification* of a language or processor in descriptive syntax may be a definition, either explicit or recursively implicit; in command syntax it may be a program or a manual of instruc-

[2] Deconcatenation or grammatical generation (the so-called 'generative ambiguities' are also considered syntactic).

tions in a "make them yourself" kit. Such specifications may also be either structural, telling how they are constructed, or behavioral, telling what they do or how they are recognized. For example, the logical diagram of a processor is a descriptive structural specification.

The most primitive processors for a language over an alphabet, so that all mechanical languages possess one or the other or both, are therefore the structural *generators*, and the behavioral *recognizers*. From a certain point of view, *every* processor may be analyzed into component generators and recognizers of suitably defined auxiliary languages, but such an analysis will more often than not be too clumsy to be useful.

A most direct generator specification of a mechanical language is given implicitly by a set of processors called 'producers' which may be applied recursively. Such a set may be called, with very little violence to standard terminology, a 'production system'. The main verb in a producer means 'substitute ($\rightarrow$)' in command context, 'implication ($\Rightarrow$)' in descriptive context, and a combination of the two ($\equiv\!\!>$) which means 'if $\cdots$ is true, do $\cdots$' in control context. This last is the primitive transformer of statements into actions; in defining *mixed languages* we assumed the existence of such a control processor.

As an example, let us assume a base alphabet of two characters, $\mathscr{D} = \{T, F\}$ and two producers to generate a language $\mathscr{P}\mathscr{D}$ over this alphabet. The first producer, called an 'ad hoc' producer, specifies that the string 'F' belongs to $\mathscr{P}\mathscr{D}$; the second producer, an 'expanding replacement producer', specifies that, if a string α over $\mathscr{D}$, belongs to $\mathscr{P}\mathscr{D}$, then so does the string '$T\alpha$' obtained by concatenating T at the head of α. The extension of $\mathscr{P}\mathscr{D}$ is clearly the sequence of strings $F, TF, TTF, \cdots \mathscr{P}\mathscr{D}$ may be used as a primitive ordinal number representation language. Like all languages generated by production systems all of whose producers are ad hoc or expanding and none of which involve auxiliary languages, $\mathscr{P}\mathscr{D}$ is decidable, possesses a recognizer, and may be used as an alphabet. It is an example of an *ordinal alphabet*, every one of which is assumed to have available a generating processor called a 'controlled counter', and a recognizer. An ordinal alphabet is assumed available in the syntax of every mechanical language. We might call the extension of $\mathscr{P}\mathscr{D}$ the 'tally code', anticipating the command meanings *tally* and *finish* for 'T' and 'F' in processors involving counting. The alphabet $\mathscr{D}$ might be used in some descriptive syntax contexts to assign meanings *true* and *false* to the same characters 'T' and 'F'. $\mathscr{P}\mathscr{D}$, besides being a primitive ordinal alphabet, is also the most primitive of the infinite class of *prefix languages*[3] about which we will have more to say.

In considering any language over an alphabet $\mathscr{A}$ we must have available processors for the language $\Sigma\mathscr{A}$ whose extension is the set of all finite strings of characters from $\mathscr{A}$. If, for example, $\mathscr{A}$ itself is a language over a prior alphabet some punctuation or means of separation of the strings of $\mathscr{A}$ must be available, at least for the recognizer of $\mathscr{A}$. It is assumed that $\Sigma\mathscr{A}$ has *registers* which keep the characters of $\mathscr{A}$ somehow distinguishable. For some alphabets, such as $\mathscr{A} = \mathscr{P}\mathscr{D}$, this is achieved without additional control devices of the punctuation type because the language is *uniquely deconcate-*

[3] I have since renamed these 'complete prefix languages'. See Saul Gorn, "An axiomatic approach to prefix languages" in *Symbolic languages in data processing*, Gordon & Breach, 1962.

nable; this is an important property of one-dimensional mechanical languages which we will discuss further in connection with syntactic ambiguity. For general linear mechanical languages, we will assume that there are available such registers for $\Sigma\mathscr{A}$ in idealized pairs for double register shifting, thereby permitting us to concatenate and deconcatenate strings over $\mathscr{A}$. . We also assume available for $\Sigma\mathscr{A}$ and hence all languages over $\mathscr{A}$ such processors as string comparators which recognize the identity of two strings, and, in particular, a generator and recognizer of the null string, i.e. a cleared register over $\mathscr{A}$; we might consider these the generator and recognizer of the null language $\Lambda\mathscr{A}$. Such processors as recognizers and generators of $\mathscr{A}$, $\Lambda\mathscr{A}$, $\Sigma\mathscr{A}$, and ordinal alphabets are the components out of which processors for a language $\mathscr{L}\mathscr{A}$ over $\mathscr{A}$ are specified. An important example would be a *substitutor* of a string β for a character γ wherever it appears in a string α. We might represent this processor for $\Sigma\mathscr{A}$ by the symbol $\alpha(\beta \to \gamma)$ in a syntax language whose alphabet includes greek characters for variables, parentheses, and the arrow symbol explained above. If, however, the substitutor is defined over a uniquely deconcatenable language such as $\mathscr{P}\mathscr{D}$, we could use a simplified syntax language to present it in the form $S\alpha\beta\gamma$. For example $STTFTFF$ would represent $TTF(TF \to F)$, in other words the command to produce $TTTF$. In fact, for the tally code $\mathscr{P}\mathscr{D}$ we would specify the processor called 'adder' as follows: $+\alpha\beta = {}_{df}S\alpha\beta F$; it is specified in a syntax language of $\mathscr{P}\mathscr{D}$ which actually contains $\mathscr{P}\mathscr{D}$ as a sublanguage.

We notice that in such cases we have avoided the use of the *control characters* for parentheses by designing the language to incorporate the Lukasciewicz parenthesis-free convention. These are further examples of the class of prefix languages, which we give the class name $\mathscr{P}$. The symbol $\mathscr{P}$ therefore represents a language function applicable to an infinite class of alphabets, one of which is $\mathscr{D}$. But we note that the symbol '$\mathscr{P}\mathscr{D}$' itself is part of a prefix language which names languages. In general, then, we will have language functions '$\mathscr{L}$' applicable to classes of alphabets '$\mathscr{A}$' to yield languages '$\mathscr{L}\mathscr{A}$' in a combinator notation; $\mathscr{L}$ is specified, generally, by a set of processors made independent of the alphabet $\mathscr{A}$ by the fact that the subprocessors dependent on $\mathscr{A}$ are simple generators and recognizers which may be replaced from alphabet to alphabet. This notation therefore permits us to express directly the recursive definition of mechanical languages; $\mathscr{L}_n\mathscr{L}_{n-1}\ldots\mathscr{L}_2\mathscr{L}_1\mathscr{A}$ would mean an nth level language over the alphabet $\mathscr{A}$, where $\mathscr{L}_1\mathscr{A}, \mathscr{L}_2\mathscr{L}_1\mathscr{A}, \cdots, \mathscr{L}_{n-1}\cdots\mathscr{L}_2\mathscr{L}_1\mathscr{A}$ are all decidable even as *English words* is a decidable language over *English characters*, *English sentences* over *English words*, *English paragraphs* over *English sentences*, etc. A more formal example is the sequence: logical alphabet, logical expressions (well-formed formulae), logical proofs, logical theories, etc.

Our language of language names will include not merely alphabet symbols, and language function symbols, but also *language operator* symbols which will produce language functions from language functions. For example, besides the *direct product* language function of two variables 'χ' which applies to two languages $\mathscr{L}_2\cdot\mathscr{A}_2$ and $\mathscr{L}_1\mathscr{A}_1$ to produce concatenated words in an alphabet which is a *union* of $\mathscr{A}_1$ and $\mathscr{A}_2$, words of the form $\alpha\beta$ where α is in the extension of $\mathscr{L}_2\mathscr{A}_2$ and β in the extension $\mathscr{L}_1\mathscr{A}_1$, there is a related operator which produces the language function $\mathscr{X}\mathscr{L}_2\mathscr{L}_1$ applicable to any alphabet $\mathscr{A}$ in the intersection of the domains of $\mathscr{L}_2$ and $\mathscr{L}_1$.

4. Stratified alphabets and the prefix languages. The complete prefix language function $\mathscr{P}$ is applicable to any alphabet which possesses a mapping into an ordinal alphabet; in such a context we call the mapping a 'tail stratification'. The generator specification of $\mathscr{P}\mathscr{A}$ calls for a recognizer of $\mathscr{A}$ and also for a generator which will produce a character of an auxiliary ordinal alphabet $\mathscr{N}$ for any character of $\mathscr{A}$. If we call the mapping n_t, then the tail stratification of a character a of $\mathscr{A}$ is represented by n_ta. The generator of $\mathscr{P}$ is then specified as follows: If $n_ta = n$ and $\alpha_1, \alpha_2, \cdots, \alpha_n$ belong to $\mathscr{P}\mathscr{A}$, then so does $a\alpha_1\alpha_2\cdots\alpha_n$.

We interpret this producer to be ad hoc when $n = 0$, i. e. we interpret it to mean that, if $n_ta = 0$, then a is a word of $\mathscr{P}\mathscr{A}$ as well as of $\mathscr{A}$.

The symbolism '$a\alpha_1\alpha_2\cdots\alpha_n$' is ambiguous. The character 'a' under a purely syntactic processor is concatenated to the left of the string designated by '$\alpha_1\alpha_2\cdots\alpha_n$'. Under a command interpreter, as exists, for example for the string '$+TTFTF$', the result is the string $TTTF$ as remarked before. The intension in the definition of $\mathscr{P}$ is the object interpretation of direct concatenation, rather than the command interpretation calling for a functional operator.

The tally code $\mathscr{P}\mathscr{D}$ results from the application of $\mathscr{P}$ to $\mathscr{D}$ from the tail-stratification $n_tT = 1$ and $n_tF = 0$.

There are a number of other processors available to all prefix languages besides generators. First of all, these languages are also all decidable and have a common recognizer. As a matter of fact, the bulk of this recognizer computes a pair of ordinal numbers we might call the 'head and tail deficiencies' of a string of characters from $\mathscr{A}$, a pair of stratifications of the alphabet $\Sigma\mathscr{A}$. Given any alphabet possessing a pair of mappings into an ordinal alphabet (let us call them head and tail stratifications, n_h and n_t), there exists an extension to a pair of stratifications, N_h and N_t, of $\Sigma\mathscr{A}$ such that:

(1) If 'a' is a character of $\mathscr{A}$, $N_ha = n_ha$ and $N_ta = n_ta$.

(2) If α and β are strings of $\Sigma\mathscr{A}$, then

$$N_h(\alpha\beta) = N_h\alpha + (N_h\beta \dot{-} N_t\alpha)$$

and

$$N_t(\alpha\beta) = N_t\beta + (N_t\alpha \dot{-} N_h\beta)$$

where

$$x \dot{-} y \text{ means } \begin{cases} x - y \text{ if } x \geqq y\,, \\ 0 \text{ otherwise}\,. \end{cases}$$

(Note the duality of h and t; taking this duality into account. N_h and N_t are generated by the same function.)

(3) (Associative invariance)

$$N_h((\alpha\beta)\gamma) = N_h(\alpha(\beta\gamma)) ,$$
$$N_t((\alpha\beta)\gamma) = N_t(\alpha(\beta\gamma)) .$$

Thus $N_h\alpha$ and $N_t\alpha$ are independent of the association of characters chosen within the string α of $\Sigma\mathscr{A}$. The deficiency computer for $\mathscr{P}$ is obtained by defining the mapping n_h to yield 1 for every character of $\mathscr{A}$. A string α is recognized as a word of $\mathscr{P}\mathscr{A}$ if and only if $N_h\alpha = 1$ and $N_t\alpha = 0$. (For other processors achieving such a purpose, see Burks, Warren, and Wright [**1**] and Rosenbloom [**9**]). As a matter of fact, the possession of the same head and tail deficiencies is an equivalence relation in $\Sigma\mathscr{A}$ whose equivalence classes are the extensions of an infinite number of decidable languages, $\mathscr{T}_{ij}\mathscr{A}$; here $\mathscr{T}_{ij}$, for $i \geqq 1$ and $j \geqq 0$, is the language function whose recognizer employs the deficiency computer and inquires whether $N_h\alpha = i$ and $N_t\alpha = j$. By far the most important of these is $\mathscr{T}_{10}$, for it is a fundamental theorem that $\mathscr{T}_{10} = \mathscr{P}$.

All prefix languages may be translated into two-dimensional graphic languages of *rooted trees* in which the nodes are named by or contain characters of the alphabet. Thus every character of every word of a prefix language is the head of a uniquely determined connected substring representing the subtree having that character as a root; in the common syntax language for prefix languages this substring would be called the 'scope' of the character. Similarly, every character of every word in a prefix language determines an ordinal called its 'depth'. Thus, various processors of varying degrees of efficiency and direction of scan are available to $\mathscr{P}$, which processors may be called 'scope analyzers' and 'depth analyzers'. The more efficient among them use counters and storage assignments known as 'push-down counters' and 'push-down storage'. Whenever the control of a processor shifts the interpreter of the characters of $\mathscr{A}$ from the object level of concatenation to the command level of operation or storage recognition (as might be done with the character '+'), a scope analyzer becomes a 'scanner'. A scope analyzer and scanner may be subject to a control processor called an 'interlocker' to cause them to share all their control instructions. This process of *interlocking* allows a linear sequential language to return to combinations of its processors a good portion of the efficiency lost because the language lacks simultaneous action features.

All the statements just made about prefix languages, together with specifications in flow-chart language of the processors mentioned, may be found in greater detail in Gorn [**4**] and *loc. cit.*

We now complete the information needed about prefix languages for the study of ambiguities by listing some of the more important alphabets to which we apply the function $\mathscr{P}$.

I. *Logical alphabets and their extensions to larger alphabets with command interpretation*:

$$\mathscr{D} = \{T, F\} ,$$
$$\mathscr{A}_0 = \mathscr{D} \cup \{|, p, q, r, \cdots\} \quad \text{where} \quad n_t| = 2 ,$$

and every other character has tail-stratification 0. In command interpretation, | calls for a truth table look-up;

$$\mathscr{A}_1 = \mathscr{A}_0 \cup \{\wedge, \vee, \supset, \equiv, \sim\},$$

where $n_t \sim = 1$, and the other new characters have tail-stratification 2, and all have truth-table look-up for command interpretation;

$$\mathscr{A}_2 = \mathscr{A}_1 \cup \{S_1, S_2, S_3, S\}$$

where

$n_tS = 3$ and S has the command meaning *substitute* described above.

$n_tS_1 = 1$ and $S_1\alpha$ has the command meaning 'produce α if α is a one character string (p, q, r), the null-string otherwise'.

$n_tS_2 = 1$ and $S_2\alpha$ has the command meaning 'produce $\sim\alpha$ if α is a good word in $\mathscr{P}\mathscr{A}_1$, the null-string otherwise'.

$n_tS_3 = 2$ and $S_3\alpha\beta$ has the command meaning 'produce $\supset\alpha\beta$ if α and β are good words in $\mathscr{P}\mathscr{A}_1$, the null-string otherwise';

$$\mathscr{A}_3 = \mathscr{A}_2 \cup \{T_1, T_2, T_3, S^*, M\},$$

where

$n_tT_1 = 3$ and T_1pqr has the command meaning 'produce $\supset\supset p\supset qr\supset\supset pq\supset pr$', i.e. an axiom.

$n_tT_2 = 2$ and T_2pq means 'produce $\supset p\supset qp$', another formation rule.

$n_tT_3 = 2$ and T_3pq means 'produce $\supset\supset\sim p\sim q\supset qp$', a last formation rule (axiom).

$n_tS^* = 3$ and $S^*\alpha\beta\gamma$ means (a transformation rule) substitution with α restricted to $\mathscr{P}\mathscr{A}_3$, β restricted to $\mathscr{P}\mathscr{A}_2$ and γ restricted to variables p, q, r, and finally,

$n_tM = 2$ and $M\alpha\beta$ means the transformation rule called modus ponens which, assuming β is of the form $S_3\alpha\gamma$, and α and γ are restricted to $\mathscr{P}\mathscr{A}_1$, produces γ (i.e. M is a 'contracting producer').

II. *Algebraic alphabets and their extensions to larger alphabets with command interpretation*:

$\mathscr{N}$-an ordinal alphabet (e.g. $\mathscr{P}\mathscr{D}$), where $n_t\alpha = 0$ for α a word of $\mathscr{N}$;

$\mathscr{A}_0 = \mathscr{N} \cup \{+, -, *, \div, \uparrow, x, y, z, a, b, c, \cdots\}$ where each of the new operational characters has $n_t = 2$,

the new *variables* have $n_t = 0$,

the command meaning of the variables is the contents of designated storage positions,

the command meaning of the operations is as defined appropriate to the syntax of $\mathscr{N}$ (e.g. the algebraic expression $y(a + b)^{c/x} - z$ would be the word of $\mathscr{P}\mathscr{A}_0$: '$-*y \uparrow +ab/cxz$');

$\mathscr{A}_1 = \mathscr{A}_0 \cup \{S, D_x\}$, where

$n_t S = 3$ and S has the command meaning *substitute* already described,
$n_t D_x = 1$ and D_x is defined by recursion to yield:

$$D_x S\alpha\beta\gamma \underset{df}{=} {}^*SD_{x\gamma}\alpha\beta\gamma D_x\beta \text{ for } \gamma \text{ a variable, } \neq x, \text{ and } \alpha \text{ containing no } \gamma,$$

$$D_x{}^*\alpha\beta \underset{df}{=} + {}^*D_x\alpha\beta{}^*\alpha D_x\beta,$$

$$D_x + \alpha\beta \underset{df}{=} +D_x\alpha D_x\beta, \text{ etc.}$$

III. *The prefix languages of tree names and their extension to a tree arithmetic*:

$\mathscr{N}$-ordinal alphabet (e.g. $\mathscr{P}\mathscr{D}$), where $n_t\alpha = \alpha$ for α a word of $\mathscr{N}$;

$\mathscr{P}\mathscr{N}$ is then a language of rooted tree names in which each node is named by an ordinal number equal to the number of emerging branches (there is one entering branch to each node), and $\alpha = 0$ designates an end-point;

$\Sigma\mathscr{N}$ may be looked upon as a language of syntactic types, where a word α of $\mathscr{T}_{ij}\mathscr{N}$ designates i disconnected trees with the last j end-points to the right missing;

$\mathscr{P}\mathscr{N}$ may be looked upon as a generalization of finite ordinal numbers to finite *ramified* ordinals. The extension to an arithmetic language has the alphabet:

$\mathscr{A} = \mathscr{X}\{+\}\mathscr{P}\mathscr{N} \cup \mathscr{P}\mathscr{N} \cup \{*\}$, where the *characters* α of $\mathscr{P}\mathscr{N}$, when not preceded by '+' have $n_t\alpha = 0$, the characters '$+\alpha$' for α in $\mathscr{P}\mathscr{N}$ have $n_t + \alpha$ equal to the number of end-points of α,

$$n_t{}^* = 2\ ;$$

the command meaning of '$+\alpha\beta_1\beta_2\cdots\beta_n$', when α has n end-points as an element of $\mathscr{P}\mathscr{N}$, and $\beta_1\beta_2\cdots\beta_n$ are n elements of $\mathscr{P}\mathscr{N}$, is the element obtained from α by substituting β_i for the ith zero of α reading from the left, for $i = 1, 2, \cdots, n$ — in other words, the tree obtained from α by hanging the trees β_1 to β_n in left to right order from the end-points of α;

the command meaning of $^*\alpha\beta$ for α and β elements of $\mathscr{P}\mathscr{N}$ is $+\alpha\beta\beta\cdots\beta$ where the number of 'β' used is the number of end-points of α;

clearly, the language $\mathscr{P}\mathscr{A}$ can be extended recursively in the same way as $\mathscr{P}\mathscr{N}$ to yield an alphabet $\mathscr{A}^*$ such that $\mathscr{P}\mathscr{A}^* = \mathscr{P}\mathscr{P}\mathscr{A}^*$ and

$$\mathscr{A}^* = \mathscr{X}\{+\}\mathscr{P}\mathscr{A}^* \cup \mathscr{P}\mathscr{A}^* \cup \{*\}$$

(e.g. $\mathscr{P}\mathscr{A}^*$ would include such expressions as '$^{**}\alpha\beta + \gamma\delta_1\cdots\delta_n$' where γ has n end-points).

IV. *Language-naming prefix languages*:

Let $\mathscr{A}_t = \{\mathscr{R}_t, \mathscr{L}\}$, $\mathscr{A}_h = \{\mathscr{R}_h, \mathscr{L}\}$, and

$$\mathscr{A} = \{\mathscr{X}, \mathscr{R}_t, \mathscr{R}_h, \mathscr{L}\}, \text{ where } n_t\mathscr{L} = 0\,,$$

$n_t\mathscr{X} = n_t\mathscr{R}_t = n_t\mathscr{R}_h = 2$ and command meanings for the characters are: a language function (variable) for '$\mathscr{L}$', the language concatenating operator for '$\mathscr{X}$', and the *head-* and *tail-residual* language operators '$\mathscr{R}_h$' and '$\mathscr{R}_t$'. The latter are defined (possibly ineffectively) as follows:

If $\alpha_1 = \rho\alpha_2$, where α_1 is a word of $\mathscr{L}_1\mathscr{A}_0$ and α_2 is a word of $\mathscr{L}_2\mathscr{A}_0$, then ρ is a word of $\mathscr{R}_h\mathscr{L}_1\mathscr{L}_2\mathscr{P}_0$; similarly for $\alpha_1 = \alpha_2\rho$ and $\mathscr{R}_t\mathscr{L}_1\mathscr{L}_2\mathscr{A}_0$. Whereas $\mathscr{X}$ is an associative operator and therefore has *powers* which are ordinal numbers, $\mathscr{R}_h$ and $\mathscr{R}_t$ are not associative and require ramification symbols (i.e. trees) as *exponents* to indicate manner of iterated association.

The language $\mathscr{P}\mathscr{A}$ provides a possibly infinite class of language functions derived from language function $\mathscr{L}$ by application of the language operators $\mathscr{X}$, $\mathscr{R}_h$, and $\mathscr{R}_t$. The language $\mathscr{P}\mathscr{A}_t$ is called the language of 'tail derivatives of $\mathscr{L}$'; $\mathscr{P}\mathscr{A}_h$ is called the language of 'head derivatives of $\mathscr{L}$'. We could extend the alphabets to include boolean symbols, $\{\cup, \cap, \sim\}$, a *dualizing* symbol, $\{d\}$, for the operator which reverses the order of characters in a string, an ordinal alphabet $\mathscr{N}$ to yield exponents of iterated associative operators by use of the prefixed exponent symbol $\{\uparrow\}$, the tree alphabet $\mathscr{P}\mathscr{N}$ to yield exponents of iterated non-associative operators, variables $\{\mathscr{L}, \mathscr{L}_1, \mathscr{L}_2, \cdots\}$ for language functions, constant language functions such as $\mathscr{T}_{ij}$, Σ, Λ, etc. The language-naming prefix names so constructed can be expected to use the set of processors of $\mathscr{P}$ in the automatic construction of appropriate processors for the languages they name.

5. Syntactic ambiguity. If we represent by $\mathscr{L}^\omega$ the union of all natural powers of $\mathscr{L}$, excluding Λ, then the extent of $\mathscr{L}^\omega\mathscr{A}$ is the set of all concatenates of finite numbers of words of $\mathscr{L}\mathscr{A}$. $\mathscr{L}\mathscr{A}$ is called 'uniquely deconcatenable' if every word of $\mathscr{L}^\omega\mathscr{A}$ is obtained by a unique concatenation of words of $\mathscr{L}\mathscr{A}$. Examples of uniquely deconcatenable languages are:

a. those possessing *control* characters appearing at, and only at, the ends of words; e.g. $\mathscr{P}\mathscr{D}$ in which the character F is a right delimiter;

b. those for which all words have the same number of characters;

c. all prefix languages, *e.g.* $\mathscr{P}\mathscr{A}$;

d. all languages which are generalized Dewey decimal codes of end-points of a fixed tree (note that $\mathscr{P}\mathscr{D}$ is again an example).

We notice that these four types are particularly simple examples of unique deconcatenability in that in each case $\mathscr{R}_t\mathscr{L}\mathscr{L} = \Lambda$, i.e. no word of $\mathscr{L}\mathscr{A}$ is a proper head of another. We call this type of language 'simply analyzable' and note that the processor called a 'deconcatenator' needs only one internal storage for a string obtained by a single left to right scan of the input word.

Whether $\mathscr{L}\mathscr{A}$ is uniquely deconcatenable or not, if it is decidable, it is easy to design a *complete deconcatenator* which will produce all valid decompositions of words of $\mathscr{L}^\omega\mathscr{A}$. How much internal storage is necessary to do the job in either the case of unique deconcatenability or under the existence of ambiguities? To answer this question we will first design a language which names patterns of syntactic ambiguities and then show an effective relationship between such names and a sublanguage of the tail-derivatives of $\mathscr{L}$.

Suppose a word of $\mathscr{L}^{\omega}\mathscr{A}$ partitions in exactly m distinct ways into words of $\mathscr{L}\mathscr{A}$. Then the union of the m partitions divides that word into, say, n connected segments belonging to $\Sigma\mathscr{A}$. We can then represent this *m-ambiguity* by an m by n incidence matrix in which we have a '1' in the ith row and jth column if the jth segment begins a word of $\mathscr{L}\mathscr{A}$ in the ith decomposition; otherwise we place a zero in that cell.

In this incidence matrix the first column is composed of ones and no column contains only zeros. We call the ambiguity 'irreducible' if no other column contains only ones, i.e. if no head of the given word is also m-ambiguous.

Such incidence matrices form a two-dimensional language naming ambiguity patterns. We can translate it into a one-dimensional language by grouping the row names i in which ones appear in each column. Thus

$$(12345)(124)(135)(234)(245)(13)(25)$$

names an irreducible 5-ambiguity pattern using 7 segments.

For our purposes we need only consider irreducible *simple ambiguities* for which $m = 2$. In this case, except for the first column, the two rows of the incidence matrix are *ones complements* of each other, so that the ambiguity is completely described by a binary word of $n - 1$ bits.

By scanning the string of segments from left to right, we find that, because of the ambiguity, certain combinations of them are recognized as belonging to appropriate tail-derivatives of $\mathscr{L}\mathscr{A}$, the tree of the derivative becoming one deeper at each step in the scan. For example, for ambiguity type $h =$ (12)1112212 with binary designator 1110010, the pattern has eight segments, the word is of the form $\alpha_1\alpha_2\alpha_3\alpha_4\alpha_5\alpha_6\alpha_7\alpha_8$, and the two decompositions into words of $\mathscr{L}\mathscr{A}$ are $\{\alpha_1, \alpha_2, \alpha_3, \alpha_4\alpha_5\alpha_6, \alpha_7\alpha_8\}$ and $\{\alpha_1\alpha_2\alpha_3\alpha_4, \alpha_5, \alpha_6\alpha_7, \alpha_8\}$; in this case $\alpha_2\alpha_3\alpha_4$ is in $\mathscr{R}_t\mathscr{L}\mathscr{L}$, $\alpha_3\alpha_4$ is in $\mathscr{R}_t\mathscr{R}_t\mathscr{L}\mathscr{L}\mathscr{L}$, α_4 is in $\mathscr{R}_t\mathscr{R}_t\mathscr{R}_t\mathscr{L}\mathscr{L}\mathscr{L}\mathscr{L}$, $\alpha_5\alpha_6$ is in $\mathscr{R}_t\mathscr{L}\mathscr{R}_t\mathscr{R}_t\mathscr{R}_t\mathscr{L}\mathscr{L}\mathscr{L}\mathscr{L}$, α_6 is in $\mathscr{R}_t\mathscr{R}_t\mathscr{L}\mathscr{R}_t\mathscr{R}_t\mathscr{R}_t\mathscr{L}\mathscr{L}\mathscr{L}\mathscr{L}\mathscr{L}$, α_7 is in $\mathscr{R}_t\mathscr{L}\mathscr{R}_t\mathscr{R}_t\mathscr{L}\mathscr{R}_t\mathscr{R}_t\mathscr{R}_t$ $\cdot\mathscr{L}\mathscr{L}\mathscr{L}\mathscr{L}\mathscr{L}$, and, finally α_8 is in $\mathscr{R}_t\mathscr{L}\mathscr{R}_t\mathscr{L}\mathscr{R}_t\mathscr{R}_t\mathscr{L}\mathscr{R}_t\mathscr{R}_t\mathscr{R}_t$ $\cdot\mathscr{L}\mathscr{L}\mathscr{L}\mathscr{L}\mathscr{L}$ as well as in $\mathscr{L}$. In short α_8 is in $\mathscr{L}\cap\mathscr{L}_t^\tau$ where $\tau=$ 202022022200000. The mechanical translation between the binary designator of the ambiguity type, 1110010, and τ is as follows: 'scan the ambiguity type from right to left and build up τ from left to right; place a '2' for each character in the ambiguity type and a '0' for each change of character ('0' in τ works like a shift-key control in a typewriter); compute the tail deficiency of τ, and place at the end of it that number of zeros, thereby rendering τ a word of $\mathscr{P}\mathscr{N}$'.

From the construction of the word $\mathscr{L}_t^\tau$, it is clear that only those binary trees will appear which have only one end-point at each depth but the last; these are called simple binary trees. There are exactly 2^{d-1} simple binary trees of depth d. The tail-derivatives of $\mathscr{L}$, $\mathscr{L}_t^\tau$, in which τ is a simple-binary tree are called the 'simple tail derivatives of $\mathscr{L}$'. The command specification of the translator to τ which we have just given makes evident the following statement in the descriptive syntax of ambiguities:

THEOREM. *To each simple ambiguity of n segments $\alpha_1, \alpha_2, \cdots, \alpha_n$ for a word of $\mathscr{L}^\omega$ there is a uniquely determined simple tree τ_{n-1} of depth $n-1$ such that the last segment α_n belongs to the simple tail derivative of $\mathscr{L}$, $\mathscr{L}_t^{\tau_{n-1}}$. Conversely, to every simple tree τ_{n-1} of depth $n-1$ there is a uniquely determined simple ambiguity in $\mathscr{L}^\omega$ of n segments such that α_n belongs to the simple tail derivative $\mathscr{L}_t^{\tau_{n-1}}$.*

The same analysis could have proceeded by use of head derivatives only; the '1' in an ambiguity matrix would then mark the tail of a word, and the new matrix is obtained from the one we defined by a left end-around shift of one column.

The following theorem, even though the criterion stated may often be undecidable, is now *obvious*.

THEOREM. *A necessary and sufficient condition that $\mathscr{L}$ be uniquely deconcatenable is that for every simple binary tree τ we have $\mathscr{L} \cap \mathscr{L}_t^\tau = \Lambda$.*

If we designate by $\mathscr{L}_t^{(n)}$ the union of all the tail derivatives of $\mathscr{L}$ at depth n, we get as an immediate corollary the condition enunciated by Patterson and Sardinas [7]:

$$\mathscr{L} \cap \mathscr{L}_t^{(n)} = \Lambda \quad \text{for every } n > 0 .$$

All simply-analyzable languages are uniquely deconcatenable because $\mathscr{L}_t^{200} = \Lambda$, whence every $\mathscr{L}_t^\tau$ is also null.

We can now recognize, both for unique-deconcatenability and for ambiguity of $\mathscr{L}^\omega$, a hierarchy of types of $\mathscr{L}$ of different levels of complication to answer our questions about the internal storage required.

To this end, let τ be a simple binary tree whose immediate predecessor is the simple binary tree τ_1. (Note that if τ had not been simple-binary, τ_1 would not have been unique.) $\mathscr{L}$ is called 'τ-ambiguous' if $\mathscr{L}_t^{\tau_1} \cap \mathscr{L}$ is null but $\mathscr{L}_t^\tau \cap \mathscr{L}$ is not. $\mathscr{L}$ is called 'τ-analyzable' if $\mathscr{L}_t^\tau$ is null but $\mathscr{L}_t^{\tau_1}$ is not. Thus 200-*analyzable* means *simply-analyzable.*

Both the τ-ambiguous and the τ-analyzable languages require an internal push-down storage of n cells to recognize their status, where n is the depth of τ. For example, if we refer to the ambiguity of depth 7 referred to before, and take an alphabet of 8 characters:

$$\mathscr{A}_0 \{a, b, c, d, e, f, g, h\} ,$$

then the language $\mathscr{L}_1 \mathscr{A}_0 = \{a, b, c, abcd, e, def, fg, gh, h\}$ is τ-ambiguous and 7-ambiguous, while the language with only one of these words missing, $\mathscr{L}_2 \mathscr{A}_0 = \{a, b, c, abcd, e, def, fg, gh\}$ is uniquely deconcatenable because it is 8-analyzable. Clearly we may take the required depth as large as we please, making the two languages as finitely extensive as we please, but differing by only one word; yet one will have ambiguities and the other not. Such is the thin line between unique interpretation and ambiguity! More formally stated:

THEOREM. *For any simple binary tree τ of depth n there are two languages,*

$\mathscr{L}_1$ and $\mathscr{L}_2$, over an alphabet, $\mathscr{A}_0$, of $n+1$ characters such that $\mathscr{L}_1\mathscr{A}_0$ has $n+2$ words and is τ-ambiguous and n-ambiguous, while $\mathscr{L}_2\mathscr{A}_0$ has $n+1$ words, all in $\mathscr{L}_1\mathscr{A}_0$, and is τ-analyzable and n-analyzable.

It is easy to give an example of a uniquely deconcatenable language which requires an infinite available storage in its deconcatenator:

Let $\mathscr{A}$ be an infinite alphabet (e.g. $\mathscr{P}\mathscr{D}$), say, $\mathscr{A} = \{a_0, a_1, a_2, \cdots\}$, and let the extension of $\mathscr{L}\mathscr{A}$ be the set of words of the form

a. '$a_0a_1 \cdots a_na_n \cdots a_1a_0$', n even,

b. 'a_ia_{i+1}' for $i =$ even,

c. '$a_{i+1}a_i$' for $i =$ even.

Then, whether the scan is from left to right or from right to left, the word from $\Sigma\mathscr{A}$, '$a_0a_1a_2 \cdots a_{n-1}a_na_na_{n-1} \cdots a_2a_1a_0$' which belongs to $\mathscr{L}^\omega\mathscr{A}$, must have $[n/2] + 1$ internal storage positions for words of $\mathscr{L}\mathscr{A}$ filled before the even-ness or odd-ness of n becomes apparent and determines the proper deconcatenation.

6. Pragmatic ambiguity. Let us now discuss the more subtle ambiguities due to the fact that the general syntax language has an unstratified control. In these ambiguities, the same expression is submitted to different processors, or to different sequencing of the same processors under call by different control processors. They are systematically introduced whenever a language is extended by a standard process we call 'syntactic definition and importation.'

For example, consider the gradual extension of $\mathscr{D} = \{T, F\}$ into the algebraic alphabets. Since $\mathscr{P}^\omega\mathscr{D}$ is characterized by having all its words ending in F, and $\mathscr{P}\mathscr{D}$ is uniquely deconcatenable, there is a simple deconcatenator applicable to every string of $\Sigma\mathscr{D}$ which ends in F. In scanning from left to right it peels off each character from the string, using 'F' as a control character to complete a recognition of a word of $\mathscr{P}\mathscr{D}$; we might ask the recognizer to signal the *yes* at each recognition by concatenating a character 'T' into a special counting register, and to signal the *no* when the null string has been reached by concatenating an 'F'. The control behavior of the recognizer has thereby given command meanings *tally* and *finish* to T and F and the processor thus produces an element in $\mathscr{P}\mathscr{D}$ which has counted the number of words of $\mathscr{P}\mathscr{D}$ in the given string. This counting word which has been produced can now be interpreted as an object in $\mathscr{P}\mathscr{D}$, say by a $\mathscr{P}\mathscr{D}$-comparator. It is thus easy to see how such deliberate shifting of the recognition and generation of T and F at object, control, and command levels can be used to design a processor which decides whether two words of $\Sigma\mathscr{D}$ deconcatenate into the same number of words of $\mathscr{P}\mathscr{D}$.

Now we have seen how, in the syntax of $\mathscr{P}\mathscr{D}$, we can define the command $+$ by a program involving the command S. This is syntactic definition. We now import the character '$+$' into the alphabet $\mathscr{P}\mathscr{D}$ to form an extended alphabet, and at the same time assign it the stratification number TTF (i.e. 2) by entry in an appropriate list, allowing every word of $\mathscr{P}\mathscr{D}$ to produce a stratification number F(i.e. 0). We have just designed a *stratifier*, but, more important, if we now apply the processors of $\mathscr{P}$ to this extended alphabet to yield a language of expressions, $\mathscr{E}_+$ the character '+' has now become as ambiguous as T and F. It has an object interpretation which would be used in a *scope analyzer* or in a *recognizer*, a command interpretation which would be used in a *scanner*, and it would shift these interpretations regularly in the control processor which interlocks these two.

Consider further the symbol '='. We could introduce it as an object and design a language whose words have the form '$=\alpha\beta$' where α and β are words of $\mathscr{E}_+$. We could call this language the 'sentences', $\mathscr{S}_=$, over the alphabet $\mathscr{E}_+ \cup \{=\}$. We could also give '=' a command meaning by having it produce 'T' or 'F' depending on whether the *yes* exit of the comparator of $\mathscr{P}\mathscr{D}$ (to be used only after all characters '+' have been given command interpretation) or the *no* exit has been effected. The resulting processor would also serve as a recognizer of the language of 'true sentences', $\mathscr{T}_=$.

The sequence of symbols '$= = + \ TFTTFTTTFF$' is now multiply ambiguous, depending on object, command, or control interpretation of the characters, and on which of the processors, *comparators for* $\mathscr{P}\mathscr{D}$, *recognizers for* $\mathscr{P}\mathscr{D}$, *stratifiers, recognizers for* $\mathscr{E}_+$, *scanners for* $\mathscr{P}$ *at various levels, recognizers for* $\mathscr{S}_=$, *or for* $\mathscr{T}_=$, are called, and in which sequence.

The lesson seems to be that every one of an important class of *extension by definition* processors systematically introduces pragmatic ambiguities; *the number of interpretations of expressions increases exponentially with the number of such definitions.*

It is now instructive to recognize some of the implications of such extension processes for logical alphabets. Here we could again begin with $\mathscr{D} = \{T, F\}$ as objects, and with the variables $p, q, r, \cdots$ as possessing command meaning, i.e., designating one bit storage positions which may contain the objects T and F. The character '|' (the Scheffer stroke function) may now be assigned command meaning, for the usual truth-table look-up, and then may be imported into the object alphabet with stratification number 2. We could now define $\sim\alpha \underset{df}{=} |\alpha\alpha$ at syntax level by a pair of replacement producers, and at object level by assignment of stratification number 1. Similarly, with stratification number 2, for

$$\wedge\alpha\beta \underset{df}{=} \sim|\alpha\beta \underset{df}{=} ||\alpha\beta|\alpha\beta\,,$$
$$\vee\alpha\beta \underset{df}{=} |\sim\alpha\sim\beta \underset{df}{=} ||\alpha\alpha|\beta\beta\,,$$
$$\supset\alpha\beta \underset{df}{=} \vee\sim\alpha\beta \underset{df}{=} ||\sim\alpha\sim\alpha|\beta\beta \underset{df}{=} |||\alpha\alpha|\alpha\alpha|\beta\beta\,,^4$$
$$\equiv\alpha\beta \underset{df}{=} \wedge\supset\alpha\beta\supset\beta\alpha \underset{df}{=} \text{etc.}^4$$

The symbol '$\underset{df}{=}$' is now an equivalence relation in the descriptive syntax which means *mutually derivable by use of the defining productions.* But there is another equivalence at command syntax level which means *truth table equivalent.* $\sim\sim p$ is equivalent to p under the second meaning, but not under the first. The inequivalence is proved as follows:

Define by recursion the function f for all strings of $\mathscr{P}\mathscr{A}$ as follows:

---

[4] The definitions in Quine's *New foundations* would be

$$\supset\alpha\beta \underset{df}{=} |\alpha|\beta\beta \quad \text{and} \quad \equiv\alpha\beta \underset{df}{=} ||\alpha\beta||\alpha\alpha|\beta\beta.$$

$$fp = fq = fr = 0\ ,$$
$$f\,|\,\alpha\beta = 1 + f\alpha + f\beta\ ,$$
$$f \sim \alpha = 1 + 2f\alpha\ ,$$
$$f \wedge \alpha\beta = 3 + 2f\alpha + 2f\beta\ ,$$
$$f \vee \alpha\beta = 3 + 2f\alpha + 2f\beta\ ,$$
$$f \supset \alpha\beta = 5 + 4f\alpha + 2f\beta\ ,^{5}$$
$$f \equiv \alpha\beta = 23 + 12f\alpha + 12f\beta\ .^{5}$$

f actually means *the number of strokes when all other operations are eliminated.* It is then easy to see that, if α is equivalent to β, $f\alpha = f\beta$. Thus $\sim\sim p$ is not equivalent to p because $f \sim\sim p = 3$ and $fp = 0$. Thus the inequivalence is proven at purely syntactic level, while the equivalence is proven at semantic level.

Continuing this type of language extension, as with the algebraic alphabets, we find that there is a sublanguage of $\mathscr{P}\mathscr{A}_3$ which designates valid proofs in $\mathscr{P}\mathscr{A}_1$. In this sublanguage the string '$MT_2pqS{*}MT_2pqS{*}T_1pqrprS_3qpq$'

a. means *a valid proof* when being processed by the recognizer of that sublanguage,

b. produces *that* proof in standard form when being processed by a *scanner* for that sublanguage.

c. produces the theorem $\supset pp$ when being processed by the scanner for $\mathscr{P}\mathscr{A}_3$ which withholds command interpretation from the strings of $\mathscr{P}\mathscr{A}_1$ which appear. The steps followed are the same as those of the proof in b.

The syntactic type of this string is a formal analysis of the proof, somewhat in the manner of Gentzen, and called a 'construction' by Curry and Feys [**3**].

In this case, the multiplicity of interpretations included a number of closely related ones; the resultant ambiguity far from being chaotic, achieved a notable economy of expression.

7. Paradoxes. An important effect of the possibility of pragmatic ambiguity, due to unstratified control, is that paradoxes become possible. It is an interesting fact that, when we translate such paradoxes from the descriptive syntax to the command syntax in our mixed language, the shifting of levels of interpretation becomes much more evident.

As an example posed in command form, every general purpose machine, where instructions and data share the same storage, is capable in one instruction of being made to cycle indefinitely. We merely place the control instruction 'do α' in storage position α. Incidentally, such a device is a useful safeguard in many programs. This command might be considered a direct translation from the statement in descriptive syntax, 'this sentence is true'. The interpretation of α shifts back and forth between control level, changing α from an object in storage into a command, to command level; the result is an infinite *fetch-execute* cycle, where the execution triggers the fetching and the fetching triggers the execution.

---

[5] With the definitions in Quine's *New foundations* these would be

$$f \supset \alpha\beta = 2f\alpha + 2f\beta$$

and

$$f \equiv \alpha\beta = 5 + 3f\alpha + 3f\beta.$$

Changing the sentence to 'this sentence is false', we have the well-known Epimenides paradox. Its translation into command syntax calls for a somewhat more complicated interpreter. We have in storage cell α the expression '$=\alpha F$'. The interpreter is supposed, first of all, to do *complete substitution* in the sense that as long as a variable is recognized the contents of the storage position it names (the command interpretation of a variable) must be substituted for the name (the object interpretation of the variable), and this process is to continue until all variables have disappeared. When this is done (which, of course, is never), the result is supposed to be processed by the scanner of $\mathscr{T}_=$ to produce a value T or F. If we represent the complete substitutor by '$A\alpha\beta\gamma$' to represent the iteration of $S\alpha\beta\gamma$ until the variable γ has disappeared, we find that the cycling command and object interpretation of α produces a recursive sequence:

$$A = \alpha F = \alpha F\alpha \rightarrow A = = \alpha FF = \alpha F\alpha$$
$$\rightarrow A = = = \alpha FFF = \alpha F\alpha, \text{ etc.}$$

A more interesting example along this same line is a producer of a recursive sequence of tautologies. Here we place in storage position α the expression '$\supset\wedge p\supset pqq$', i.e. the tautology whose corresponding command interpretation would be $M\alpha\supset\alpha\beta\rightarrow\gamma$ to produce 'β' in storage position γ. We also place in α_0 the expression '$\wedge p\supset pq$', and in β we place the pair of commands

$$\text{'}S(\alpha)(\alpha_0)p \rightarrow \alpha, \text{do } \beta\text{'},$$

which mean 'substitute the contents of α_0 for p into the contents of α, place the result back into α, and then do β'.

This will clearly cycle indefinitely, producing a bigger and bigger expression in α at each iteration of β. If we let α_i be the contents of α at the beginning of the ith application of β, then $S(\alpha_0)(\alpha_0)p = \wedge(\alpha_0)\alpha_1$,

$$\alpha_1 = \text{'}\supset\wedge p\supset pqq\text{'} = \supset(\alpha_0)q\ ,$$
$$\alpha_2 = \text{'}\supset\wedge\wedge p\supset pq\supset\wedge p\supset pqqq = \supset\wedge(\alpha_0)\alpha_1 q\ ,$$

and, recursively,

$$\alpha_n = S\alpha_{n-1}(\alpha_0)p = \supset\underbrace{\wedge\cdots\wedge}_{n-1}(\alpha_0)\alpha_1\cdots\alpha_{n-1}q\ .$$

This is precisely the recursive sequence of tautologies discussed in Lewis Carroll's paradox of the tortoise and Achilles. If, at any time, the contents of α were shifted to a command interpretation, whereby the first character '$\supset$' is interpreted as M (modus ponens) and the last character 'q' is ignored, 'q' would nevertheless be produced. But this control shift is never performed, and this, of course, is the whole point of the paradox.

We will now complete our examples of paradoxes translated into command form by giving a miniature program which simultaneously simulates the Russell paradox, the "autologous-heterologous" paradox, and the paradox of "the catalogue of all catalogues which do not catalogue themselves" (note that in the last cited, the shifting interpretation of the word 'catalogue' is quite apparent).

The paradoxical processor is most simply specified by a flow chart, but let us first explain some of the notation. There will be six variable instructions in storage cells $\tau, \nu_1, \nu_2, i_1, i_2$, and i_3. At various positions within these storage cells will be a variable a with interpretation depending upon the instruction and the position within that instruction. τ and i_3 are control instructions of the form '$x \in y$? if yes do α, if no do β'. Here x is automatically interpreted as an object, y is interpreted as the name of a list of storage cells each of which must have its contents compared with x; α and β are instructions indicated in the flow chart by arrows leading from one box to another. The operational commands are of the form '$x \rightarrow y$', which we have already explained, 'erase x in y', which means 'compare the contents of x with the successive contents of list y and remove from y whenever x appears'. Finally, $a \oplus 1$ means the symmetric sum (boolean) of the contents of a and 1, and "'x' $\rightarrow y$ in α' means $S(\alpha)xy$. The flow chart then is:

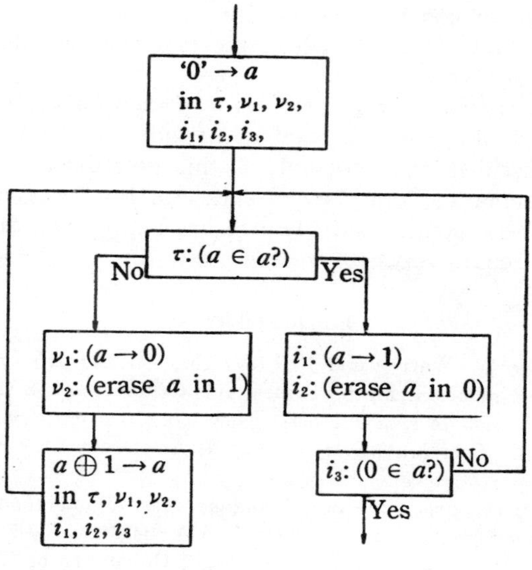

The processor never completes its action. It cycles the operations of placing 0 in list 0, erasing 0 from list 1, placing 0 in list 1, and erasing 0 in list 0.

8. Conclusion. We have now the dangerous task of interpreting what we have just discussed.

Just as we would not consider losing the power of a general purpose machine by removing its loop control, we should not reject a mechanical language with unstratified control, descriptive or command. The possibility of paradox is a necessary condition for self-referencing and universality. If we accept the control shifting of context and interpretation we must also accept the responsibility of making the resulting ambiguities work for us and not against us. By having a mixed language we can have theory about processors and processing of theories and formal systems, but even more important, we can use a handful of control processors and language function processors recursively to handle very many mechanical languages. Many of the classical proofs concerning algorithms turn into trivially simple and natural ones by mere translation into command form.

Looking into some of the past methods in the foundations of mathematics, specifically into the logistic program, we recognize that its bootstrapping aim of proving theorems which can in turn be reinterpreted as methods of proving theorems was a natural one, that the appearance of paradoxes was also a natural result of the unstratified control, and that the attempt to avoid them by making operations meaningless which led to them was in effect throwing the baby out with the bath. The artificial stratification imposed by the theory of types had to be weakened by a still too weak control axiom, the axiom of reducibility. Attempts like that of Quine [8] to permit a descriptive language of unstratified expressions were most reasonable, not in spite of, but because of the fact that a decision method to determine whether any expression is stratifiable is impossible, and because of the fact that paradoxes became possible.

It is the characteristically pragmatic flavor of the Computer and Information Sciences which, I think, generates such an attitude as I have been expressing. If you feel it is justified, then not only is this new discipline an application of logic and mathematics, but they in turn are applications of it. If you agree with me on this point, then the pleasure I expressed at being heard by you is increased,—recursively.

BIBLIOGRAPHY

1. A. W. Burks, D. W. Warren and J. B. Wright, *An analysis of a logical machine using parenthesis-free notation*, Math. Tables Aids Comput. vol. 8 (1954) pp. 53–57.

2. Carnap, *Introduction to symbolic logic*, New York, Dover, 1958.

3. Curry and Feys, *Combinatory logic*. Vol. I, Amsterdam, the Netherlands, North Holland, 1958.

4. S. Gorn, *Common programming language task*, Final reports AD59UR1 and AD60UR1, U. S. Army Signal Corps Contract No. DA-36-039-SC-75047, Part I, 1959, 1960.

5. R. M. Martin, *Truth and denotation*, Chicago, University of Chicago Press, 1958.

6. ———, *Toward a systematic pragmatics*, Amsterdam, the Netherland, North-Holland, 1959.

7. G. W. Patterson, and A. A. Sardinas, *A necessary and sufficient condition for unique decomposition of coded messages*, Convention Record of IRE, Part 8, 1953.

8. W. V. Quine, *A system of logistic*, Cambridge, Cambridge University Press, 1934.

9. P. Rosenbloom, *The elements of mathematical logic*, New York, Dover, 1950.

UNIVERSITY OF PENNSYLVANIA,
PHILADELPHIA, PENNSYLVANIA

COMPUTER PROGRAMS FOR CHECKING MATHEMATICAL PROOFS

BY

JOHN MCCARTHY

1. **Introduction.** Checking mathematical proofs is potentially one of the most interesting and useful applications of automatic computers. Computers can check not only the proofs of new mathematical theorems but also proofs that complex engineering systems and computer programs meet their specifications. Proofs to be checked by computer may be briefer and easier to write than the informal proofs acceptable to mathematicians. This is because the computer can be asked to do much more work to check each step than a human is willing to do, and this permits longer and fewer steps. Detection of *nonsequiturs* by the computer will catch errors in engineering system designs at an early stage. The combination of proof-checking techniques with proof-finding heuristics will permit mathematicians to try out ideas for proofs that are still quite vague and may speed up mathematical research.

There is much to do before computer proof-checking becomes practically useful. We need improved techniques for computing with symbolic expressions [**10**; **12**], new kinds of formal systems that permit brief proofs, the formal expression of the basic parts of mathematics in such systems, computer systems that are continuously available to individual users [**8**; **13**], and not least the development of specific proof-checking techniques.

The remaining sections of this paper deal with the following topics: the potential applications of computer proof-checking, formal systems admitting brief proofs, and a specific proof-checking system based on LISP, the language for describing computations with symbolic expressions devised by the author and his colleagues in the M.I.T. Artificial Intelligence Group.

2. **Potential applications to engineering and mathematics.** Increasingly, engineering has to deal with complicated systems. These systems are sometimes designed by first determining performance specifications for the system as a whole, then making a design in terms of subsystems whose performance specifications are then determined. The specifications for the subsystems are then given to the groups that have the job of designing and building them.

It is intended that if the subsystems meet their specifications and are connected in the specified way, then the system as a whole will meet its specifications. The more complicated the system, the less likely it is that this will be the case. Complicated systems have to be debugged, which means testing the system on a variety of cases and fixing any deficiencies that are discovered. However, debugging is an uncertain process, and there is no assurance at

Received by the editors June 13, 1961

any given time that all the bugs have been fixed.

One of our objectives is to make it possible to prove that systems will meet their specifications provided that subsystems meet theirs and are connected in the specified way. Such proofs may replace debugging.

In order to replace debugging by proofs that systems work, two requirements must be met. First the proofs must be checkable by computer; otherwise there will be too much opportunity for wishful thinking. Secondly proving the system meets specifications must not be a larger task than designing the system. There is good reason to hope that these requirements can be met, and that the main burden of the work can be put on the computer. The feasibility of computer proof-checking in engineering will depend on the formalization of the branches of engineering concerned and on engineers becoming familiar with formal methods.

We envisage the use of computer proof-checking in mathematics as follows: The mathematician already has formalizations of his branch of mathematics and the computer system has stored in it the theorems that have previously been proved. In addition, there are a number of techniques embodied in programs for generating proofs. The mathematician expresses his ideas of how a proof may be found by combining these techniques into a program. The computer carries out the program which may prove the theorem, may generate information that will guide another try, may indicate an elementary misconception, or may be of no help whatsoever.

Both the mathematical and the engineering applications depend on the existence of formal systems covering the subject matter of interest in which proofs can be expressed briefly and naturally.

3. Formal systems that admit brief proofs. A formal mathematical system M has two parts. The morphological part may be regarded as a computable predicate g whose domain is the set of strings of symbols in a certain alphabet, and which is such that $g(s)$ is true if the string of symbols s is a sentence of M. Usually, the set G of sentences of M is described recursively in terms of elementary expressions and certain operations, possibly with the help of auxiliary classes of expressions that are also recursively defined.

The syntactic part of a formal system may be regarded as a two-place predicate v such that $v(s, p)$ is true if the expression p is to be regarded as a proof of the sentence s. In the usual cases v is computable. We say that s is a theorem of M if $th(\mathrm{s}) = (\exists p)v(s, p)$ is true. In the most interesting systems $th(s)$ is not decidable.

In most well known formal systems v (the rule for verifying a proof) has a very special form [**1**; **2**; **18**]. Namely, p is a list of sentences ending with the sentence s and v checks that each sentence of p is either an *axiom* of the formal system M or else is an immediate consequence of one more of or its predecessors in p according to one of the *rules of inference* of M. This special kind of formal system is good for investigating the foundations of mathematics and logic because it is easy to see what each step accomplishes, but if we want systems in which formal proofs are as brief as possible, it seems better to take the generalized notion of proof based on the verification predicate. We can explain the generalized notion as follows: in general the set of theorems is not decidable so that we cannot determine directly whether a sentence s is a theorem. However, if we have the right additional information, the information that a mathematician tries to discover when he investigates a conjecture, we can determine that s is a theorem. Our generalization is to say that a proof p is just the additional information that enables us to determine mechanically that a sentence s is a theorem.

The following are the ways we have found for reducing the work involved in writing a proof at the expense of making the computer work harder in checking it.

1. *Decision methods.* The more extensive and interesting branches of mathematics including those with the most applications do not and cannot have decision methods. Moreover, in some cases where decision methods exist, e.g. for certain classes of formulas of the predicate calculus, they may involve impossible amounts of calculation. Nevertheless, important subdomains of the interesting branches of mathematics have feasible decision procedures. Propositional calculus [**1**], some parts of predicate calculus [**1**], elementary algebra [**20**], and the elementary theory of conditional expressions [**11**] (our formalization of case analysis) are examples. Other decidable parts of mathematics will be isolated once it becomes worthwhile to do so.

In mathematics papers one encounters phrases like "The reader will easily verify that ..." and "It follows from equations (1) and (3) by an algebraic calculation that ...". We would like to regard such phrases as appeals to decision procedures which the author assumes the reader to have available. If we take this view, then the proofs can be regarded as rigorous, and even formal, once it is stated what decision procedures the reader is assumed to know.

If we take full advantage of decision procedures for elementary algebra and case analysis, it may be possible to isolate the references to the axioms, e.g. induction, that extend the decidable subsystem to an undecidable system. A proof that consisted of appeals to decision procedures for elementary though complicated formulas, interspersed with a few applications of rules of inference like mathematical induction, might be quite brief.

The form we have adapted for rules of inference requires no distinction between application of a decision procedure and more elementary rules of inference.

2. *Proofs as programmed dialogs.* A proof is normally considered to be a sequence of sentences each of which follows from the preceding sentences by one of the rules of inference belonging to the mathematical system. (Systems of natural deduction are an exception but are a generalization in a different direction than concerns us now.) We propose a generalization of this notion that may make proofs easier to write. Namely, we shall consider the proof-checker to be able to check steps, but also to be able to carry out procedures that purport to generate steps. This allows the prover to tell how to generate a proof rather than give the proof itself. Many informal proofs may be regarded as directions for constructing proofs.

3. *Heuristic proof generators.* Instead of requiring that the proof program generate only correct steps in the proper order, we can allow trial and error procedures. For example, suppose the statement to be proved is

$$\int_{-\infty}^{\infty} e^{-x^2} dx = \pi^{1/2} .$$

The solution is to be obtained by transforming the integral on the left. We may have heuristic rules that suggest transformations to be tried, and these may lead to new possible transformation and so forth. Thus we have a tree of possible transformations and we may have a heuristic program that explores the tree in an attempt to evaluate the integral [**19**]. Such a program if successful may serve as the proof since the proof-checker will accept the correct steps it finally generates. Suppose that our heuristic program is not bright enough to take the hint from the occurrence of $\pi^{1/2}$ that it should square the integral, regard the result as a double integral and then change to polar co-ordinates. Then the user may have to furnish this information. All this goes to show that the line between a proof that the computer checks and the use of a computer program that generates a proof is a blurred one. Any programs for proof generation that can be devised can be incorporated into the proof-checker. See references [**3**; **4**; **5**; **6**; **7**; **14**; **15**; **16**; **17**; **19**; **21**; **22**; **23**]. [**14**] and [**15**] are a survey article and a bibliography respectively on artificial intelligence.

4. *Metamathematical methods.* The symbolic calculations involved in proof checking are themselves subject to mathematical analysis [**10**; **12**]. In particular it may be possible to establish a result of the form

$$(\forall p's)v'(p', s) \supset (\exists p)v(p, s)$$

for some computable predicate v'. This allows us to use v' as a new proof verifier. It is natural to consider including in M sentences about its own proof procedures and means for proving them. We also need a rule that permits us to infer a sentence s from a sentence that asserts the existence of a proof of s.

4. The proof-checking system PC. PC is a proof-checking system based on LISP [**10** ; **12**] functions and predicates.

The form of a rule of inference in PC is a computable partial predicate *rule* taking three symbolic expressions as arguments. If *rule*[*premisses*; *conclusions*;*parameter*] has the value T, we shall say that each of the statements in the list *conclusions* follows from the conjunction of the statements in the list *premisses* in a manner specified by the symbolic expression *parameter*. If

$$rule\,[premisses;\ conclusions;\ parameter] = F$$

or if its value is undefined because the computation does not terminate, then we shall say that *conclusions* do not follow from *premisses* according to *parameter* although they might follow if a different value of *parameter* were used or if a different rule were used.

The rules we shall use will have the property that if every element of the list p_1 is an element of the list p_2 and if every element of the list c_2 is an element of the list c_1, then $rule[p_2; c_2; para] = T$ whenever $rule[p_1; c_1; para] = T$.

A logical system within PC is characterized by a collection of rules of inference and a collection of formulas serving as axioms. As usual, the theorems are those formulas which can be derived from the axioms by successive applications of the rules of inference. However, as we shall see, the proofs are not just records of successive applications of the rules of inference. In fact, proofs are more to be regarded as programs for generating applications of the rules of inference, and the proof-checker is to be regarded as an arbiter that keeps track of what has been admitted as proved, checks inferences suggested to it by the *proof program* and checks whether it has admitted the result to be proved when the *proof program* announces that it is done.

The form of rule of inference we have just described has a property that requires explanation.

Namely, the rule is not required to be a total function. This has the effect that if a proposed proof is valid, PC will tell us so, but if the proposed proof is not valid, the calculation may not terminate. Admitting this kind of rule does not change the classes of formulas that may be theorems. This can be shown as follows.

Suppose *rule* is only a partial function; we can define *rule*2 which admits the same inferences as *rule* but which is a total function. The parameter of *rule*2 for a given inference is a pair consisting of the parameter of *rule* for the same inference and a symbolic expression n which measures the number of steps of *rule* to be simulated by *rule*2. *Rule*2 carries out n steps of *rule* and gives the answer T of F according to whether *rule* gives T in this number of steps. The existence of *rule*2 follows from the existence of a universal function which can simulate any computable function given a symbolic description of it. Thus, any *conclusions* which follow from *rule* also follow from *rule*2 with a suitable n.

In spite of the fact that we could restrict ourselves to total functions, we shall admit partial functions as rules of inference. This is because the effect of the process described above for making total rules out of partial rules can be achieved by stating a bound on the computer time taken to verify a proof. If we regard such a bound, or the equivalent in terms of the number of elementary list processing operations executed as part of the proof, the process of verification becomes definite. The advantages are that there is no need to bound subprocesses and, more important, that tentative procedures for generating inferences can be allowed in proofs.

As examples of rules of inference, we shall give the rules for modus ponens and substitution in the propositional calculus. These rules do not require much of the power of the system.

Modus ponens is the rule that permits us to deduce q given p and $p \supset q$. In the LISP notation we shall write (IMPLIES, p, q) for $p \supset q$, and the rule is

$$modusponens\ [prem; concl; para] = contained\ [list\ [car\ [para];$$

cons [*IMPLIES*; *para*]]; *prem*] ∧ *contained* [*concl*; *list* [*cadr* [*para*]]] .

The only function used which is not a basic LISP function is *contained*. *Contained* [*u*; *v*] is true if every member of the list *u* is also a member of the list *v*. It may be defined by the formulas

$$contained[u; v] = null[u] \vee among[car[u]; v] \wedge contained[cdr[u]; v]\,]$$

where

$$among[x; v] = \sim null[v] \wedge [equal[x; car[v]\,] \vee among[x; cdr[v]\,]\,]$$

The parameter *para* in the deduction of q from p and $p \supset q$ is the two element list (p, q). For example, *modusponens*[*prem*; *concl*; *para*] is true in the case

prem: ((*IMPLIES*, *P*, (*IMPLIES*, *Q*, *P*)), (*IMPLIES*, (*IMPLIES*, *P*, (*IMPLIES*, *Q*, *P*)), (*IMPLIES*, *P*, *P*)))

concl: ((*IMPLIES*, *P*, *P*))

para: ((*IMPLIES*, *P*, (*IMPLIES*, *Q*, *P*)), (*IMPLIES*, *P*, *P*) .

The rule of substitution is given by

$$substit[prem; concl; para] = among[car[para]; prem] \wedge andlis[cadr[para]; \lambda[\,[j]; var[car[j]\,] \wedge formula[cadr[j]\,]\,]\,]\,] \wedge contained[concl; list[sublis[cadr[para]; car[para]\,]\,]\,]$$

The auxiliary functions here are

$$andlis[u; \pi] = null[u] \vee [\pi[car[u]] \wedge andlis[cdr[u]; \pi]]$$

$$var[x] = atom[x] \wedge \sim eq[x; IMPLIES] \wedge \sim eq[x; AND] \wedge \sim eq[x; OR] \wedge \sim eq[x; NOT] \wedge \sim eq[x; T] \wedge \sim eq[x; F]$$

$$formula[x] = var[x] \vee eq[x; T] \vee eq[x; F] \vee \sim atom[x] \wedge [eq[car[x]; NOT] \wedge one[cdr[x]] \wedge formula[cadr[x]] \vee [eq[car[x]; IMPLIES] \vee eq[car[x]; AND] \vee eq[car[x]; OR]] \wedge two[cdr[x]] \wedge formula[cadr[x]] \wedge formula[caddr[x]]$$

$$one[x] = \sim null[x] \wedge null[cdr[x]]$$

$$two[x] = \sim null[x] \wedge \sim null[cdr[x]] \wedge null[cddr[x]]\,.$$

The parameter *para* in the substitution rule has the form $(kn, ((v_1, e_1), \cdots, (v_n, e_n)))$ where *kn* is a formula among the premisses containing the variables $v_1, \cdots, v_n$, and $e_1, \cdots, e_n$ are formulas to be substituted for the variables in *kn*. For example, in the deduction of $r \supset (r \vee s \supset r)$ from $p \supset (q \supset p)$ we may have

prem: ((*IMPLIES*, *P*, (*IMPLIES*, *Q*, *P*)), (*OR*, *R*, (*NOT*, *R*)))

concl: ((*IMPLIES*, *R*, (*IMPLIES*, (*OR*, *R*, *S*), *R*)))

para: ((*IMPLIES*, *P*, (*IMPLIES*, *Q*, *P*)), ((*P*, *R*), (*Q*, (*OR*, *R*, *S*))))

Now we shall describe how proofs are checked. First we have a predicate *stepcheck* defined by *stepcheck*[*prem*; *concl*; *para*; *rule*] = *isrule*[*rule*] ∧ *apply*[*rule*; *list*[*prem*; *concl*; *para*]; *NIL*] .

Here *isrule* is a predicate that has value *T* provided its argument is an *S*-expression for a rule of inference in the system. In the simplest case *isrule* simply checks whether its argument is on a list, but in systems permitting metamathematical argument rules of inference may be derived. In the latter case *isrule* also checks the possibility that the expression (RULE, *isrule*) is among the members of *prem*. *apply* is the universal function of LISP, **[10]**.

Next, we have the proof-checking predicate itself. This can have a number of forms and it is not yet clear which is the most convenient. One possibility is *proofcheck* defined by

proofcheck[*prem*; *concl*; *proof*] = *null*[*concl*] ∨ λ[[*prem*1; *concl*1; *para*1; *rule*1; *state*1; *proof*1]; *stepcheck*[*prem*1; *concl*1; *para*1; *rule*1] ∧ *contained*[*prem*1; *prem*] ∧ *proofcheck*[*union*[*prem*; *concl*1]; *less*[*concl*; *concl*1]; *proof*1]] [*apply*[*cadr*[*proof*]; *list*[*car*[*proof*]]; *NIL*]; *apply*[*caddr*[*proof*]; *list*[*car*[*proof*]]; *NIL*]; *apply*[*cadddr*[*proof*]; *list*[*car*[*proof*]]; *NIL*]; *apply*[*caddddr*[*proof*]; *list*[*car*[*proof*]]; *NIL*]; *apply*[*cadddddr*[*proof*]; *list*[*car*[*proof*]]; *NIL*]; *cons*[*apply*[*caddddddr*[*proof*]; *list*[*car*[*proof*]; *NIL*]; *cdr*[*proof*]] .

In this the auxiliary functions *union* and *less* are defined as follows;

union[*u*; *v*] = [*null*[*u*] → *v*; *among*[*car*[*u*]; *v*] → *union*[*cdr*[*u*]; *v*]; *T* → *union*[*cdr*[*u*]; *cons*[*car*[*u*]; *v*]]]

less[*u*; *v*] = [*null*[*v*] → *u*; *among*[*car*[*v*]; *u*] → *less*[*delete*[*car*[*v*]; *u*]; *cdr*[*v*]]; *T* → *less*[*u*; *cdr*[*v*]]]

where

delete[*x*; *u*] = [*null*[*u*] → *NIL*; *equal*[*x*; *car*[*u*]] → *cdr*[*u*]; *T* → *cons*[*car*[*u*]; *delete*[*x*; *cdr*[*u*]]]] .

We now explain the somewhat long formula for *proofcheck*. A proof is a list of six items and has the form

(*state*, *premgen*, *conclgen*, *paragen*, *rulegen*, *stategen*)

where all but the first are S-expressions which may be applied to the *S*-expression *state*. If *concl* is null, there is nothing to prove and this is the termination condition. Otherwise, *proofcheck* applies *premgen*, *conclgen*, *paragen* and *rulegen* to *state* to generate quantities *prem*1, *concl*1, *para*1 and *rule*1 *stepcheck* is then used to verify that the immediate inference with these four arguments is valid. Then it must also be verified that the premisses of *prem*1 are contained in *prem*. We then refer to proofcheck again but with the *concl*1 added to the premisses and deleted from the conclusions to be proved and with a new *proof* which has a new *state* generated by applying *stategen* to the old state.

The generality permitted by this form of *proofcheck* comes from the fact that each step in the proof is computed from the *state* and the new state is also computed. Because we allow the use of arbitrary computation rules for generating the steps of the proof, there is no way of insuring that the computations will always terminate. However, as we explained before, imposing a time limit on the proof-checking process makes it definite.

We need the usual form of proof which is a sequence of steps as well as the general form just described. Consider a proof having the form $(s_1, \ldots s_n)$ where each s_i has the form (*conclusions,parameter,rule*). A proof-checker for checking such proofs is *listcheck* given by

listcheck[*prem*; *concl*; *steplist*] = *null*[*concl*] ∨ [*stepdheck*[*prem*; *caar*[*steplist*]; *cadar*[*steplist*]; *caddar*[*steplist*]] ∧ *listcheck*[*union*[*prem*; *caar*[*steplist*]]; *less*[*concl*; *caar*[*steplist*]]; *cdr*[*steplist*]]

Notice that *proofcheck* and *listcheck* each have the form of a rule of inference, and indeed we propose to admit them as simple rules of inference. This allows either kind of proof to be incorporated as a step in the other.

A form of natural deduction can be incorporated in our system by adding the following rule:

natural[*prem*; *concl*; *para*] = *equal*[*concl*; *list*[*list*[*IMPLIES*; *car*[*para*]; *cadr*[*para*]]] ∧ *proofcheck*[*cons*[*car*[*para*]; *prem*]; *list*[*cadr*[*para*]]; *caddr*[*para*]]

This allows us to infer $p \supset q$ from given premisses provided that adjoining p to the premisses allows us to infer q.

References

1. Alonzo Church, *Introduction to mathematical logic*, Princeton, 1958.

2. M. Davis, *Computability and unsolvability*, New York, McGraw-Hill 1958.

3. M. Davis and H. Putnam, *A computational proof procedure*, Troy, N.Y., Rensselaer Polytech. Inst. AFOSR TR 59-124; 1959.

4. ———, *A computing procedure for quantification theory*, J. Assoc. Compat. Mach. vol. 7 (1960) pp. 201-215.

5. H. Gelernter, *Realization of a geometry theorem-proving machine*, Proc. International Conference on Information Processing (ICIP), Paris, France, UNESCO 1959.

6. P. C. Gilmore, *A program for the production of proofs for theorems derivable within the first order predicate calculus from axioms*, Proc. International Conference on Information Processing (ICIP), Paris, France, UNESCO House, 1959.

7. ———, *A proof method for quantification theory: its justification and realization*, IBM J. Res. Develop. vol. 4 (1960) pp. 28-35.

8. J. C. R. Licklider, *Man-computer symbiosis*, I. R. E. Trans. on Human Factors in Electronics, vol. 1 (1960) pp. 4-11.

9. J. McCarthy, *Programs with common sense*, Proceedings of Symposium on Mechanization of Thought Processes, Teddington, England, National Physical Laboratory, 1959.

10. ———, *Recursive functions of symbolic expressions*, Comm. ACM vol. 3 (1960) pp. 184-195.

11. ———, *A basis for a mathematical theory of computation*, Proc. Western Joint Computer Conference, May 1960.

12. ———, *LISP programmer's manual*, March, 1960.

13. ———, *Time sharing*. Computer Systems in "Management and the Computer of the Future," M.I.T. Press, 1962, pp. 220-236.

14. M. L. Minsky, *Steps toward artificial intelligence*, Special Computer Issue, Proc.

I. R. E. vol 49 (1961) pp. 8-30.

15. ———, *A selected descriptor-indexed bibliography to the literature on artificial intelligence*, PGHFE, December, 1960.

16. A. Newell and H. A. Simon, *The logic theory machine*, IRE Trans. on Information Theory, vol. IT-2 (1956) pp. 61-79.

17. D. Prawitz, H. Prawitz, and N. Vogera, *A mechanical proof procedure and its realization in an electronic computer*, J. Assoc. Comput. Mach. vol. 7 (1960) pp. 102-128.

18. P. C. Rosenbloom, *The elements of mathematical logic*, New York, Dover Publications, Inc., 1950.

19. J. Slagle, *A heuristic program that solves symbolic integration problems in freshman calculus*, M.I.T. Ph. D. Thesis, 1961.

20. A. Tarski, *A decision procedure for elementary algebra and geometry*.

21. H. Wang, *Toward Mechanical Mathematics*, IBM J. Res. and Dev., vol. 4, no. 1, pp. 2-22; January, 1960.

22. ———, *Proving Theorems by Pattern Recognition, I*, Communications of the Association for Computing Machinery, vol. 3, April, 1960, pp. 220-234.

23. ———, *Proving Theorems by Pattern Recogniton, II*, Bell Telephone Laboratories Inc., Murray Hill, N.J. unpublished memorandum.

Massachusetts Institute of Technology
Cambridge, Massachusetts

SIZE AND STRUCTURE OF UNIVERSAL TURING MACHINES USING TAG SYSTEMS[1]

BY

M. L. MINSKY

One of the most interesting and basic results of modern work on the foundations of mathematics was the demonstration, by Turing and others, that there exist "Universal" machines. A Universal Turing machine is a finite automaton with (infinite) external tape which can "simulate" any other such machine's behavior given a description or encoding of the other machine.

The existence of Universal machines is remarkable enough and it is startling to find that such machines can be quite simple in structure. One may ask just how small they can be; one needs then some measure of size and Shannon [**3**] has suggested that the product of the numbers of states and symbols is appropriate for this since, as he has shown, this product has a certain invariance—one can exchange states and symbols without greatly increasing the product. One could also measure the numbers of "quintuples" (we use Turing's formulation rather than Davis' "quadruples") or the number of conditional transfers, etc.,

In this paper we discuss (1) the relation between Post Tag systems and Turing machines, (2) a particular Universal machine believed to be the smallest known, and (3) a few remarks on the structure of such machines.

1. Tag systems and Turing machines

A "Tag" system consists of an alphabet of symbols $a_1, \cdots, a_m$, and a set of transformations associated with the symbols. With each symbol a_i is associated a certain sequence $a_{i,1}, \cdots, a_{i,n_i} = P_i$ of symbols from the same alphabet. The sequence P_i contains n_i symbols, and this number may vary from one symbol to another. The Tag system is characterized also by an integer P and defines a process to be applied to arbitrary strings of symbols, as follows.

Suppose that a certain string of symbols is initially given. Read the first (leftmost) symbol. If it is a_i, concatenate P_i at the right-hand end of the string. Then delete the first P symbols of the string, and iterate the process. The process continues until either the string vanishes in the deletion process or a certain special "halting symbol" a_H is encountered in the reading operation.

We give as an example a simple 2-symbol Tag system with $P=3$; this is the smallest apparently nontrivial system encountered by E. Post

[1] Received by the editor May 22, 1961.

(see [2]). The two symbols are 0 and 1, the associated "Production" sequences are 00 and 1101, respectively. Thus if one begins with the sequence 1010111, one will obtain next 01111101 and then 1110100; 01001101; 0110100; 010000; 00000; 0000; 000; 00; 0; Halt. It is not known whether, for this particular system, there exists a finite initial sequence which yields an infinite set of different successor strings. It is known [1] that there do exist Tag systems for which the problem of deciding whether an arbitrary string is infinite-generating is recursively unsolvable.

Tag systems representing Turing machine calculations. Theorem II of [1] states essentially that we can find, for each partial recursive function $T(k)$ a Tag system which is equivalent to $T(k)$ in the following sense. The Tag system alphabet contains, among others, four distinguished symbols, A, a, B, b with the property: If the system is applied to the string Aa^{2^k} then, if the value of $T(k)$ is defined, the Tag system will eventually generate the string $Bb^{2^{T(k)}}$ and furthermore it will generate no other string beginning with the symbol B. If we identify this symbol B with the "halting" symbol A_H this will terminate the process, so that the Tag system and the partial function $T(k)$ are equivalent in a straightforward sense.

We use this result to replace the problem of constructing a Universal Turing machine by the equivalent problem of constructing a Turing machine which can simulate the performance of an arbitrary Tag system (given an initial string together with a description of the productions P_i of the Tag system). The next section of the paper describes such a machine with 6 symbols and 6 states; this is believed to be the smallest known Universal Turing Machine.[2] This section is concerned with explaining the encoding for the machine, and how it is interpreted. As such, the discussion can be regarded as concerned with questions of digital computer programming. As a matter of "programming appreciation" it seems interesting that such a small structure can be so complicated.

2. Structure of the 6×6 machine

SYMBOL READ	STATE q_1	q_2	q_3	q_4	q_5	q_6
O	XL	XRq_3	ALq_4	XRq_5	YLq_4	$HAIT$
I	BR	BRq_1	BR	ILq_2	BR	IR
A	XLq_2	BR	AR	AL	AR	ORq_1
B	IL	IL	—	IL	—	AR
X	YR	BLq_6	XR	XL	XR	OR
Y	XL	XL	YR	YL	YR	OR

Each entry is (Symbol written, direction, new state if different).

[2] Except for the (4, 7) machine described at the end of the paper.

2.1. Outline of the encoding. The tape for the universal machine is divided into seven regions. Starting at the left end of the tape we have

I. *Termination region.*

END:

O	X	I	A	A	O

This region is involved in the restoral of conditions after erasure of words at the front of the Tag string, and in the final halting of the machine when a special "halting symbol" is encountered in the Tag process.

II. *Erasure region.*

I	I	I	I	$\cdots$	I	I	I	

This region contains a very large number, Q, of I's. It is used in the erasure process.

III. *Production regions.* This region contains an encoding of the

$$(P_m)(P_{m-1})\cdots(P_1)$$

productions of the Tag system as strings of O's and I's. To each symbol a_i of the Tag system there will correspond a certain number N_i whose value is defined below. Let $a_{i,1}a_{i,2}\cdots a_{i,n_i}$ be the sequence corresponding to a_i in the Tag system. Then the string (P_i) occurring in this production region will have the form

$$(P_i) = IIO^{N_{i,n_i}}OI\cdots IO^{N_{i,2}}OIO^{N_{i,1}}O.$$

Thus the production region contains a representation of the productions of the system. Note that (P_i) contains $n_i + 1$ I's.

IV. *Spacing region.* This region contains a large number, R, of I's.

$$I\ I\ I\cdots I\ I\ I = I^R.$$

It is used to make the numbers N_i large enough, as explained below.

V. *Erased region.* This region is composed of O's, representing the places where the Tag system has erased symbols from the beginning of its string.

VI. *Working string region.* This region contains a symbolic representation of the current string $a_\alpha\cdots a_\omega$ on which the Tag system is operating. It has the form

$$Y^{N_\alpha}AY^{N_\beta}AY^{N_\gamma}\cdots AY^{N_\omega}$$

VII. *Empty space.* This is the infinite right-hand remainder of the tape, containing O's.

The entire tape then has the form

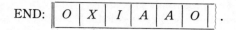

$OXIAAO$	I^Q	$(P_m)\cdots(P_1)$	I^R	$OOO\cdots OOO$	$Y^{N_\alpha}A\cdots AY$	$OOO\cdots$
I	II	III	IV	V	VI	VII

2.2. Outline of the machine's procedure. In each cycle of operation of the Tag system, the machine will (1) read the first symbol of the working string, and locate the corresponding (P_i) region. It will then (2) make a copy of the production string (P_i) at the right of the working string, i.e., in the left end of the empty region. Then (3) it must erase a certain number, P, of terms from the beginning of the working string and (4) restore conditions for the next cycle.

Phase (1) is accomplished by removing Y's from the beginning of region VI and at the same time moving a location marker along region III; the number N_α will tell how many I's to mark in region III. This phase stops as soon as the process encounters the first A in region VI.

Phase (2), copying, scans the (P_i) zone located in the previous phase, and copies it in the empty region VII. The first O of a block yields an A in the empty zone; the further O's yield Y's. Copying repeats when interrupted by a single I; and terminates when the scanning encounters the two successive I's which mark the end of the (P_i) zone.

When copying of a production is completed, the system has to erase P symbols from the left end of region VI. Actually one such symbol is already erased by phase 1. Erasure is accomplished by a trick using the same machinery already used in phase (1); the machine attempts to make $P-1$ further copies. But in these subsequent attempts to locate and copy, the marking symbols will have entered region II, which contains no O's. Hence *no symbols will be copied*. The system will reiterate the process, erasing more symbols, and moving the location marker along region II. The lengths of regions II and IV are so chosen that this cycle will repeat precisely P times, and the process will terminate when the location markers finally break into region I.

Breaking into region I causes the machine to restore the tape to the standard format. Then, the entire program is repeated. A halt can occur only if the erasing phase ends with the location marker landing precisely on the O to the right of region I. We shall arrange this event to occur only as a consequence of encountering a particular symbol a_H of the Tag system; this in turn will depend on the halting of the original Turing machine T being simulated.

2.3. Details of the encoding. The production string (P_i) contains $(n_i + 1)$ I's; one more then the number of symbols in the string associated with the symbol a_i of the Tag system in use. Now let us define N_i to be

$$N_i = \sum_{j=1}^{i-1} (n_j + 1) + R$$

where R will be defined below. Note that if one counts N_i I's from right to left, starting at the right of region IV, one will have located the beginning of the zone (P_i). We will therefore use the sequence Y^{N_i} to correspond to a_i in region VI, and the string (P_i) will be represented in region III as noted in (2.1). Define S to be the total number of 1's in the entire region III:

$$S = \sum_{i=1}^{m} (n_i + 1) \qquad (m = \text{number of symbols} \geqq 2)$$

and define $R = PS$, where P is the Tag number. Note that for all i we will have $R \leqq N_i < R + S$.

Observe that for any sequence of P symbols $N_{i,1}, N_{i,2}, \cdots, N_{i,P}$ we have the relation:

$$n_i + 1 + \sum_{j=1}^{P-1} N_{i,j} < (P-1)\cdot(R+S) + S = PR ,$$

while on the other hand we have

$$n_i + 1 + \sum_{j=1}^{P} N_{i,j} > PR .$$

The left-hand terms are the numbers of I's that would be marked in the course of dealing with, respectively, $P-1$ and P terms of the arbitrary sequence.

Finally we set $Q = (P-1)R - S$, so that the total number of I's in regions IV, III, and II is just PR. The two inequalities then assure us that region I will be reached by the marking process in the course of dealing with any set of P symbols, but that this cannot happen in the course of dealing with any set of $P-1$ symbols. This is the device that insures that precisely P symbols will be erased in each cycle.

2.3.1. The Halting symbol. In order to terminate the process, we need a special symbol which, when encountered at the beginning of the working string, will cause the machine to halt. Examination of the state diagram shows only one condition yielding a halt; this condition will occur if the machine can be made to arrive at the right hand O of region I, in state q_2. This condition will occur if a location operation terminates precisely with the marking (replacing with B) of the extreme left-hand I of region II; i.e., if the machine has marked precisely $PR\, I_s$ since the beginning of the previous operation cycle. But the symbol representation lengths, N_i, were chosen precisely so as to prevent this from occurring in normal operation; this is why we arranged the strong inequalities

$$\sum^{P-1} < PR < \sum^{P} .$$

To cause a stop, we have to force marking to terminate precisely at the end of region II. This is done by using, for a special termination sequence, the production string

$$(P_H) = \boxed{I \ I \ O^{PR} \ O \ I \ O^{PR} \ O}.$$

Suppose this production has been copied, yielding $A\ Y^{PR}\ A\ Y^{PR}$ in the working region. When state q_1 processes the first Y^{PR} block, it will either terminate at the left-hand end of region II, meeting the condition for Halting, or it will cause state q_1 to run into the region I, causing clearing and restoration of the tape condition. In this latter case, the very next location phase, operating on the second Y^{PR} block, will certainly terminate in precisely the desired location, and Halting will ensue. Thus (P_H) will serve as the production string for a special Halting symbol a_H.

2.4. Details of the machine's performance. The system starts in state q_1 on the first Y in region VI; this Y is changed to X and the machine moves to the left (changing O's to X's) until it encounters an I, which it changes to a B. It then moves to the right (changing X's to Y's) until it encounters another Y. It changes that Y to an X and again runs to the left looking for another I. Whenever it runs to the left it changes Y's and O's to X's and changes B's to I's; whenever it runs to the right it changes X's to Y's and I's to B's. This trick makes it possible to use a single state to perform the location function. One might work out the example $I I O O I O O I I I Y \dot{Y} Y A\ Y$. If started at Y, in state q_1, this will yield $B B Y Y B Y Y B B B Y Y Y Y Y X Y A\ Y$ (where we have stopped the machine at the marked Y, in state q_2, on its way to the left to attempt another copy operation). The machine converted the production string $I I O O I O O$ into the string $A\ Y A\ Y$ at the end of the working region, but has already started to eat into this new material in the course of a new locate cycle.

The copying phase, which begins with a transfer out of state q_1 (because an A was encountered), uses states q_2, q_3, q_4, and q_5. States q_3, q_4, and q_5 do the copying proper, while state q_2 is used mainly as a decision element to decide when copying should be terminated. When state q_2 encounters an O to the left, the machine enters state q_3 which causes an A to be entered in the empty region. State q_4 returns to the left, looking for an O or an I. If an O is seen first, state q_5 is entered; this writes a Y in the empty region and returns to state q_4, so that the pair of states (q_4, q_5) will copy any number of O's. When state q_4 sees an I, it returns the system to state q_2, again moving left. If an O is next encountered, we write an A and continue copying, but if an I is encountered the machine returns to state q_1. Thus the return from state q_4 to state q_1 is contingent upon seeing two successive I's, i.e., the end-of-production marker.

No copying can occur from the region II, because there are no O's to cause the transfer into state q_3. Thus phase (3), "erase", is subsumed in phase (1). The restoring phase is entered when, the machine runs into the first A of region I. The X that it writes here is detected two steps later, when in state

q_2, and this transfers the machine to state q_6. The B's that have just been written are replaced by the proper A's, all O's and I's are restored, and all unused Y's at the beginning of region VI are erased. The terminal A is also destroyed, incidentally transferring the machine back to the initial state q_1. State q_6 can encounter an O only under a special condition, since regions II through VI have had all O's replaced by X's during the previous location cycles. In fact, the only O that state q_6 can ever see is the initial O of region I, and this can be reached only under the special circumstance of entering that region in state q_2, as noted in 2.3.1.

3. On the structure of Universal machines

One may ask whether the property that a Turing machine is universal implies some interesting structural condition about the state-symbol transition diagram of the machine's finite automaton part. In this section we describe a variety of Universal machine whose structure is so simple that it would seem nothing interesting of this nature can be said. It will be sufficient to exhibit a single example, for the simple Tag system

$$\begin{Bmatrix} 0 \to 0\,0 \\ 1 \to 1\,1\,0\,1 \end{Bmatrix}$$

discussed earlier; construction of similar machines for arbitrary Tag systems (including Universal Tag systems) will be equally simple.

Our machine will use five symbols. "0" and "1" are the symbols of the Tag system in question, "A" and "B" are the symbols of a corresponding auxiliary alphabet, and X is an extra symbol corresponding to blank tape squares. An initial string for the Tag system will be given as a tape sequence

$$\cdots XXX \cdots XXX {A\,0\,0 \atop B\,1\,1} \cdots {0\,0 \atop 1\,1} XXX \cdots XXX \cdots$$

where ${A \atop B}$ etc. means a choice of one of the two symbols indicated. Assume that the machine is started at the position marked X. We want it to move to the left until it finds the ${A \atop B}$ symbol, read that symbol, erase it and the next two symbols, move to the right and write in the X region the appropriate string (00) or (1101). Location to the left is done by a state Q_L. When it hits A or B it branches to the states Q_1 or Q_1', depending on whether it read A or B. The states $Q_1 \cdots Q_4$ and $Q_1' \cdots Q_4'$ erase the first three symbols (there are four of them because the machine moves left after reading and has to return). The states Q_5 or Q_5' convert the next symbol to the AB alphabet for use in the next cycle. Q_6 or Q_6' move the machine over to the right-hand blank region and a few more states inscribe the appropriate production sequence there. Finally the machine re-enters state Q_L and repeats the process.

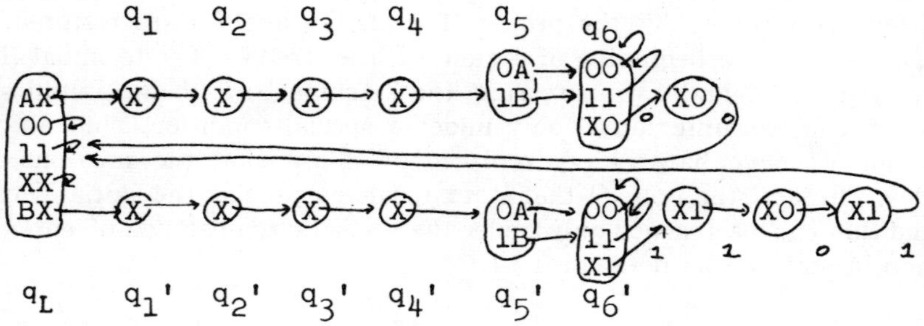

q_L moves left. Other states move right.
Pairs mean (symbol read, symbol written).
EXAMPLE OF TURING MACHINE DIAGRAM FOR TAG SYSTEM

All but Q_L always move right. An X state always writes "X". We could introduce an X-Halt condition in Q_5 to detect that the Tag string has become too short, and other halt conditions are easy to insert. To make a similar machine for some other Tag process, one just gives Q_L more branches, and for each of them adds a linear string of states similar to the two in the present example.

3.1. Observations. (i) There is only state which can go to more than one other state, (ii) there is only one state reached by more than one other state, (iii) all states are "directionally pure"—they either always move left or always move right, (iv) there is only one state which moves left, (v) the machine can change direction only by changing a symbol, (vi) the branch states of (i) and (ii) and (iv) can be a single unique state.

By sacrificing (v) we can make the unique state a non writing state, e.g., by changing AX and BX to AA and BB in its table. The isolation of the left-going mode to a single state reflects the monogenic nature of the Tag process in which no memory is required between application of productions. The extra AB alphabet can be eliminated easily by adding a few more states, since the leftward motion can as well be interrupted encountering an X; this entails the loss of property (vi). If we use Y's for the marks at the left, instead of X's, we get a sort of nonerasing machine; each tape square is changed at most twice, either from X to O to Y or from X to 1 to Y; it is quite simple to convert this to a 2-symbol nonerasing machine.

REMARKS. The first really small Universal machine seems to be that of N. Ikeno [4]. His 6-symbol 10 state machine (6,10) uses a "relative addressing" in which the location of the subsequent state is described in terms of the distance of its representation from that of the present state on the tape, rather than in terms of the distance from the beginning of the description block as in earlier (e.g., Turing's) constructions. The author [5] constructed a (6,7) machine based on Tag systems shortly after obtaining the results in [1]. Following Ikeno, Watanabe [6] constructed a (6,8) machine; the innovation was a sort of "indirect addressing" which eliminates machinery used by Ikeno to remember the direction of motion to the next instruction. After hearing of our (6,7) result, Watanabe [7] refined his method to produce a (5,8) machine.[3] The author, who had supposed that his use of the Tag theorem gave an advantage, was then driven to produce the present (6,6) machine with its monstrous encoding. The greatest difficulty was encountered in providing for explicit Halting; this was done by using our "location inequality" method. A. Tritter has recently constructed a (4,9) machine along these lines, but using the end-of-tape for a halting symbol. The Tag system that Post mentions in [2], which we used for the example in part 3, has a (3,7) Turing machine, and a similar system has a (3,6) machine. It is still not known whether the halting problems for these systems is solvable, and if they turn out to be Universal, these would be Universal machines. Actually, no one seems to have ruled out the unlikely possibility that there is a universal (2,2) machine.[3] Watanabe [7] has advanced arguments that with certain restrictions on the manner of computation the minimum product must be $\geqq 30$, but these restrictions may not be essential; our (6,6) machine does not obey them. A very small machine might turn out to be useful in construction of a very simple unsolvable decision problem, e.g., one in elementary number theory.

Added in proof.

The author has constructed a machine with a simpler encoding and a smaller state-symbol product, (4,7). This is based on the discovery, by John Cocke and the author, that there are Universal Tag systems with deletion number $P = 2$. The encoding eliminates regions I, II, and IV on the tape and is otherwiset he same except that each P_i ends with an I, N_i is just $1+\sum_{j=1}^{i-1}(n_i+1)$, and (P_H) is *IIOIOI*. The machine structure is

	q_1	q_2	q_3	q_4	q_5	q_6	q_7
Y	OL	OLq_1	YL	YL	YR	YR	OR
O	OL	YR	$HALT$	YRq_5	YLq_3	ALq_3	YRq_6
I	ILq_2	AR	AL	ILq_7	AR	AR	IR
A	IL	YLq_3	ILq_4	IL	IR	IR	ORq_2

In (8) Watanabe published his (5,8) machine and also his interesting (5,6) machine which uses an infinite periodic representation of the instructions. This is equivalent to including an extra circular tape.

[3] D. Bobrow and the writer have recently ruled these out.

REFERENCES

1. M. Minsky, *Recursive unsolvability of Post's problem of "Tag"*, MIT Lincoln Laboratory Group Report 54G-0023, June, 1960. Ann. of Math. vol. 74 (1961). pp. 437–455.

2. E. Post, *Formal reductions of the General combinatorial decision problem*, Amer. J. Math. vol. 65 (1943).

3. C. E. Shannon, *A Universal Turing Machine with two internal states*, Automata Studies, Princeton, 1956 (Annals of Mathematics Study, no. 34).

4. N. Ikeno, *A 6-Symbol 10-state Universal Turing machine*, Proceedings, Institute of Electrical Communications, Tokyo, 1958.

5. M. Minsky, *A 6-symbol 7-state Universal Turing machine*, MIT Lincoln Laboratory Group Report 54G-0027, August 1960. (Contains a description of Ikeno's machine.)

6. Shigeru Watanabe, *On a minimal Universal Turing machine*, MCB Report, Tokyo, 1960.

7. ———, personal communication.

8. S. Watanabe, J. A. C. M., vol. 8, no. 4, pp 476–483, October, 1961.

MASSACHUSETTS INSTITUTE OF TECHNOLOGY,
CAMBRIDGE, MASSACHUSETTS

AUTHOR INDEX

Italic numbers refer to pages on which a complete reference to a work by the author is given.

Roman numbers refer to pages on which a reference is made to a work of the author. For example, under Kreisel would be the page on which a statement like the following occurs: "This theorem was proved earlier by Kreisel [7, §2] in the following manner..."

Boldface numbers indicate the first page of articles in this volume.

SUBJECT INDEX